TRUMAN AND THE STEEL SEIZURE CASE
The Limits of Presidential Power

Contemporary American History Series
WILLIAM E. LEUCHTENBURG, GENERAL EDITOR

Maeva Marcus

TRUMAN AND THE STEEL SEIZURE CASE

the
limits
of
presidential
power

COLUMBIA UNIVERSITY PRESS   NEW YORK   1977

Library of Congress Cataloging in Publication Data

Marcus, Maeva, 1941–
    Truman and the steel seizure case.
    Bibliography: p.
    Includes index.
    1. War and emergency powers—United States.
2. Steel industry and trade—Law and legislation—
United States.   3. Truman, Harry S., Pres. U.S., 1884–
1972.   4. Sawyer, Charles, 1887–   5. Youngstown
Sheet and Tube Company.   I. Title.
KF5060.M37   343'.73'078   77-4095
ISBN 0-231-04126-8

Columbia University Press
New York   Guildford, Surrey
Copyright © 1977 Columbia University Press
Printed in the United States of America

for Danny

contents

preface

On April 8, 1952, President Harry S. Truman announced that, to avert a strike, the federal government was seizing the steel mills of all the major companies involved in a labor dispute with the United Steelworkers of America. Response to the President's action was uniformly one of shock. Presidents had seized plants before, but never had the government, in what was technically a time of peace, taken over the major portion of an industry as basic to the American economy as steel. Nor had any President invoked his inherent executive powers under Article II of the Constitution to justify a seizure when a statute—in this case, the Taft-Hartley Act—offered an alternate, unquestionably legal method for preventing the strike.

Truman's seizure of the steel mills created a political and constitutional crisis that raised fundamental questions about the role of the President and the nature of presidential power in the American scheme of government. In theory the Constitution sets up three separate branches, the executive, the legislative, and the judicial, each knowing the metes and bounds of its powers as well as having the authority to check the acts of the other two. In practice the limits of each branch's powers have proved to be indefinite, and the methods of exercising the checking authority frequently have been difficult to determine. The relative influence of the President, the Congress, and the courts has fluctuated throughout American history. The steel seizure provided the country with a unique opportunity for reassessing, in the aftermath of two decades of unprecedented employment of executive power, the balance of authority among the branches.

Since the birth of the nation, Americans have taken a particular interest in separation-of-powers disputes. This is hardly surprising, for, as the late Justice John Marshall Harlan noted, the separation-of-powers concept is basic to a society based on individual freedom:

We are accustomed to speak of the Bill of Rights and the Fourteenth Amendment as the principal guarantees of personal liberty. Yet it would surely be shallow not to recognize that the structure of our political system accounts no less for the free society we have. Indeed, it was upon the structure of government that the founders primarily focused in writing the Constitution. Out of bitter experience they were suspicious of every form of all-powerful central authority and they sought to assure that such a government would never exist in this country by structuring the federal establishment so as to diffuse power between the executive, legislative, and judicial branches.[1]

President Truman's seizure of the steel mills triggered a classic constitutional and institutional debate. Had the President acted without lawful authority? If so, what was the role of Congress in passing judgment on the propriety of the President's action? Or should the responsibility for answering these questions devolve on the Supreme Court?

By congressional default, the resolution of the crisis waited upon the progress of a lawsuit through the federal courts. In an attempt to retrieve its property, the steel industry brought suit against Secretary of Commerce Charles Sawyer, the official nominally in charge of running the steel mills. That litigation culminated in the Supreme Court's landmark decision in *Youngstown Sheet & Tube Co.* v. *Sawyer*,[2] which invalidated President Truman's action.

This study will examine the history and significance of the remarkable *Steel Seizure* case. The political, economic, and legal setting in the spring of 1952 will be explored to determine what elements were crucial in shaping the Supreme Court's unexpected decision. The case arose against a backdrop of twenty years in which presidential power had steadily expanded—a phenomenon which led Truman to believe that he had the authority to take over the mills. Moreover, Truman thought the seizure was necessary to ensure the flow of adequate supplies to American troops in Korea and to keep the economy healthy at home. Given these substantial reasons for the President's act, why did the New Deal–Fair Deal Court, usually receptive to the exercise of presidential power, take the occasion of the steel seizure to curb the use of inherent executive authority? Could the Court, as a matter of constitutional theory and precedent, as easily have upheld the President? If so, what actions and attitudes—of the parties, the President, the individual Justices, the public—were critical in shifting the result?

From a historical standpoint, the impact of the *Youngstown* decision is as important as, if not more important than, the reasons why it was made. The Court's judgment was generally interpreted as a rebuke to President Truman and as an effort to halt the accretion of power in the executive branch. Closer examination of the individual opinions of the Justices, however, reveals that the Court's holding was neither as simple nor as sweeping as the broad language of the opinion of the Court suggested. In analyzing the opinions and the use to which *Youngstown* has subsequently been put, this study assesses the influence of the case on the theory and practice of presidential power and on the doctrine of separation of powers.

acknowledgments

There is one person to whom I owe an immeasurable debt of gratitude. Professor William E. Leuchtenburg guided my work from the early stages of a dissertation through publication. His trenchant criticism, wide-ranging knowledge, and unflagging interest made this experience profitable and satisfying for me.

I am grateful to Professor Stuart Bruchey, who made numerous suggestions which improved the manuscript, and to the members of my dissertation defense committee, Professors Walter Metzger, Louis Henkin, and Donald Dewey, for their perceptive comments. Alonzo Hamby, after a thorough reading of the manuscript, offered much helpful advice. The editors at Columbia University Press, Evelyn Katrack, Joan McQuary, and Bernard Gronert, gave valuable counsel too.

John Pickering, Stanley Temko, Judge Harold Leventhal, William Peirce, Robert Peabody, David Becker, James R. Perry, Peter Rousselot, E. Barrett Prettyman, Jr., Peter Strauss, Sol Lindenbaum, Philip Lacovara, Richard Darilek, David Rosenbloom, Alice Edgerton, Stephen Breyer, Robert Kopp, George Vogt, and Karen Richardson all gave generous assistance. I also appreciated the opportunity to interview Elliot Bredhoff, Bruce Bromley, Judge Oscar Davis, Judge Charles Fahy, Justice Arthur Goldberg, Ellis Lyons, Herman Marcuse, Charles S. Murphy, David Stowe, Leon Ulman, and especially Charles Sawyer who graciously permitted me to examine his papers at his home.

The Harry S. Truman Library Institute provided the funds which enabled me to use the valuable manuscript collections at the Truman Library. The staff of the Library, and especially Philip Lagerquist, made working there a delight.

For five years the Library of Congress was my second home. Without the books, manuscripts, and desk space it provided, and without the helpfulness of so many members of its staff, I could never have completed this book.

Mrs. Gladys Shimasaki did a meticulous job typing the final copy of this lengthy manuscript and was a pleasure to work with. Linda Merritt, over several years, generously typed drafts of many, many chapters. Susan Kenney, in the last hectic months, also volunteered to help with the typing. I greatly appreciate their assistance.

Throughout the years in graduate school, my parents have given aid and comfort in innumerable ways. They know how much this has meant to me. My children, Stephanie and Jonathan, have borne, with remarkable cheerfulness, the many demands made on them by a working mother. To Frances Shaw, who made the burden of these last five years ever so much lighter, my family and I are deeply grateful.

And finally, to my husband, Daniel Marcus, I owe an incalculable debt of thanks. His expert editing greatly strengthened the manuscript. But beyond that, his constant faith and support sustained me throughout.

MAEVA MARCUS

TRUMAN AND THE STEEL SEIZURE CASE
The Limits of Presidential Power

one

LIMITED WAR AND THE ECONOMY

The steel seizure crisis originated in a specific set of historical circumstances which affected both the course of President Harry S. Truman's actions and the decision of the Supreme Court. In the spring of 1952, the Korean War, the problems of the economy, and the political imperatives of an election year provided the framework within which the Truman administration had to act. The most fundamental of these factors was the cold war, suddenly turned hot in Korea.

Despite years of talk about the threat of Communist aggression, the nation was shocked to learn that on June 24, 1950, the North Koreans had invaded the Republic of Korea to the south. Although President Truman and Secretary of State Dean Acheson assumed that the Soviet Union had instigated the North Korean attack, they were wary of action that might enlarge the area of conflict. The administration believed that the Communists had to be repelled in Korea, however, in order to discourage them from using force elsewhere. Hence the United States chose to employ its power, but in a manner limited so as to avert the outbreak of a third world war.[1] In a matter of days, the Truman administration had resolved to support the Republic of Korea and to do so under the aegis of the United Nations if that organization would take the responsibility. To avoid any delay in sending help to the South Koreans, the President decided not to consult with Congress.[2] Thus the United States became involved in a protracted armed conflict, to which a large number of American ground troops would eventually be committed, without a formal declaration of war.

Although Truman did not think it necessary to obtain the explicit approval of Congress for his resolves, on June 27 he met with congressional leaders to explain the development of American policy in Korea. According to Democratic Senator Tom Connally, the reaction of the congressmen was favorable: they agreed that the United States had to assist in the defense of South Korea and that the United Nations was the proper authority to direct a cooperative effort. Democrats and Republicans appeared to be satisfied with the measures Truman had taken thus far.[3]

Anticipating a quick end to hostilities, the President sought to minimize the seriousness of the course upon which he had embarked. At a press conference soon after the North Korean invasion, Truman stated emphatically, "We are not at war."[4] He described the conduct of the United Nations members as "going to the relief of the Korean Republic to suppress a bandit raid on the Republic of Korea."[5] It was only in answer to a reporter's question as to whether the United Nations operation might be characterized as a "police action" that Truman acceded to the use of that term.[6] As the Korean conflict turned into a war in every practical sense of the word, it became obvious that "police action" was an infelicitous description. One congressman's reaction to the use of the term was typical:

Our countrymen are fighting and dying every hour of the day and the night. If we are not technically at war, pray tell me by what right doth the President of this country, the Commander in Chief of our Armed Forces, send the youth of this Nation 7,000 miles from their homes and firesides to fight, to bleed, and to die in a foreign land? . . .

Yet we sit here and seem to think that we are engaged in a kind of "police action." Certainly my son is not a policeman nor is he a member of a police force. Neither is your son, and sons of other Americans who are now getting ready for the eventualities of a war which is being waged seven or eight thousand miles from home. . . . Let us face up to the fact: This is no police action. It is nothing short of war.[7]

When the gravity of the Korean situation became evident, a number of Senators and Representatives asked why the President had not sought congressional ratification of his actions. But few went as far as Senator Robert Taft in categorically stating that the war was illegal.[8] As military circumstances in Korea worsened, however, more congressmen joined Senator Taft in charging the administration with improperly neglecting to seek the consent of Congress to American participation in the defense of Korea. Republican Senator William Knowland, who immediately after the outbreak of hostilities had asserted that the President had legitimately used his powers to take "the necessary police action," [9] reversed himself a year later and accused the administration of leading the country into war, "but without a declaration of the Congress of the United States." [10]

Throughout the course of the Korean War, the Truman administration was to be plagued by this charge.[11] The lack of a congressional declaration of war put the administration in the difficult position of having to fight a full-fledged war without acknowledging its total dimensions. When President Truman requested legislation to implement the decision to fight in Korea, some congressmen questioned the need for such laws when technically the United States was not at war. The administration thus was deprived of the whole-hearted cooperation of Congress, many of whose members were unwilling to act as if the United States were involved in an acute wartime emergency.[12] In the *Steel Seizure* case, the President would face the consequences of his determination to proceed with the war without formal legislative approval.

When the Chinese Communists intervened in support of the North Koreans in November 1950, the Truman administration realized that it would have to plan for a much lengthier conflict. Greater numbers of men and more materials would be needed. But the President still thought that total mobilization was unnecessary. Thus, on December 16, 1950, Truman declared a "limited" national emergency [13] in order to trigger the application of a number of statutory provisions giving him additional powers to face the crisis in Korea.[14]

With the prospect of greatly increasing defense expenditures, the Truman administration had to reevaluate the economic policy it had propounded soon after the United States entered the Korean conflict. At the outset of the war, the President did not ask Congress for direct controls over the economy: the administration's expectations as to the nature of the Korean hostilities led it to believe that the economy would not be seriously dislocated and that inflation could best be handled by indirect controls. Moreover, Truman feared that the nation, and the world, might interpret strict economic controls as preparation for another global war.[15] Within the administration, Leon Keyserling, chairman of the Council of Economic Advisers, was the strongest advocate of this indirect controls policy.[16]

Keyserling proposed to meet the demands of the Korean campaign by fostering a tremendous growth in production, by giving priority to defense and essential civilian requirements, and by reducing the pressures of inflation. He emphasized industrial expansion as the major goal of the administration's economic policy as well as the best means of achieving the other two objectives.[17] To implement this plan, Keyserling recommended a system of government priorities and allocations for necessary materials and facilities, accelerated amortization and long-term loans as incentives to enlarge industrial capacity, reduction of federal expenditures in nondefense areas, tax increases, and credit restrictions to inhibit the inflationary trend.[18] Keyserling thought that short-run inflation would occur regardless of whether direct or indirect controls were imposed, but he believed that direct price and wage controls might hinder the more important aim of economic expansion. He convinced the President that, in the interest of increased production, the administration should fashion a policy based on indirect controls to reduce inflation.[19]

As a result, when Truman requested congressional action to enable the government to put its economic program into effect, he did not ask for power to impose price and wage controls. Indirect restraints, coupled with the voluntary cooperation of citizens, would be sufficient to curtail inflation, the President advised. Truman indicated, however, that if prices rose excessively, he would be prepared to recommend more "drastic" measures.[20] At a press conference several days later, the President reiterated his intention to avoid wage and price restrictions "because there is no necessity for it." [21]

There were, undoubtedly, some political considerations that influenced Truman's acceptance of Keyserling's economic theories. Price control and the Office of Price Administration (OPA) had been very unpopular during World War II. In the years 1947–1949, when the administration had considered imposing price controls again, Congress had shown that it was not receptive to such a move. Hence Truman decided that it would be easier for a defense production bill to be passed if it contained no provision for price and wage restrictions.[22]

Much to his surprise, however, Congress and the public did not share the President's confidence that inflation could be effectively curbed without direct controls. They were obviously disturbed by reports of greatly increased prices. Immediately after the North Korean invasion of South Korea, a round of scare buying occurred, in which consumers and businessmen stocked up on goods and materials that they expected soon to be in short supply. This led to a sharp rise in the Wholesale Price Index, from 157.3 in June 1950, to 162.9 in July, and to 166.4 in August. The Consumers' Price Index also rose, although at a slower pace: in June 1950, the Index stood at 170.2; by August it had reached 173.0.[23] At Senate hearings on the administration-supported Defense Production Act late in July, Bernard Baruch, a highly respected authority on economic controls, testified that the administration bill—which contained no price and wage controls—was self-defeating. The former chairman of the War Industries Board claimed that no system of priorities could work effectively or for long without price control. "That was learned during World War I. It was forgotten and had to be learned anew, at what bitter cost, in World War II. Must we persist in repeating the mistakes of the past, even to inviting disaster?" [24] Set against the rise in prices, Baruch's testimony struck a responsive chord both in Congress and among the public. Speeches were made on the floor of the Senate and the House advocating, at the least, standby authority for the President to impose price and wage controls, and congressmen were inundated with constituent mail urging the passage of such legislation. [25]

In answer to the clamor for controls, the President submitted to the congressional committees considering the Defense Production Act a draft of price and wage control legislation that would be acceptable to the administration. [26] But, in a letter to the chairmen of these committees, Truman made it clear that he still did not favor direct controls. If the Congress found it necessary to pass such legislation, he stated, the authority granted should be available for use at the discretion of the President. He must be allowed to decide when, where, and how price and wage controls would be instituted. [27]

The Defense Production Act of 1950 was passed by overwhelming majorities [28] and was signed into law on September 8, 1950. [29] It contained all the provisions the President had requested—power to establish priorities and to allocate strategic materials, inducements for the expansion of productive capacity and supply, power to control consumer and real estate credit and authority to requisition—and some additional ones as well. Most important of these were Titles IV and V, dealing, respectively, with price and wage stabilization and the settlement of labor disputes. To understand the choices of action open to the administration when the steel dispute arose in 1951–1952, Titles IV and V together with Title II, the authority to requisition, must be briefly examined.

Title II empowered the President to requisition property, or the use thereof, whenever he determined "that the use of any equipment, supplies, or component parts thereof, or materials or facilities necessary for the manufacture, servicing, or operation of such equipment, supplies, or component parts, is needed for the national defense." [30] This broad power was limited only by a requirement of payment of just compensation to the property owner. [31] Although Title II appeared straightforward, there were conflicting interpretations of what it authorized the President to do. During the Senate hearings on the Defense Production Act, administration representatives asserted that the main purpose of that section was to allow the President to obtain equipment and materials needed for the national defense. In addition, Title II permitted him to take over a plant, but the administration was unsure whether this authority entitled the government to operate such a facility.[32] Some committee members were dubious about this distinction. Senator Homer Capehart, for example, asked incredulously, "What good is Title II? Why do you want the authority to take over every business in America if you do not have the authority to operate it?" [33] In the Senate, Republican Senator John J. Williams indicated agreement with Capehart's view: "Under this bill as it now stands, the President could authorize the nationalization of our great steel industry, along with the nationalization of such of our other large industries, as he saw fit." [34] In the House, Brent Spence, Democrat of Kentucky and chairman of the House Banking and Currency Committee which considered the Defense Production Act, asserted, in response to a question about whether the government could operate a plant it had seized, "Of necessity I think he [the President] could operate it. There would not be any point in seizing it if he could not operate." [35]

Title IV of the Act, dealing with price and wage stabilization, allowed the President discretion in the imposition of controls. Congress declared its intention "to prevent inflation and preserve the value of the national currency; to assure that defense appropriations are not dissipated by excessive costs and prices"; and "to stabilize the cost of living for workers and other consumers." [36] But it permitted the President, in Section 402, to choose the appropriate means of achieving these objectives.[37] In the first instance, he was authorized to encourage voluntary action by industry, labor, agriculture, and consumers to impede the inflationary spiral. If voluntary means were unsuccessful, the President could then issue regulations establishing price ceilings on individual goods or services; at the same time, he had to stabilize wages in the industry producing the material or providing the service. When price ceilings had been imposed on a substantial part of the economy, the President was required to institute overall controls.[38] If this became necessary, the President was directed by Section 403 of the act to create a new independent agency to administer the task of economic stabilization.[39] The Senate committee report on the Defense Production Act noted, "It was felt that this provision was important because of the conviction that price and wage controls directed toward stabilization are so closely related that responsibility for both must be invested in a single agency responsible to the President." [40]

In Title V, Congress asserted its desire that "effective procedures for the settlement of labor disputes affecting national defense" [41] be established. It recognized that price and wage stabilization would put unusual strains on the normal processes of collective bargaining; [42] Congress therefore authorized the President, in Section 502, "to initiate voluntary conferences between management, labor, and such persons as the President may designate to represent government and the public, and . . . to take such action as may be agreed upon in any such conference and appropriate to carry out the provisions of this title." [43] Apparently it was contemplated that the President might create a board similar to the War Labor Board of World War II, which, after the War Labor Disputes Act of 1943 was passed, had the statutory power to compel settlements.[44] Title V did not specifically establish a board, because Congress wanted labor and management to cooperate in setting up the procedures such a board would follow. Any arbitration board would be dependent on the goodwill of labor and management for its success, and asking these groups themselves to establish the procedures appeared to be one way of obtaining it. Moreover, Congress undoubtedly hoped that when labor and management got together they would agree upon a "no strike, no lockout" pledge as they had in World War II. The only specific limitation on executing Title V was contained in Section 503, which stipulated that no measures inconsistent with the Taft-Hartley Act of 1947 and other Federal labor standards statutes could be taken.

To some congressmen, Sections 502 and 503 appeared inherently contradictory: if the President had to follow the procedures set up by existing labor laws, what was the purpose of providing authority for the creation of new ones? Senator Taft was especially concerned with the meaning of these provisions:

> Either it [Title V] is a reaffirmation of the statutes we have, or it goes very much further and authorizes the President to impose practically any terms he wants to impose in the settlement of labor disputes. If it is desired to impose compulsory arbitration in time of war—and it is perfectly possible that such a desire exists—then it seems to me that purpose ought to be spelled out. But the language is completely ambiguous, and appears to me to have no possible justification for being placed in the bill. [45]

The official report of the Senate Banking and Currency Committee indicated that the two sections would give the President a choice of actions when a labor dispute erupted in industry necessary to the national defense: he could use the procedures available to him in existing statutes or rely on those provided in Title V. [46] But another Representative thought that the explicit purpose of Title V was to give the President additional power, over and above the Taft-Hartley Act, to deal with labor disputes during the Korean crisis:

> Of course title V is not in keeping with our labor laws which we have enacted for peacetime, because our labor laws all have as their purpose, whether they have that effect, to encourage collective bargaining and agreement. Whereas this bill proposes to substitute the authority of the Government for voluntary agreement in the interest of continued production. It is the antithesis of our way of life. [47]

The ambiguity of Title V thus made it a poor source of authority for the development of procedures to deal with labor disputes during the Korean War. [48]

By Executive Order 10161, issued September 9, 1950, the President set up the machinery to implement the Defense Production Act. [49] In preparation for the possible imposition of direct controls, he created the Economic Stabilization Agency (ESA) to comply with Title IV's directive that a new independent agency administer all stabilization functions. But Truman and his advisers disagreed with the congressional belief that a *single* agency was the best vehicle to regulate prices and wages.[50] Influenced by the successful World War II experience, the administration thought that if overall controls had to be imposed, two separate bodies could more properly deal with the problems of price and wage stabilization. This reflected the White House conviction that wage increases should be awarded in accordance with wage stabilization rules; they should not depend on the ability of the employer to absorb the additional labor costs without a rise in the price of his product.[51] Specifically, if, after application of the government's wage guidelines, the workers of a particular company deserved a raise, it should be granted. Then, if the new labor rates caused the employer hardship, he could apply to the price control agency for relief. In the administration's mind the criterion for a wage increase, at least during wartime, should not be the company's ability to show a profit. Congress, however, clearly intended that the impact of a wage increase on the employer's prices for his products ought to be a major consideration in determining whether that increase should be granted; this was the only way to stop the inflationary spiral.

As a compromise, the President established the ESA to meet the letter of the law but, in Sections 402 and 403 of the executive order, provided for the appointment of a director of price stabilization and a wage stabilization board, each responsible to the administrator of the ESA. The Wage Stabilization Board (WSB), modeled after the War Labor Board of World War II, would be a tripartite body composed of nine members: three to represent the public, three to represent labor, and three to represent business. One of the public members would be designated by the President as chairman of the board.[52] President Truman appointed Alan Valentine, former president of the University of Rochester, as the first administrator of the ESA. Cyrus Ching, director of the Federal Mediation and Conciliation Service, was named chairman of the Wage Stabilization Board. The position of director of price stabilization remained unfilled for three months, because the administration had difficulty in finding a qualified individual who would accept the job. There was little pressure to make the appointment, however, for the administration, at this point, was merely preparing for the possibility that a full-fledged controls program might have to be implemented.

At the outset, the wage stabilization agencies were relatively inactive. This was attributable to the administration's antipathy for mandatory controls, coupled with the stated policy of the Defense Production Act, that voluntary action should be promoted before any resort to direct controls.[53] In late September 1950, moreover, the military situation in Korea improved dramatically. Indeed, it seemed possible that the conflict might soon be ended. With the prospect of huge defense expenditures subsiding the scare buying decreasing, there was a lull in the buildup of inflationary pressures. Since the ESA and the WSB might shortly lose their raison d'être, gathering a large staff in preparation for the imposition of across-the-board controls seemed a futile exercise.[54]

The intervention, in full force, of the Chinese Communists in Korea at the end of November abruptly altered military and economic conditions. Day by day the position of American troops in Korea grew worse. In the United States, a second and more intense round of panic buying occurred. The Truman administration had to revise upward its estimates of military spending in the light of the probability of a much longer conflict. All economic indices showed that inflationary pressures were again increasing: the Wholesale Price Index rose almost four points in the one month from November 15 to December 15, 1950;[55] the Consumers' Price Index went up by more than two points in that same month;[56] wage adjustments in the last quarter of 1950 averaged 10 percent.[57] In view of these developments, President Truman decided that more vigorous measures were needed to halt inflation.

The first indication that a change in the government's economic policy would be forthcoming was the appointment, in early December, of Michael V. DiSalle as director of price stabilization and the establishment under him of an Office of Price Stabilization (OPS). An energetic administrator with much experience in government, DiSalle immediately turned his attention to the staffing of a price stabilization organization fully capable of planning for and enforcing price controls.[58]

Then, in a speech to the nation on December 15, 1950, explaining the need for a proclamation of national emergency, President Truman revealed the particular steps the administration would take to keep the economy functioning efficiently. Still adhering to the basic Keyserling philosophy, Truman emphasized the importance of expanding production: "We must produce more—more steel, more copper, more aluminum, more electric power, more food, more cotton, more of many other things." But, he continued, an increased defense effort would inevitably push prices upward unless the administration took action to prevent this. Therefore, the government immediately would impose price controls on a variety of materials and products critical to defense production and the cost of living. As required by law, the President observed, the government also would stabilize wages in those fields where price controls were instituted.[59] On December 22, passenger automobiles became the first target of the government's new selective controls policy.[60]

The President also announced in his December 15 speech a modification in the structure of the executive agencies concerned with mobilization and stabilization. Citing the necessity for more centralized control, Truman informed the nation that he was creating an Office of Defense Mobilization (ODM), which would be in charge of all mobilization and stabilization activities and agencies. The administrator of ESA would now be responsible to the director of ODM, instead of having direct access to the President. To the sensitive post of director of defense mobilization, the President appointed Charles E. Wilson, president of the General Electric Company, a Republican, and a nationally respected business executive. Originally, Truman had thought of the director of defense mobilization as a "czar" in the mobilization area only. But Wilson would not accept the appointment unless a substantial amount of power was delegated to him in the stabilization field also.[61] Wilson was the recipient, according to members of the executive office, of "the broadest grant of power ever held by anyone other than the President himself." [62]

The choice of Wilson, the administration thought, carried several advantages. An executive of demonstrated competence, with experience as vice-chairman of the War Production Board during World War II,[63] Wilson could—it was hoped—more easily gain the cooperation of industrialists. Moreover, the presence of Wilson at the head of the government's mobilization and stabilization establishment would inspire bipartisan congressional confidence. Congress admired Wilson's abilities, and it soon became apparent that Congress wanted him to use all the powers that had been delegated to him.[64] But Wilson also brought liabilities. Union leaders viewed him as very unsympathetic to organized labor; their hostility to Wilson led to many difficulties. Furthermore, Wilson had the reputation, stemming from his World War II government post, of being self-confident to the point of arrogance, a trait that complicated his relationships with his colleagues, both inferior and superior, in the government.[65]

Though Wilson's appointment signaled a change in the administration's policy from voluntary to selective mandatory controls, the executive branch still debated whether or not to impose across-the-board controls immediately. The wage-price spiral was continuing its upward rush, as businessmen and labor unions hurried to raise their prices and wages before the controls that most believed were inevitable were in fact imposed.[66] In the two months after the Chinese intervention, the Consumers' Price Index rose more than 5 percent.[67] The need for firm action seemed imperative. Nevertheless, Alan Valentine, the ESA administrator, opposed overall controls. Price Stabilization Director DiSalle, however, strongly advocated a prompt general freeze, and he persuaded Wilson to join forces with him. Valentine, realizing he would be overruled, resigned as ESA administrator on January 18, 1951. An aide to Wilson, Eric Johnston, head of the Motion Picture Association of America and former president of the Chamber of Commerce, was quickly appointed to fill the post.[68]

Wilson, Johnston, and DiSalle convinced President Truman that a freeze on wages and prices was necessary. On January 26, 1951, therefore, the administration abandoned the selective controls phase of its economic policy and began the period of across-the-board mandatory controls with a declaration of a general freeze.[69]

The Truman administration, however, was sensitive to the inequities that a total freeze perpetuated. Upon institution of the freeze, the stabilization agencies immediately set to work to rectify injustices to those businessmen and laborers who had been hurt unfairly by the freeze order. The OPS developed the Industry Earnings Standard as a yardstick to measure the need for a price increase: if an industry's earnings fell below 85 percent of the average of its three best years in the four-year period 1946–1949, price relief would be warranted. The Wage Stabilization Board, in Wage Regulation 6, decided to allow an increase of up to 10 percent in wages over those paid on January 15, 1950, to those workers who had not yet negotiated contracts reflecting Korean War economic conditions. This would enable them to catch up with workers who had received significant raises in the latter part of 1950. But the WSB's decision was not a unanimous one: the public and industry members, who made up the majority, favored this increase; the labor members dissented, because they thought the catchup formula gave workers too little.

To show how serious was their opposition to the board's policy, the labor members, on February 16, 1951, withdrew entirely from the WSB. Labor representatives in all other mobilization and stabilization agencies similarly resigned.[70] As the labor members explained afterward, their objection to Wage Regulation 6 was not the only reason for their walkout from the wage board and other defense agencies. Well before the imposition of a general freeze, the United Labor Policy Committee—which spoke for a broad spectrum of organized labor [71]—had warned the President and the public that "wage stabilization must not become wage freezing"; any wage policy "must permit raises to compensate for increases in the cost of living, correction of sub-standard wages, and adjustment of existing wage rates within or between industries." Furthermore, the government's wage policy "must embody the 'now well-recognized principle' that wage earners should share in the profits from industrial progress and increased productivity." [72] Believing that the administration's mobilization and stabilization program was controlled by businessmen, the United Labor Policy Committee demanded greater representation and leadership in the relevant agencies. It was the Committee's decision that triggered the mass exodus of labor representatives from government bodies. [73] Philip Murray, head of the CIO, president of the United Steelworkers of America, and leader of the CIO delegates to the United Labor Policy Committee, told a gathering of union people why labor had withdrawn: Labor was upset about the rising cost of living while wages were being frozen; labor wanted rent control to be extended; labor believed the administration's tax program hurt moderate-income families more than wealthy ones. But, most important of all, labor felt excluded from the decision-making hierarchy in the mobilization and stabilization program. As Murray observed, "The direction of this defense mobilization program . . . has slipped into the hands of a clique of men who represent only one attitude: the attitude of the top executive offices of Big Business." [74] Thus, barely three weeks after the imposition of mandatory controls, the administration faced a crisis in its wage stabilization program.

The demands of the United Labor Policy Committee highlighted a difficult problem for the administration's economic policy. Because of the ambiguous nature of the Korean conflict and the limited scope of both the national emergency and the mobilization, no segment of the public was willing to make the sacrifices required to halt inflation. Labor wanted compensation for every rise in the cost of living, as well as for increased productivity. Businessmen insisted that their profit margins be protected. And Congress, as will be seen, succumbed to the view that major sacrifices were not needed, thus weakening the controls program it had approved in 1950. [75]

As one disheartened stabilization official noted,

I feared then, and I know now, that it is next to impossible to make a success of anything such as a controls program at a time such as we attempted it. It was difficult enough to do even in World War II when the Japanese were threatening our West Coast and there were rumors of Nazi planes en route to bomb the East. In the fall and winter of 1950 and 1951, some of our boys were fighting in Korea, and certainly their families and friends were well aware of the seriousness of the situation, but still we were not in an all-out war. No state of unlimited emergency had been declared. There was no awareness of any immediate or great danger on the part of the people as a whole. The willingness to make great sacrifices just didn't exist.[76]

In the spring of 1951, the Truman administration had to devise a solution for the wage stabilization crisis that would enable the government to continue to prosecute the Korean War successfully. To do this, the administration needed the cooperation of industry and labor. But the nature of limited war left the government with few inducements or threats with which to gain the support of these interest groups. To elicit labor participation in the administration's mobilization and stabilization program without alienating business proved to be an exacting task.

two

TRUMAN, LABOR, AND THE POLITICS OF LIMITED WAR

President Truman by 1951 had amassed considerable experience in dealing with the conflicting objectives of labor and management during a period of inflation. While he consistently maintained that the public interest was paramount, Truman had rarely found the key to making the unions and industry see things his way and frequently had been forced to compromise his economic policy in deference to the demands of these pressure groups. With the United States at war, however, the President was more determined than ever that his economic policy must not be destroyed.

In seeking labor cooperation, Truman could draw on a long and close political alliance with organized labor. The contours of his relationship with the labor movement emerged during the first years of his career. He supported the right of workers to join unions and the unions' right to government protection of their existence. He sympathized with labor's economic and social aspirations. But Truman never hesitated to put the public welfare—as he saw it—first when the demands of unions clashed with it. In the Senate he voted for legislation favorable to labor but was always conscious of labor's obligations to the public as well. During Truman's 1940 Senate campaign, the support of unions was instrumental in his reelection. This marked the beginning of a firm association between Truman and the labor movement.[1]

The relationship was strained, however, when Truman, as President, had to deal with major strikes during the reconversion period. The demands of various unions during 1946 were not consonant with the general welfare as the Truman administration saw it. Union leaders, restless after years of denying the wishes of rank and file members in order to cooperate with the government's war production and stabilization program, had to assert themselves with a view to proving their effectiveness. Anticipating a shorter work week and, therefore, less take-home pay, labor leaders had wanted a general wage rate increase immediately after the end of World War II. The administration, realizing that inflation still would be a major problem, proposed gradual wage and price adjustments and only where necessary to encourage production of items in scant supply in the civilian economy.[2] Thus the President had to resist rewarding the most important interest group in the Democratic coalition in order to maintain the goals of his economic policy. But worse than just the denial of labor's demands, according to union leaders, were the tactics the President used to bring labor into line. Truman's threat to draft strikers during the railroad strike of 1946 shocked the labor movement, and many union leaders declared that they would seek Truman's defeat in 1948.[3]

However, when the Republican 80th Congress passed the Taft-Hartley Act and then overrode the President's veto of it, these same labor leaders rapidly returned to the Democratic fold. Described by its proponents as an attempt to equalize the positions of labor and management vis-à-vis the government, the Taft-Hartley Act (Labor Management Relations Act of 1947)[4] went too far in one direction, the unions thought, by making basic changes in the national labor law that put unions and the process of collective bargaining in a "strait-jacket."[5] The five titles of the Taft-Hartley Act covered the National Labor Relations Board (NLRB), rights of employees, unfair labor practices, representatives and elections, prevention of unfair labor practices, conciliation of labor disputes in industries affecting commerce, national emergencies, suits by and against labor organizations, restrictions on payments to employee representatives, boycotts and other unlawful combinations, restriction on political contributions, strikes by government employees, and creation of a joint committee to study and report on basic problems affecting friendly labor relations and productivity. Particularly reprehensible, according to the unions, were the provisions of the act that made possible increased use of the injunction.[6] The rash of postwar strikes had convinced Congress and the public that unions had become too powerful. Even the administration believed that some labor legislation was required.[7] But President Truman, sympathetic to the unions' view, thought that the Taft-Hartley Act was too punitive and he determined to make political capital out of his veto of it.[8]

Truman's veto message was staunchly prolabor. It showed a keen awareness of the rights of unions, which the labor movement appreciated. The Taft-Hartley Act thus was the catalyst that propelled the whole labor movement, the AFL as well as the CIO, into partisan political activity.[9] In the 1948 campaign, President Truman made effective use of his Taft-Hartley veto to win labor's support for his candidacy: his Labor Day speech in Detroit was a rousing repudiation of the actions of the Republican 80th Congress.[10] Workers responded by voting for Truman in huge numbers. One study of the election notes that "Although support from labor was not enough to produce Truman victories in the industrial states of Michigan, Indiana, Pennsylvania, New York, New Jersey, and Connecticut, it was a major factor in his victories in Rhode Island, Illinois, Ohio, and West Virginia." [11] Many groups contributed to the Democratic majority, but union leaders and the President believed that labor was responsible for his triumph.[12] From 1948 on, the alliance between organized labor and the Truman administration was firmly fixed in the public mind.

Helping to maintain the generally good relationship between the Truman administration and the labor movement was the White House staff, which was keenly aware of the importance of retaining labor's goodwill. Toward this end, members of the staff made special efforts to stay in touch with union officials and to be well informed of their desires.[13] Directing the staff in these activities were John R. Steelman and Charles S. Murphy.

Steelman, as The Assistant to the President, was the most influential person in regard to labor matters on the White House staff. In 1945, President Truman had asked Steelman to become his special consultant on labor-management problems. Labor unrest was growing, and Truman's Secretary of Labor, Lewis Schwellenbach, had no experience in the labor relations field. Steelman, who had been head of the Federal Conciliation Service for many years, seemed eminently suitable for this position. He had been a successful mediator because he had the ability to get along with labor as well as with management and was impartial in his dealings with them.[14] A prominent journalist described Steelman as "a horse of a man weighing well over two hundred pounds and with energy to match, [who] had the scrubbed, bright-eyed appearance and the brisk geniality of an Eagle Scout. He knew the leaders of the labor movement and the devious maneuverings of the bargaining table in a way that Schwellenbach could never match."[15] Steelman soon won the President's confidence and was appointed The Assistant to the President. Although he took on many more duties with his new title, Steelman continued to be in charge of labor-management affairs for the administration.[16]

Charles S. Murphy, a lawyer from North Carolina, became special counsel to the President when Clark Clifford resigned the position in 1950. Before coming to the White House he had worked for many years on Capitol Hill, where he became an expert in drafting legislation. As special counsel, Murphy was responsible for policy planning and legislation,[17] and in this capacity he frequently became involved in interpreting the powers of the President. Another member of the White House staff noted that "on a number of occasions—particularly concerning controversial legislation and executive orders—Murphy's views were, in fact, decisive."[18] A steady, "genial" lawyer, Murphy is credited with making the White House staff function smoothly, with little of the backbiting common among those who work close to the source of great power. "Everything began to mesh beautifully," Margaret Truman has commented, "from the day that Charlie arrived on the scene. . . . He shared Dad's calm, persistent approach to problems, and his awareness that they could not be solved by a slogan. . . . Charlie, who is a very modest man, insists that the President deserves most of the credit for the way the White House was run." [19]

Murphy and Steelman were theoretically at the head of two separate teams with distinct duties, and each had a staff suited to those responsibilities. The key men on Steelman's side were David Stowe and Harold Enarson, both well versed in industrial relations matters. Assisting Murphy were David Lloyd, David Bell, and Richard Neustadt, all better prepared for the legislative and speech-writing functions assigned to Murphy. In practice, however, the division of labor was not so rigid; both Steelman and Murphy participated in most major White House projects. [20] When the union representatives resigned from the Wage Stabilization Board and other government mobilization agencies, the energies of most of these staff members were directed toward finding a solution to the crisis.

At the time of organized labor's walkout, the unions suffered from a sense of being shabbily treated by the administration. In a statement issued after the mass resignations, the United Labor Policy Committee announced "that big business was dominating the [mobilization] program, that the interests of the plain people of this country were being ignored," that Wage Regulation 6 "denies justice and fair play to every American who works for wages," and that "there is absolutely no desire on the part of Mobilization Director Charles E. Wilson to give labor a real voice in the formulation of defense policy." [21] Labor publications echoed the same themes. *Steel Labor,* for example, noted that the price freeze had been announced "at a moment when prices were higher than ever before in history. And, although 'frozen,' prices are rising steadily." Profits were being protected and "flexibility, which has been denied in the wage stabilization order, is being guaranteed by the price stabilization orders." [22] Otis Brubaker, director of the research department of the United Steelworkers of America, argued that labor's position deserved sympathy. Unions had agreed that price and wage controls were necessary during the Korean mobilization and they had offered to cooperate with the government "if the conditions were created which would permit some kind of equality of sharing in the price which we are forced to pay for the kind of program of mobilization on which we are embarking." These conditions included effective price controls, a reasonable tax program, and a sound rent control law.[23] But, as the stabilization program stood in February 1951, none of them was present.[24]

Furthermore, labor, as Brubaker indicated, was advancing an additional requirement as the price for its willingness to participate in the government's program: the creation of a board that would handle the settlement of disputes.[25] The desire for a disputes settling agency was not new. The United Labor Policy Committee had from the first favored the establishment of such a board. But the committee did not make an issue of this until labor had withdrawn from the stabilization agencies. In March 1951, the committee urged that the Wage Stabilization Board be reconstituted, with authority over wage stabilization regulations and the settlement of labor disputes.[26]

From the standpoint of organized labor, the need for such a board seemed imperative. Union leaders understood the administration's desire to have continuous production during the Korean conflict, but the emergency did not seem serious enough to warrant a "no strike" pledge. The labor dispute provisions of the Taft-Hartley Act merely allowed for a delay of a strike while negotiations continued; the act did not furnish a method for the actual settlement of disputes.[27] Moreover, because of its use of the injunction, this section of the act was anathema to the labor movement. A tripartite wage stabilization board with the power to suggest settlement terms, on the other hand, appeared to be a sensible means of dealing with labor disputes during the Korean hostilities. Such a board would consider a dispute in the context of the wage stabilization program with which it was thoroughly familiar; thus its recommendations would be realistic. Furthermore, the settlement it suggested would be formulated by all concerned parties—labor, management, and government. Consequently, its recommendations probably would be acceptable to the disputants.

The business community opposed the creation of a labor

disputes board.[28] In a letter to the economic stabilization administrator, the Joint Committee for the Business Advisory Council, the
National Association of Manufacturers, and the Chamber of Commerce of the United States, indicated that its opposition was
founded on several grounds: Existing legal machinery was sufficient to deal with labor disputes; normal collective bargaining
relationships would be disrupted by government intervention; decisions of such a board would approach compulsory arbitration, for,
although they might not be enforceable in the courts, they could
be effected through presidential sanctions; and the Taft-Hartley
Act's emergency disputes provisions would be circumvented.[29]
The Joint Committee proposed instead that a reconstituted wage
stabilization board "might hear interested parties on matters growing out of or directly related to the stabilization program and involving economic or monetary issues only in order to interpret and
rule on the application of existing policies, to change existing
policies, or adopt new policies." The board would not be permitted to consider any dispute in which a strike was already in progress, and it would not be allowed to make special exceptions to
previously adopted policies.[30] The United Labor Policy Committee rejected this proposal.

To find a way out of this impasse, the President, on March
15, appointed a National Advisory Board on Mobilization Policy
consisting of sixteen members drawn from various backgrounds. Its
first assignment was to make recommendations to the President on
the reestablishment of a wage stabilization board. Those members
of the advisory board with business experience immediately announced that they were participating on the board as individuals
and that their actions should in no way be interpreted as committing specific business organizations to whatever policies the board
might suggest to the President. The National Advisory Board
meetings were not, they asserted, to be considered a labor-
management conference.[31] On April 17, the National Advisory
Board, by a vote of 12 to 4 (those with business backgrounds dissenting), recommended that the President set up an enlarged wage
stabilization board with authority to handle labor disputes.

Accordingly, on April 21, 1951, President Truman, in Executive Order 10233,[32] reconstituted the Wage Stabilization Board, giving it a new structure and new functions. He increased the number of members from nine to eighteen: six representing labor, six representing business, and six representing the public. To facilitate the board's wage stabilization function, and to ease the threat of strikes, the board was given the power, where labor disputes threatened the national defense, to make recommendations on economic and noneconomic issues. Unless they had voluntarily agreed in advance to comply with the WSB's decision, the parties to the dispute would not be bound by the board's recommendations. On disputed matters the board was ordered to report directly to the President. Thus, while on wage stabilization policies the WSB had to submit its regulations to the economic stabilization administrator who in turn checked with the director of defense mobilization, who ultimately was responsible to the President, on disputes issues the board was virtually autonomous.[33] Although these arrangements met with the approval of the labor movement, which was delighted that Defense Mobilizer Wilson would be circumvented in the new disputes procedure, industry representatives were not enthusiastic, and the administration had to use no small amount of persuasion to convince businessmen to serve on the new board.[34]

The Wage Stabilization Board resumed functioning on May 8, 1951, with George W. Taylor as its chairman. Taylor, who had accepted the post on condition that he be permitted to return to the University of Pennsylvania in September, was succeeded on August 30, 1951, by Nathan Feinsinger who continued as WSB chairman through the time of the steel dispute. A law professor with many years of industrial relations experience, Feinsinger had been used as a mediator by the Roosevelt and Truman administrations, frequently with great success.[35] Appointed as a public member of the newly reconstituted WSB, he quickly became its vice-chairman and then its chairman. The new wage board was immediately faced with a huge backlog of wage policy questions because for three months there had been no agency to handle wage stabilization problems. There were no disputes for it to consider, however.[36]

Despite the fact that no disputes had come before it, the authority of the WSB to deal with labor-management controversies became a source of contention. Section 502 of the Defense Production Act authorized the President to call a conference at which representatives of labor, management, and the government would decide the best methods for settling labor disputes; the President would then appoint the persons and agencies necessary to carry out this policy. But President Truman, believing that a business-labor consensus could not be reached, never called such a conference. Hence, in Executive Order 10233, the President based his authority to set up the new WSB on the general powers given to him "by the Constitution and statutes, including the Defense Production Act of 1950"; [37] he did not mention specifically Section 502. A member of the White House staff elaborated on the President's position: "Legally, the disputes role of WSB was in the tradition of the Steel Industry Board of 1949 [38] and similar fact-finding and recommendatory bodies going back to the turn of the century." [39]

Section 503 of the Defense Production Act, which stipulated that no action inconsistent with the Taft-Hartley Act of 1947 could be taken, further confused the issue. How the emergency disputes provisions of the Taft-Hartley Act could be reconciled with the Defense Production Act's contemplation of the new procedures remained obscure.

Congressional committees, in an attempt to unravel the mystery, studied the disputes functions of the new WSB. The report of the Senate Committee on Labor and Public Welfare did little to clarify the inherent contradiction between Section 502 of the Defense Production Act and the Taft-Hartley law. The disputes settlement procedure suggested in Section 502 (i.e., a board modeled on the War Labor Disputes Board of World War II) was the antithesis of the procedures stipulated in the Taft-Hartley Act. Taft-Hartley relied on collective bargaining to achieve a settlement; Section 502 contemplated compulsory arbitration. Ignoring this, the committee's report simply concluded that Section 502 was not being used by the President inasmuch as the President had not followed the procedure outlined in that section, nor had he set up machinery for the *compulsory* settlement of labor disputes as intended by Section 502. Furthermore, the committee found that the new Wage Stabilization Board, concerned only with the *voluntary* settlement of labor disputes, had been created by a valid exercise of the President's constitutional authority.[40] The new board was solely an additional tool to be used by the President to prevent strikes during the period of mobilization. The committee recognized that the same dispute might be ripe for action under both the Taft-Hartley Act and Executive Order 10233. Its report concluded that there was no such situation in sight, but that, should it arise, it would be the President's responsibility to decide which route would dispose of the dispute most effectively—or he might use both routes, depending upon the facts in the case. In other words, the disputes-handling functions of the Wage Stabilization Board were "an addition to, rather than a substitute for, the emergency disputes provisions of the Taft-Hartley law." [41]

In the summer of 1951, during the debate on the renewal of the Defense Production Act of 1950 (which terminated on June 30, 1951), Congress defeated an attempt to strip the WSB of its power to make recommendations in labor disputes. Representative Wingate Lucas, a Democrat from Texas, offered an amendment that would have removed disputes from the jurisdiction of the WSB. He maintained that the wage board as established by Executive Order 10233 clearly was in violation of the intent of Congress in Title V and that its power to investigate labor disputes, propose settlements, and render arbitration awards, cut across the field of many agencies created by legislation, including the National Labor Relations Board, the Federal Mediation and Conciliation Service, the National Mediation Board, and the special boards of inquiry provided for under Title II of the Labor-Management Relations (Taft-Hartley) Act.[42] Opponents of the amendment emphasized the need for a special agency to deal with labor disputes during the mobilization period and claimed that the amendment was aimed at reducing the constitutional power of the President.[43]

The amendment was rejected 217 to 113.[44] Truman considered this vote of confidence in the work of the WSB.[45] Subsequently, at the time of the steel dispute, the President would act in the belief that Congress had approved the WSB procedure as an alternate to the Taft-Hartley process and had deliberately left the choice of remedies to the President.[46]

While the vote on the WSB was a victory for the administration, other amendments to the renewed Defense Production Act of 1950 were clearly defeats. Chief among these was the Capehart amendment to the price and wage stabilization provisions of the Act.[47]

In the spring of 1951 the economy was experiencing a respite from inflation. Administration economists believed that this was a temporary condition attributable to the abatement of scare buying. Prices, which in the first six months of the Korean War had risen much faster than the impact of the defense program appeared to warrant, now were finding their natural levels. But, these economists warned, greater military expenditures planned by the administration would significantly intensify inflationary pressures. The expansion in defense production would be accompanied by the growth of individual incomes, and the increase in the manufacture of consumer goods would not be sufficient to absorb the additional income. Consequently, prices would tend to rise.[48]

Guided by this prognosis, and hurt by criticism of its slowness in imposing mandatory controls,[49] the administration decided to ask Congress to strengthen the anti-inflationary measures contained in the Defense Production Act.[50] But Congress was no longer receptive to such a plea. The reasons for this change in the attitude of Congress are not readily apparent. Indeed, polls taken in April, 1951, revealed that the public was still very concerned about inflation. It was dissatisfied with the government's stabilization program and expected that prices would rise in the next six months.[51] But there is no evidence that this public feeling was effectively mobilized by the administration. Without the counterpressure of a great public outcry, Congress, secure in the knowledge that prices were not increasing at the moment, yielded to the lobbying efforts of the various business groups that wanted controls eased. Moreover, one may speculate that the deep antipathy between the conservative Republican/Southern Democratic congressional coalition and President Truman, submerged after the outbreak of hostilities in Korea, reasserted itself and was reflected in Congress' unwillingness to cooperate with Truman. Thus, in the 1951 battle over inflation controls, the President and Congress reversed positions: the President was now the champion of a stringent anti-inflation program; Congress did all it could to reduce the effectiveness of the measures it had enacted into law in the Defense Production Act of 1950. The Capehart amendment was the major means by which the renewed Defense Production Act was weakened.

The Capehart amendment authorized price increases to compensate companies for all additional costs incurred between June 1950 and July 26, 1951. During the hearings on the extension of the Defense Production Act of 1950 businessmen from a great variety of enterprises testified that in principle they favored price and wage controls but urged a special exception for their particular businesses, which were unfairly hurt by such controls. Each house of Congress responded by passing an amendment that in some way safeguarded profits. The language of the Capehart amendment came out of the conference committee, which advocated its passage by stating that it "will protect the fair and reasonable profit of those who have merely added to their prewar prices the necessary and unavoidable costs of doing business which they have since incurred." [52] The Senate passed the conference bill with little debate and no recorded vote; the House endorsed the conference bill 294 to 80.[53]

President Truman, incensed by the action of Congress, announced when he signed the bill, that he was totally displeased with the inflation control measures of the Defense Production Act Amendments of 1951. He had to sign the bill, however, to prevent the expiration of other provisions of the act, which were necessary for continuation of the defense program. But since he viewed the inflation controls as entirely inadequate he intended to ask

Congress to amend the act again.[54] Specifically attacking the Capehart amendment, the President declared, "This complicated amendment will force price ceilings up on thousands of commodities, clear across the board. It is like a bulldozer, crashing aimlessly through existing pricing formulas, leaving havoc in its wake." [55] Several weeks later, in a message to Congress recommending repeal of the Capehart amendment, Truman, despite his previous efforts to win the support of business for his mobilization and stabilization programs, placed himself squarely on the side of the common man and showed no sympathy for the complaints of businessmen:

All along the line, under the Capehart amendment, business is protected. Business is told that it need not absorb rising costs. But no such assurance is extended to the consumer, the wage earner, and the people living on pensions and other fixed incomes. They stand at the end of the line, and the effect of the Capehart amendment is to take all rising costs—the cost of materials, of labor, overhead, advertising, corporate salaries—everything—out of their pockets.[56]

The administration, however, was unsuccessful in its attempt to have the Capehart amendment repealed or even modified. The price lull still continued and Congress had no interest in tough controls. As Senator Irving Ives remarked, "Politics and controls do not mix well—this the Administration has taken almost 1 year to discover." [57]

Because the Capehart amendment made the task of price control infinitely more difficult, many stabilization officials, exhausted by the job of trying to enforce the administration's policy with little support from Congress, resigned.[58] Foremost among those who left the government were Eric Johnston, who resigned on November 30, 1951, after less than a year as economic stabilization administrator, and Michael V. DiSalle, director of price stabilization, who departed on February 15, 1952.

Johnston was replaced by Roger L. Putnam, a Democratic politician and industrialist from Springfield, Massachusetts, who had had a successful career in both fields. The offspring of two prominent Massachusetts families, the Putnams and the Lowells, he was the recipient of a typical upper-class education: preparatory school, Harvard College (from which he graduated in 1915 magna cum laude), and a postgraduate year at the Massachusetts Institute of Technology studying mechanical engineering. Putnam joined the navy during World War I. When discharged he started working as a salesman for the Package Machinery Company of which his father was president. Eight years later he became the company's president. The Package Machinery Company was one of the first (in 1926) to offer its employees profit-sharing and life insurance plans. Throughout his business career, Putnam continued to show an unusual concern for the working class. During the depression, the Package Machinery Company switched to a five-day week in order to keep employment high, and Putnam served on the Massachusetts commission that formulated that state's unemployment laws. In 1932, Putnam registered as a Democrat, and he subsequently characterized himself as a "consistent, liberal Democrat." [59]

A stocky man with gray eyes and brown hair, Putnam was popular with the voters of Springfield, Massachusetts, who elected him mayor three consecutive times between 1937 and 1941. In 1942, however, Putnam lost the election for governor of Massachusetts to the incumbent Republican, Leverett Saltonstall. Subsequently, Putnam served in the navy and in various government agencies in Washington. [60] When he accepted the position of economic stabilization administrator, Putnam did not expect to remain at that post for long, recognizing the challenge that faced him. "Everybody may deplore inflation, but everyone is very happy with a little of it happening in his behalf," he observed soon after assuming the post. "The job of this office is to sit on the lid and stop people from getting all they want." [61]

To succeed DiSalle, President Truman appointed Ellis G. Arnall, a former governor of Georgia (1943–1947) who, in 1952, was president of the Dixie Insurance Company and the Society of Motion Picture Producers. [62] Despite an unprepossessing appearance (he was short, heavy-set and balding), Arnall had achieved spectacular political success, winning, at the age of twenty-six, his first elective office, a seat in the Georgia House of Representatives, where he was immediately elected speaker pro tem. Exactly ten years later, after serving as assistant attorney general and attorney general of Georgia, Arnall fought the Democratic machine of Governor Eugene Talmadge and won the primary by a decisive 45,158 votes to become the youngest governor in the United States.[63] Campaigning on a liberal reform platform, Arnall sought to turn Georgia into a modern, economically prosperous, and politically democratic state.[64] During his term of office, Georgia adopted a new constitution, the voting age was lowered to eighteen and a new act was passed making it easier for Georgia's citizens in the armed services to vote, the legislature repealed the poll tax, the penal system was updated, the state education system was made independent of political influence,[65] and measures were passed to create a more hospitable atmosphere in the state for industry. Believing that economic opportunity was the key to prosperity for whites and blacks alike, Arnall professed a strong distaste for monopolistic corporations and a positive admiration for unions, which he believed had greatly improved the condition of the working class.[66]

When called upon to become director of price stabilization in 1952, Arnall announced that price controls were in the interest of the nation and that he hoped to administer them effectively, energetically, and fairly. Thus, as the steel dispute began to take shape, two of the four highest stabilization officials were relatively new to their jobs and approached the task of administering economic controls with a firm commitment toward holding inflation down.

The refusal of Congress to cooperate with the President in strengthening the anti-inflation measures of the Defense Production Act was indicative of a mood that had been developing in Congress and the country since the Chinese Communists intervened in Korea. Initially, congressional and public opinion fully supported President Truman's decision to fight in Korea. Joseph C. Harsch, in an article in the *Christian Science Monitor*, declared that in the twenty years in which he had lived and worked in and out of Washington, never before had he felt such a sense of relief and unity as passed through the city when the administration announced its intentions. While the decision to act had already been taken, almost everyone assumed that there would be no action—that the administration would "miss the boat and do something idle or specious." When action came, "there was a sense first of astonishment and then of relief. Mr. Truman obviously did much more than he was expected to do, and almost exactly what most individuals seemed to wish he would do. I have never seen such a large part of Washington so nearly satisfied with a decision of the government." [67] The Senate minority leader, Kenneth Wherry of Nebraska, was one of many congressmen who applauded Truman's action: "A sigh of relief has swept across the country that, at last, at long last, the President has accepted the suggestion of some of us that he draw a line, that he stop his vacillation in the Pacific. . . . At last, at long last, he says to the Communist hordes, 'thus far and no further.' " [68]

A public opinion survey taken in July 1950 revealed that 81 percent of the people polled approved the sending of troops to Korea. [69] Moreover, for the first time since the beginning of his second term, President Truman's popularity increased. It rose from 37 percent before the conflict erupted in Korea to 46 percent after American troops were sent to repulse the invasion. [70] And, like Congress, the public, after experiencing a few weeks of inflation, favored the imposition of price controls. A slight majority even thought that all prices and wages should be frozen at their levels of June 25. [71]

Public opinion changed quickly, however, after the Chinese Communists entered the war. The public's faith in the Truman administration was undermined with every military defeat and with each rise in prices. By early 1951 the public was disillusioned with the administration's Korean policy and with inflation, which the administration seemed slow to check. [72] One-half of the people polled in March 1951 thought the United States had made a mistake by fighting in Korea and a large majority advocated pulling American troops out of the war. [73] "Despite considerable effort," a student of the period remarked, "the administration had not convinced the public of Korea's strategic importance. It had created a crisis of ambivalence between support for American troops and doubts about the war, between a militant anti-communism and a sense of caution based upon a fear of escalation." [74]

Truman's personality may have contributed to his inability to sustain public opinion in his favor. Commentators noted his "humility, unassuming manners, and down-to-earth qualities," his "humanity, determination, and fighting spirit." But they saw unattractive traits too: "At times the President seemed impetuous, ill-informed, cocky, and quarrelsome." [75] Moreover, Truman always suffered from unfavorable comparison with Roosevelt. Lacking Roosevelt's breadth of vision and his charisma, Truman, as a war leader, did not often inspire the public.

It may have been impossible, however, given the nature of limited war, to avoid an equivocal reaction on the part of the public. In a total war, as in World War II, the enemy is identified, the goal is to defeat him, and all the resources of the country are employed to that end. A war psychology is easily created. But a limited war is very different: force must be used in a restricted manner to obtain a specific objective which may be far short of "victory." In the Korean conflict, the administration, while trying to build public understanding and support for its aims, also took care not to arouse a war psychology that might be interpreted as a prelude to World War III. It required a higher degree of sophistication than the American public possessed to sustain enthusiasm for an administration that walked a seemingly endless tightrope.

The military stalemate after the Chinese Communist intervention caused the administration to revise its strategy to bring the war to a conclusion. It became apparent that a termination of hostilities could not rapidly be achieved by force; the Truman administration therefore sought to end the war by negotiation. Truce talks began on July 10, 1951. [76]

But armistice discussions brought no cease-fire, and, while the negotiators haggled endlessly over conflicting demands, American troops continued to die on the battlefield in apparently pointless fights for numbered hills. Americans became increasingly frustrated, as time passed, by the events in Korea. They realized that there would be no victory, that the United States would have to settle for the status quo ante bellum. Moreover, in Korea the United States was not even fighting its most feared enemy, the Soviet Union. The American public believed that Moscow was behind every move the Communists made in Korea, yet while thousands of American lives were lost the Russians remained unscathed.[77] Influenced by the frequent Republican attacks on the Truman administration's policies, Americans began to think of the Korean War as unnecessary, and became incensed that men were dying in such a useless cause. Truman's popularity—at a peak when the United States entered the war—steadily declined. It reached its lowest point in November 1951, when the Communists broke off truce talks, as they had done once before. A meager 23 percent of the electorate approved of Truman's performance in office.[78]

Despite the fact that negotiations were resumed, a particularly trying period for the administration ensued. The public fretted because only one issue—the forced repatriation of Communist prisoners held by the UN forces—seemed to be keeping the UN and the Communists from agreement. This issue remained a sticking point to the end of Truman's presidency. General Matthew Ridgway urged Americans to be resolute while his staff dealt with the Communists: "Patience is a necessity. The stakes in this case are not only the lives of allied prisoners in communist hands, but world peace itself." [79] Patience wore thin, however, in the face of mounting casualties and fruitless negotiations; people were unwilling to make the sacrifices asked of them by an administration they considered inept.

The low political standing of the Truman administration at the time of the steel seizure, caused by the difficulties with the Korean War and impatience with economic controls, was compounded by accusations that the government was riddled with communism and corruption. Senator Joseph McCarthy's charges of Communist infiltration of the government made it difficult for the Truman administration to govern effectively; the civil service was shaken by charges of disloyalty, and public trust in the administration was sapped.[80]

Repeated disclosures of corruption in the government also embarrassed the Truman administration. Though people did not accuse President Truman himself of dishonesty, it was thought that he was aware of what was going on and yet did little to correct the situation.[81] When various charges of improper and illegal activities were made against some of his closest associates (Harry Vaughn, Matthew Connelly, Wallace Graham) the President accepted their explanations at face value and none was asked to resign. And he was slow in reacting to revelations of scandalous behavior on the part of some officials in the Bureau of Internal Revenue and the Department of Justice. In November 1951, Truman finally fired T. Lamar Caudle, assistant attorney general in charge of the tax division of the Department of Justice. But it was January 1952 before the President ordered a general investigation of corruption in the government, and then he chose Attorney General J. Howard McGrath, himself under a cloud of suspicion, to head the inquiry.[82] The House of Representatives, disgusted with the President's action, designated a subcommittee of its Committee on the Judiciary to undertake its own investigation of corruption in the Department of Justice.[83] McGrath then appointed Newbold Morris, a Republican, to be special assistant to the Attorney General in his corruption probe. Weeks later Attorney General McGrath fired Morris because of a disagreement over whether McGrath himself was required to submit to Morris' investigation. Truman then asked for and received McGrath's resignation [84]—just a few days before the President seized the steel mills.

Truman immediately nominated a new Attorney General, U.S. District Judge James McGranery. The Senate, however, wished to make sure that there was nothing unsavory in McGranery's background and took the time for a thorough review of his record and beliefs. McGranery was not confirmed until May 20, 1952.[85] Thus, the Department of Justice had no Attorney General to give it leadership and direction during the entire course of the *Steel Seizure* case. According to Richard Neustadt, a White House staff member, the absence of an Attorney General probably made little difference in the quality of advice the President received from the Justice Department. During the Truman administration that department "tended to be evasive, sometimes downright unresponsive, in providing the Executive Office with forthright legal guidance on legislative or operational issues," Neustadt noted.[86]

The "mess" in Washington reflected poorly on the Truman administration's image. As a result confidence in the administration's ability to deal with the serious problems of the Korean War and the economy was further undermined. One seasoned Washington reporter observed, on April 8, 1952: "Under a parliamentary system of government, the Truman Administration would at this point of near collapse go out of office on a vote of no confidence. That would in all probability have happened months ago as the inability of the Administration to carry out even rudimentary policy was demonstrated over and over again."[87]

Moreover, 1952 was an election year, and Congress, preoccupied with politics, did not wish to give the President many legislative victories. Despite the numerical majority of Democrats, the 82d Congress enacted few measures to implement the President's Fair Deal program.[88] Congress was deeply divided ideologically and regionally; the group of Senators and Representatives loyal to the President was not strong enough to withstand the frequent coalition of Republicans and Southern Democrats.[89]

With his political popularity at a low ebb, Truman announced, on March 29, 1952, that he would not run for reelection. His lame-duck status, however, did not deter Truman from using all the powers he could command to achieve a desired result. From his study of history and from his experience in both the legislative and executive branches of government, Truman had emerged a firm believer in the strong Presidency. He thought the explicit powers granted to the President in the Constitution were insufficient for a successful administration. History had changed the character of the Presidency; not only was the chief executive the head of state but he was a political leader as well. As Truman noted in a speech in 1954, "Out of the struggle and tumult of the political arena a new and different President emerged—the man who led a political party to victory and retained in his hands the power of party leadership. That is, he retained it, like the sword Excalibur, if he could wrest it from the scabbard and wield it." [90] The separation of powers stipulated by the Constitution made a certain amount of political conflict inevitable; consequently, in Truman's opinion, a strong President was even more of a necessity. As guardian of the whole people, Truman averred, the President must use all his powers to protect their welfare. "These powers," Truman commented,

which are not explicitly written into the Constitution are the powers which no President can pass on to his successor. They go only to him who can take and use them. However, it is these powers, quite as much as those enumerated in Article II of the Constitution, which make the presidential system unique. . . .

For it is through the use of these great powers that leadership arises, events are molded, and administrations take on their character. Their use can make a Jefferson or a Lincoln administration; their nonuse can make a Buchanan or a Grant administration.[91]

When the steel dispute arose in the last year of his Presidency, Truman would have the opportunity once again to translate theory into practice. But this time his actions would undergo the ultimate test, examination by the Supreme Court of the United States.

three

THE STEEL INDUSTRY AND THE UNITED STEELWORKERS OF AMERICA

From the first indications of conflict between the Steelworkers and the steel companies, President Truman perceived the urgency of bringing the 1952 contract negotiations to a successful conclusion. Without a constant supply of steel, the Korean War mobilization program would be severely handicapped. And, because steel held a central position in the economy, the terms of the settlement would provide a major test of the administration's stabilization policy. But the task of imposing a reasonable agreement and avoiding a strike was a formidable one, for the administration had to deal with two powerful and antagonistic parties—the steel industry and the United Steelworkers of America.

When the United States Steel Corporation was formed in 1901, the configuration of the modern steel industry was forecast. The merger between the Carnegie Steel Company and a number of smaller steel firms, engineered by the House of Morgan, created an immense corporation that controlled

213 different manufacturing plants and transportation companies; 41 mines; 1,000 miles of railroad; 112 ore vessels, one-third the tonnage of

the Great Lakes; 78 blast furnaces, more than one-third the total in the United States; the great coal and coke holdings of H. C. Frick, and the giant Rockefeller ore reserves in the Mesabi. It controlled 43.2 percent of the nation's pig iron capacity, and roughly 60 percent of the basic and finished steel production.[1]

Carnegie Steel had been the most successful company in the latter part of the nineteenth century, not only as a result of technical efficiency but also because it owned all of the means and materials necessary for each step in the manufacture of steel. Since integrated production involved a huge outlay of capital, most other steel companies manufactured finished goods and relied on Carnegie Steel for their supply of basic steel. The founders of U.S. Steel planned to follow and extend the Carnegie example: the parent corporation would possess all of the facilities necessary to engage in each stage of steel production. Increased efficiency, greater stability, lower prices, and larger profits would be the fruits of the new order to be imposed on the industry.[2]

Few steel plants of significant size and capacity remained outside the U.S. Steel orbit. Among these were Jones & Laughlin, Republic, Youngstown Sheet & Tube, and Bethlehem. They prospered too, however, as the demand for steel grew during the new century. Indeed, mergers and technical innovations enabled Bethlehem to become the second largest steel producer in the nation. Bethlehem's growth was guided by Charles Schwab, a Carnegie protégé who had played a prominent role in the formation of U.S. Steel but who left it in 1903 to take over the management of Bethlehem.[3]

As a result of the economic difficulties experienced by the steel companies during the period following the financial panic of 1907, a change occurred in industry practices. In the quest for business in a depressed economy, fierce competition, which included price-cutting, had not produced profits. Hence the steel companies decided to take a different tack: cooperation, not competition, became the dominant ethic. At the behest of the directors of U.S. Steel, industry executives agreed to maintain the same prices in the future and to consult one another freely on their common problems.[4] The pattern was set for operation of the steel industry in the twentieth century.

The largest companies, increasing their capacity mainly by mergers and improved technology, kept the greatest share of steel production in their own hands. In 1950 the U.S. Steel Corporation remained the leader of the industry, although its share of the market had decreased from 60 to 30 percent. Despite the fact that more than 250 firms were in the steel business, the four largest companies, U.S. Steel, Bethlehem, Republic, and Jones & Laughlin, supplied 60 percent of the total industry capacity, and the ten biggest furnished 80 percent. The entire industry was capable of producing more than 100.5 million tons of steel in 1950 and actually manufactured 96.8 tons.[5] To make this tremendous amount of steel, the industry employed almost 600,000 people.[6] The geographical center of the industry continued to be in Pennsylvania and Ohio, where 45 percent of the nation's steel was produced. Other states in which a substantial quantity of steel was manufactured included California, Indiana, Illinois, Maryland, Michigan, New York, Alabama, and West Virginia.[7]

In view of the strength of the United Steelworkers of America in 1952—it could shut down most of the industry with a strike—one tends to forget that the steel union was a relatively recent arrival on the industrial scene. A tradition of antiunionism, stemming from the individualism of the founders of the various steel companies, coupled with the entrepreneur's usual concern to keep his costs as low as possible, had made the steel industry particularly unreceptive to union organization. Not until the government provided support were any advances achieved by the union.

The history of labor relations in the steel industry is bitter and violent. Unionism first became a significant factor in the industry in the latter half of the nineteenth century. A number of smaller unions merged to form the Amalgamated Association of Iron, Steel, and Tin Workers of North America, a craft union whose membership reached a peak of 24,000 in 1891. At that time it was considered by many to be the most powerful labor organization in the country. The Amalgamated had organized about one quarter of the eligible iron- and steelworkers [8] and had been more successful in collective bargaining than any other union. In steel, however, the union made little headway in those plants that had started as steel mills; it was better organized in plants that had begun as iron producers and then added steel converters. As a result, only one of the three principal Carnegie plants built as steel mills was unionized. That plant was Homestead, located on the Monongahela River about ten miles east of Pittsburgh. [9]

The early steel industry operated in an extremely competitive atmosphere, which "produced the distinguishing quality of American steelmen: the economizing temper. The embodiment of this spirit was Andrew Carnegie." [10] Carnegie was preeminently cost conscious; in Henry Clay Frick he found a like-minded man with the ability to implement the Carnegie philosophy. Frick became chairman of the Carnegie business and manager of the Homestead plant in 1889. By 1892 he had concluded that the Homestead mills were not producing as efficiently as they might because of the terms of employment negotiated with the Amalgamated in 1889. Carnegie and Frick therefore decided that when the 1889 contract terminated on June 30, 1892, they would no longer deal with the union.

The Carnegie Steel Company believed it had the wherewithal to break the union permanently. Its nonunion plants would continue to function throughout the duration of a strike at Homestead; hence, the company would have a ready, and more than sufficient source of funds. Even more important, Frick was confident that he could replace striking union men with nonunion labor and that production could be resumed in a short time at Homestead. Mechanization of the steel manufacturing process had progressed to the point where the skilled craftsmen of the Amalgamated were no longer needed; untrained men could man the steel mills— which proved to be the undoing of the union. [11]

Events at Homestead foretold the demise of the Amalgamated. Long remembered for the bloody confrontation between striking workers and strikebreakers, Homestead was a defeat for the union that it vowed to avenge. The failure of the strike affected the course of unionism in the steel industry for many years thereafter. Thousands of members withdrew from the Amalgamated, and its treasury shrank. At the Carnegie Company, management undertook a zealous antiunion program: members of the Amalgamated were fired. The company discontinued publication of its wage scales, additional money for working on Sunday was no longer offered, and grievance procedures were terminated. When Carnegie Steel became part of the U.S. Steel Corporation, the corporation inherited a group of executives whose forceful antiunionism "had been hammered out on the anvil of the strike," and a mass of workers "who knew the meaning of industrial absolutism." [12]

In a last-ditch attempt to establish the steel union within the industry, the Amalgamated, in 1901, demanded that three subsidiaries of the newly formed United States Steel Corporation sign contracts with the union for all their plants regardless of whether they had been previously organized. When the companies refused, the Amalgamated called a strike. The advantage gained by taking the offensive before U.S. Steel's position was solidly entrenched was squandered by ill-advised tactics. [13] As a result of the Amalgamated's actions during this strike, J. P. Morgan, who initially did not have the antiunion bias prevalent among the Carnegie men in the corporation, turned wholly against the union and effectively ensured its demise. Two months later the union surrendered. It lost fourteen of the mills it had already organized and had to promise to make no further attempts actively to encourage unionization at U.S. Steel. [14]

Thus, from its very beginning, the United States Steel Corporation pursued an antiunion policy. And, because of U.S. Steel's predominant position in the industry, its competitors tended to follow the leader in labor relations as in price and wage matters. By use of blacklists, dismissals, discrimination, and espionage, the entire industry aimed to drive the union out of its plants and, aided by the ineptitude of the Amalgamated's leaders, was largely successful. [15] The Amalgamated continued to exist, but it was essentially helpless against the combined power of the steel companies.

Other factors worked to ensure the failure of the few feeble attempts at organization made by the union before World War I. The Amalgamated was an association of skilled craftsmen which, although it liberalized its constitution to allow unskilled laborers to join, never totally committed itself to organizing the mass of workers in the steel industry. Moreover, this mass of unskilled workers, mainly immigrants from southern and eastern Europe who were interested only in earning money and going back home, provided the mills with a constant source of labor. In a period of labor shortage, as during World War I, the worker has some bargaining power because his services are of value to his employer. But when a worker can be easily replaced, he cannot afford to take any action that might jeopardize his employment. In the pre–World War I period, with job security the foremost concern of the workers, an "each man for himself" attitude resulted. Consequently, under the guidance of industry leaders who totally opposed unions, the steel industry became a bastion of the open shop.[16]

America's entry into World War I altered the balance of power in the steel industry. At a time when the demand for steel was increasing sharply, the supply of immigrant workers was cut off and the draft was taking large numbers of workers out of the mills. The labor shortage made management more aware of the need to improve the lot of the workers in order to maintain their loyalty. In addition, the Wilson administration, anxious to maintain continued high steel production for the war, courted organized labor. Union men were given prominent positions in war agencies concerned with labor problems. And in March 1918, the War Labor Conference Board announced the principle that was to guide wartime labor relations: "the right of workmen to organize in trade-unions" without hindrance "in any manner whatsoever."

Though steel management was more aware of the need for contented employees, it was not amenable to bargaining with unions or to having employment terms dictated by the government. The efforts of the National War Labor Board (NWLB) to give practical effect to the principle enunciated by the War Labor Conference Board met with stiff resistance from the steel industry. Before an official decision in a wage dispute involving the Bethlehem Steel Company was handed down, government investigators privately advised Eugene Grace, the company's president, to talk with a committee of workers. Grace flatly refused: "We do not employ a committee, we employ a particular workman and naturally, we are always ready to listen to what he has to say and make corrections." [17] On July 31, 1918, the NWLB ordered Bethlehem Steel to adjust its minimum hourly wage rates in accordance with the scale used by the War and Navy Department and to compensate work performed beyond eight hours at time-and-a-half and work on Sundays and holidays at double time. The board also directed the company to meet with shop committees to be chosen in an election supervised by the board, and it informed Bethlehem that charges of past discrimination against the union would be investigated.

The company refused to comply with the NWLB orders, and the workers threatened to strike. With a strike imminent, Eugene Grace appeared before the NWLB on September 15 and explained that his company had not implemented the board's decision because the new rates were too financially burdensome; Bethlehem would need a price increase to cover the cost of the award. After promising to refer the price request to the War Industries Board, the NWLB obtained Grace's acquiescence to its orders concerning the organization of workers. Shop committees, the majority of whose members were union men, were finally elected at the Bethlehem Steel Company. To management, however, this did not represent union recognition. Steel industry leaders hoped to return to their former labor policies once the war ended. [18]

With the knowledge that the government supported its organizational efforts, the American Federation of Labor (AFL) embarked on a campaign to unionize the steel industry in August of 1918. The jurisdictional setup of the AFL, however, precluded unionizing the steelworkers under the banner of one steel union. Craft workers in steel belonged to twenty-four unions, while the unskilled workers were divided between two of the traditional craft unions: the Mine, Mill, and Smelter Workers, to which belonged the blast furnace workers, and the Amalgamated Association of

Iron, Steel, and Tin Workers whose members worked in the steel-making and finishing branches. A National Committee for Organizing Iron and Steel Workers managed the campaign and by the war's end had made good progress.[19]

But the armistice on November 11, 1918, abruptly reversed the fortunes of the steel unionizing endeavor. Even during the war, industry leaders had maintained the fiction that the shop committees with which they dealt were not connected with unions and that treating with them in no way constituted union recognition. Free of any obligation to the government, steel executives reverted with a vengeance to their old habits. Bethlehem reneged on the NWLB award; the company would not meet with the shop committees. With Judge Elbert Gary of U.S. Steel as its spokesman, the industry determined to win back whatever management prerogatives it had lost during the war. Steel executives were fully prepared to use all the repressive measures at their command to ensure the perpetuation of the open shop.[20] At the same time, improvements would be made in terms of employment so that unions would appear to be unnecessary. As production cutbacks occurred and the work force was reduced, labor activists were the first to be fired. Steel town officials, at the behest of industry executives, forbade public assemblies of workers. In the Pittsburgh area, labor organizers were driven out of company towns.[21]

Surprisingly, steelworkers continued to join unions, neither lulled by the welfare measures instituted by their employers nor deterred by the repressive tactics. Their experience during World War I gave them confidence and made them responsive to the rhetoric of union agents. Once they became union members, the steelworkers were impatient for results. They demanded action that would lead to union recognition by employers. The National Committee for Organizing Iron and Steel Workers wanted strong unions behind it before forcing the recognition issue, but the men in the mills clamored for a strike vote. The balloting acted as a powerful stimulus to organization; large numbers of men joined the unions while the vote was being tabulated. Those polled included union members and nonunion workers who chose to vote.

Approximately 98 percent favored a strike if the demands of the National Committee were not met. The committee had compiled a list of twelve items to be negotiated, but the key issue was the right to bargain collectively. Since it was apparent that the steel companies would not agree to bargain with the unions, a strike was a certainty. [22]

The strike, considered premature by experienced labor organizers, began on September 22, 1919. The workers' response in the first weeks was tremendous. But more potent forces favored an industry victory. A conservative reaction had set in after the war ended. The seeds of the Red Scare had begun to be sown, and charges that the strike was led by radicals hurt the union image in the public mind. Attempts to use the federal government to support the strikers' cause failed. President Woodrow Wilson was too involved in the fight for ratification of the peace treaty to expend his credit in the dispute. The steel companies, with the support of local community officials, were free to use whatever tactics they thought appropriate to break the strike. And, in the last analysis, the judgment of the labor organizers was correct: the unions were not ready. They had neither the resources nor the discipline necessary for an extended strike. The National Committee held out until January 8, 1920, but some unions, including the Amalgamated, had deserted its ranks much earlier. Steelworkers returned to the mills resentful and apathetic, not likely to respond to a renewed effort at organization. [23]

Thus the steel industry emerged triumphant from its war and postwar experiences. On the production side, the manufacture of ever-increasing quantities of steel was proof, to the government and to the public that bigness coupled with cooperation among the companies led to greater efficiency and growing output. The antitrust fervor of the progressive era was dissipated by the impressive production records achieved by big business during the war. Even the Supreme Court succumbed in the face of the steel industry's fine performance: The Court upheld a district court decision declaring that the U.S. Steel Corporation was not an unlawful monopoly under the antitrust laws. [24] Thus, the government's strongest challenge to concentration in the steel industry came to naught and the companies were free to perpetuate their pattern of growth. Consolidation increased during the 1920s; following Bethlehem Steel's merger with Midvale, Bethlehem and U.S. Steel between them controlled almost two-thirds of the industry.[25] In the area of labor relations, steel executives, despite some government opposition during the war, had succeeded in retaining total power over their work force. In the 1920s the steel industry remained a bulwark of the open shop and became a flourishing example of welfare capitalism in action.

"Our job primarily is to make steel," Charles Schwab, chairman of the board of Bethlehem Steel, asserted in December 1927, "but it is being made under a system which must be justified. If . . . this system does not enable men to live on an increasingly higher plane, if it does not allow them to fulfill their desires and satisfy their reasonable wants, then it is natural that the system itself should fail." [26] A belief in the paternalistic duty of employers to care for their workers underlay the practice of welfare capitalism. But in the 1920s, industry leaders came to believe that it served their own interests as well—that efficiency would increase if companies treated their workers well.[27] Judge Gary explained to U.S. Steel's stockholders that expenditures of $10 million a year for employee welfare were necessary "because it is the way men ought to be treated, and secondly because it pays to treat men in that way." [28] The industry assumed the responsibility for assuring workmen a decent living, and with this most employees seemed content. Wage rates were comparable to those in other manufacturing industries.[29] There were improvements in the fringe areas— pension plans; savings, stock-purchasing, and home-ownership programs; insurance to protect against losses due to accident, illness, old age, and death; improved safety measures and medical services. But only after tremendous pressure from the public and from the Harding administration did U.S. Steel abolish the twelve-hour day in 1923. Employee representation plans were instituted, but no one believed them to be independent of company control.[30] Nonetheless, these developments served to keep workers away from the trade union movement.[31]

The depression, however, destroyed the industry's ability to fulfill the obligations it had taken upon itself and ultimately brought about the failure of welfare capitalism. The onset of the depression severely curtailed the resources available to the companies for welfare purposes. Steel executives tried hard to ease the plight of their workers; prices were maintained so that wage rates would not be reduced, and share-the-work programs proliferated. But by 1932 the steel industry was operating at less than 20 percent of capacity. Prices had to fall. Wages were cut and welfare programs expired. When management was unable to make good on the responsibility it had claimed for itself, steel labor had to look elsewhere for help. The abandonment of the welfare philosophy, brought about by the worsening economic situation, paved the way for unionism in steel.[32]

In June 1936 a serious attempt to establish a union in the steel industry began. Passage of the National Labor Relations Act of 1935 assured the organizational movement some chance of permanent success. The act guaranteed to workers the right to join unions, to bargain collectively through representatives of their own choosing, and to participate in all activities necessary to the accomplishment of these ends. John L. Lewis, president of the United Mine Workers of America and chairman of the Committee for Industrial Organization (CIO) had a special interest in the creation of a strong steelworkers' union, which he believed crucial to the efforts of his own union to entrench itself in U.S. Steel's captive coal mines. Lewis was instrumental in the formation of the Steel Workers' Organizing Committee (SWOC) and contributed funds and talent from his own United Mine Workers (UMW).[33] Perhaps the most valuable asset Lewis bequeathed to the SWOC was Philip Murray, then vice-president of the UMW.

Born in Scotland in 1886, Murray, the son of a coal miner who was president of a local union, learned from an early age both the hardships of work in the mines and the pains of labor organization. In 1902 the Murray family emigrated to America and settled in western Pennsylvania, where young Phil soon obtained employment in the coal mines. Two years later, he accused the weighmaster at the mines of short-weighting his coal, a charge that was promptly denied. A fight ensued. As Murray describes it, "The weight boss took a shot at me with a balance weight. . . . I hit him with a stool over the head. I happened to get the best of the argument, but I was discharged." Some 550 men went out in support of Murray. Mine guards were stationed around the mine, and Murray's family was evicted from the company house. "That was the beginning of these things that I am interested in," explains Murray.[34]

The miners lost the strike, but the incident launched Murray's trade union career in the United States. He was immediately elected president of his local union. To prepare himself for a leadership role in the labor movement, Murray took correspondence courses. His investment appeared worthwhile, for, as he rose through the ranks of the United Mine Workers, he was recognized for his prodigious store of economic facts and figures that supported the union's position. After becoming a citizen of the United States in 1911, he was elected to the executive board of the UMW in 1912. Subsequently, he became president of District 5 (the Pittsburgh area) of the UMW and in 1919 was chosen vice-president of the international union.[35]

Tall, and balding, Murray, in appearance and personality, was the antithesis of John L. Lewis. Moderate and soft-spoken, but a born orator, Murray was a "warm and sensitive person who had good rapport with people high and low," [36] and he enjoyed a reputation as a skilled negotiator and peacemaker. Although Murray believed in the need for organization and collective bargaining, he also took a great interest in the social reforms necessary to improve the workers' standard of living. In the 1920s and 1930s Murray, as vice-president of the United Mine Workers, aided the union in its quest for advancement. He also helped the Roosevelt administration draft legislation on bituminous coal mining.

As chairman of the Steel Workers' Organizing Committee, Murray is credited with directing one of the best organizing campaigns in labor's history.[37] His years of experience in the labor movement, together with his intelligence and appealing personality, made him ideally suited for leadership of the steel union. Although in the first years of the SWOC's existence Murray operated in the shadow of Lewis, the subsequent development of the United Steelworkers of America reflected Murray's rather than Lewis' style. Lewis won the first and biggest victory for the SWOC, but it was Murray's patient nurturing of the union and its relationship with the steel companies that accounted for its secure position within the industry by 1950.

On March 2, 1937, barely nine months after the SWOC was created, Carnegie-Illinois, a subsidiary of U.S. Steel, signed a contract recognizing the SWOC as bargaining agent. The agreement, which surprised many observers, had been reached as a result of private conversations between John L. Lewis and Myron Taylor, chairman of the board of U.S. Steel. Other U.S. Steel subsidiaries soon signed similar contracts. A number of reasons led to the peaceful surrender of the stronghold of antiunionism, chief among them being U.S. Steel's desire to maintain continued production when the economy at last seemed to be recovering after years of depression. The SWOC had successfully infiltrated the company unions, but when it had no more to gain from them the SWOC declared its independence and filed charges against U.S. Steel with the National Labor Relations Board (NLRB), accusing the firm of unlawful control of employee representation plans. Evidently it was the unfavorable publicity resulting from the NLRB proceedings and the strong appeal of the SWOC that caused the Carnegie-Illinois management to succumb, rather than risk a strike for recognition.[38]

This was a momentous occasion in the history of labor relations in the steel industry. Robert R. R. Brooks has written of the contract signing:

As the pen was being handed around there was an understandable tension in the atmosphere. One of the signatories, attempting to make conversation, nodded to the man across the table from him and asked, "Who is that in the oil portrait behind you?" Without turning his head, the person addressed replied in a hoarse whisper, "He wasn't there yesterday." "Is that so? Whose picture was there yesterday?" "Old H. C. Frick. They took him out. Didn't think he could stand it." [39]

But U.S. Steel's action could not eliminate all the vestiges of a half century of opposition to unions. Other steel companies, which had already instituted many features of the Carnegie-Illinois contract, were slow to follow U.S. Steel's recognition of the SWOC. Jones & Laughlin fell into line after a brief strike and an NLRB election, but other companies proved much harder to crack. In the spring of 1937, strikes were called against Bethlehem, Republic, Youngstown Sheet & Tube, and Inland, all of whom had refused to sign contracts with the SWOC. Led by Tom Girdler, whose antiunion bias was notorious, the "Little Steel" companies determined to resist the SWOC. The strikes were punctuated by violence, and the companies and local authorities showed a callous disregard for the lives of workers.[40] A Chicago plant of the Republic Steel Corporation was the site of the most shocking incident, the "Memorial Day Massacre": As a peaceful parade of strikers seeking to set up mass picket lines approached the plant, the police opened fire. The workers turned and fled; the police continued to shoot. Ten marchers were killed and 125 people, including 35 police, were injured. The Little Steel strikes collapsed soon after. Thus, both successful NLRB elections and bloody strikes dotted the career of the incipient steel union before World War II, and its continued existence was by no means assured. The SWOC persisted in its organizational activities at the Little Steel companies, however, and a combination of NLRB orders, favorable court decisions, and the rearmament program which required sustained steel production, succeeded in convincing these firms that their interest lay in recognizing the union. By 1941 the Little Steel companies had begun negotiations with the SWOC and by the end of 1942 the union had obtained written agreements giving it exclusive bargaining status with all the important steel producers.[41]

Murray believed that the way for the SWOC to achieve security within the industry was to show that a responsible union, one that kept its contractual obligations, fostered stable labor relations which, in turn, contributed to the economic well-being of the company. Murray, who commanded the total respect of the

SWOC's membership, was able to guide the union along this path. During World War II he cooperated with the government—in contrast to his mentor, John L. Lewis—and built a solid reputation for his union. By making the United Steelworkers of America (SWOC was formally transformed into the United Steelworkers of America (USA) in May 1942) a model of democracy, honesty, and responsibility, Murray gave the steel companies little opportunity to attack the union.[42]

At the same time that it was learning to live with the union, the industry gained its first experience of operating under economic controls for a prolonged length of time. The companies early got a taste of what was in store when, in April 1941, U.S. Steel and the other major firms granted a 10-cents-an-hour wage increase only to have the Office of Price Administration (OPA) immediately freeze steel prices at first quarter of 1941 levels.[43] Although the OPA freeze order allowed for review, steel prices were held to that level until 1945. From the government's point of view price stabilization in the industry was very successful. Steel executives thought otherwise. Despite the tremendous growth in steel output, wartime profits reached a high point in 1941, declined slightly in 1942 and dropped significantly in 1943. Industry leaders sought to raise the price levels after 1943 but OPA was unreceptive.[44] The steel companies emerged from the period of wartime controls feeling that the government had not permitted them fair compensation for their increased costs.[45]

One further development of the World War II years helped mold the character of collective bargaining in the steel industry. The War Labor Board, in the interest of efficiency, treated the steel industry as a single unit in its dealings with the union. It therefore became the practice for uniform agreements to be reached between the Steelworkers and the companies. Although, after the war, the union and the industry reverted to bargaining on a company-by-company basis, the settlements continued to be virtually the same for all companies. U.S. Steel, because of its important economic position within the industry and because it had been the first to recognize the union, remained the leader in contract negotiations and set the pattern for the rest of the industry.[46]

During the postwar years the United Steelworkers of America continued to work toward an amicable and mutually beneficial relationship with the industry and most especially with the United States Steel Corporation. It appeared that the good faith efforts of Philip Murray and of John A. Stephens, director of industrial relations and negotiator for U.S. Steel, had done much toward promoting a durable and peaceful relationship between the industry and the union. Stephens evidently had been able to convince his colleagues at U.S. Steel that such a relationship was in the economic interest of the company.[47] Moreover, the operating heads of the various steel companies—Benjamin Fairless of U.S. Steel, Arthur Homer of Bethlehem, Admiral Ben Moreell of Jones & Laughlin, Charles M. White of Republic, J. Lester Mauthe of Youngstown, and Clarence Randall of Inland—were a new generation of steel executives who were not as infected with the antiunion animus as their predecessors and who had to accept collective bargaining as a fait accompli. The strikes of 1946 and 1949 bore no resemblance to the bloody encounters of the prewar years; disagreements developed over economic issues, not over the existence of the union.

Collective bargaining was always more difficult, however, when the freedom of the parties was restrained by the exigencies of government controls or by the intervention of the government to prevent a strike. This was particularly the case in the steel industry during the Truman years. The presence of the government as a third party to a dispute exacerbated the conflict between the union and the companies, because the steel executives believed that the government did not understand the industry. In the post–World War II years the government had been urging the steel companies to expand their capacity. But industry leaders, haunted by the specter of unused steelmaking facilities during the depression, had refused to comply. An extended debate had occurred resulting in strained relations between steel executives and the administration.[48] Moreover, the industry thought that the special relationship that existed between the Steelworkers and the Truman administration inevitably caused the government to be partial to the union's demands.[49]

The industry learned an important lesson in its first dealings with the Truman administration during the reconversion crisis, when economic controls were theoretically still in effect. In the 1945–1946 contract negotiations, the union asked for an across-the-board wage increase of 25 cents an hour. The industry said that in order to grant such a raise it would have to charge $7 a ton more for steel. The OPA advised that $2.50 per ton would be sufficient to cover increased industry costs, and the price control organization maintained this position throughout the controversy. Truman, wishing to avoid a strike, appointed a fact-finding board to help settle the dispute. Complicated and lengthy negotiations ensued: the union bargained with the industry and with the government for its wage increase; the companies dickered with the government for a price increase. No agreement was reached and the union went out on strike.

The stumbling block to a settlement was the price issue. The three parties had already decided that 18½ cents an hour would be the wage increase. After a three-week strike, the White House overruled OPA and granted the industry a $5 per ton price rise. Steel executives did not soon forget that they had achieved better results from the White House than from other government agencies. Moreover, the industry realized that this was the way to obtain its demands from an administration that was sympathetic to the labor movement.

Above and beyond the close alliance between the Truman administration and organized labor was the special esteem in which the President held the steel union. Philip Murray, whom Truman considered a man of great integrity, had always cooperated with the administration by agreeing to postpone a strike when the President requested such action in the public interest. Murray's reputation for moderation and responsibility formed the basis for Truman's particular regard for the Steelworkers.[50] Furthermore, it was especially important for Truman to maintain good relations with Murray who, as president of the CIO (he had been elected to this position after John L. Lewis resigned in 1940) as well as of the Steelworkers wielded tremendous influence throughout the labor movement. "In an election year," a journalist commented, "President Truman would let a lot of unfortunate things happen before he would deliberately alienate his old friend Phil Murray, the champion of four million [sic] C.I.O. members and one of the country's most respected labor leaders." [51]

The friendly association between Murray and President Truman was reinforced by the good rapport between representatives of the steel union and the White House staff. Arthur Goldberg, general counsel of the United Steelworkers of America as well as of the CIO, maintained frequent contact with several of the top men of the President's staff, John R. Steelman, Charles Murphy, and David Stowe. Noted for his powers of persuasion, Goldberg was admired by the staff for his intellectual ability, sober judgment, and honesty. Born in Chicago in 1908 into a poor family, Goldberg began working, at the age of twelve, as a delivery boy for a shoe factory. He was educated in Chicago public schools and at Northwestern University, where he received his law degree. He had a brilliant career at law school (Goldberg had had to receive special dispensation to take the Illinois bar exam before he had reached the age of twenty-one), all the while working nights at the post office and vacations on a construction crew. In 1933, Goldberg started his own law practice. It was not until 1938, however, after his political efforts in behalf of Franklin D. Roosevelt had put him in touch with labor leaders, that Goldberg began to specialize in labor law. He soon represented a variety of unions, including the United Steelworkers of America, and gained a reputation as an authority in the labor-management field. In March 1948, Goldberg was appointed general counsel of the United Steelworkers and of the CIO. Philip Murray is reported to have said that recommending Goldberg for the CIO position was his own biggest contribution to the success of that organization. [52]

Given the congenial relations between the Steelworkers and the White House, it is not surprising that the steel companies resented any intervention by the government in a steel labor-management dispute. In 1951, administration policies once again interfered with collective bargaining in steel, and the industry determined to protect its interests with vigor.

Enjoying the most profitable period since World War I,[53] the steel companies hoped to maintain their high rates of return throughout the Korean conflict. The industry was willing to give wage increases as long as it could raise prices and keep its profits high. But as soon as price control went into effect, the companies recognized that their profit margins would be threatened by any additional costs. Remembering the lean years of the depression, steel executives wished to do everything possible to improve profits during a time of prosperity and to avoid increasing costs to such an extent that their companies would be hurt if production fell after the war. New contracts with the Steelworkers were to be negotiated by December 31, 1951, and the union would surely request sizable wage increases. Because of the excellent profit situation in the steel industry, the Industry Earnings Standard promulgated by the OPS probably would not yield any price relief even if substantial raises were awarded to the union. Hence, steel company executives began a campaign against the Industry Earnings Standard. They argued that the steel industry, because of its critical position in a mobilization economy, should be given special treatment. Ben Moreell summed up the industry's case in a note to the economic stabilization administrator. He pointed out that the price controllers, in their zeal to hold down prices, were likely to overlook the primary objective of the Defense Production Act—to ensure production, not only currently but at a time when the country might be engaged in a full-scale war. The amount of "financial nourishment" needed for this purpose could not be determined by reference to "earnings standards achieved during a base period chosen arbitrarily for another purpose" and based on the revenue needs of the average industry. While government was urging the steel industry to expand its capacity to meet prospective war needs, it was failing to take into account that in a heavy industry such as steel the investment per dollar of sales was far greater than that of other industries.[54]

The Steelworkers, on the other hand, conscious of the extraordinary profits of the companies during this period, believed that the workers should be rewarded for their part in helping the steel industry achieve such a favorable position. David J. McDonald, secretary-treasurer of the union, in a speech soon after the labor members had withdrawn from the WSB, claimed, "There is no [effective] control on prices. There is no control on profits. There is control on wages only." [55] According to McDonald, there was no reason for labor to be singled out to carry the burden of stopping inflation, and he vowed, "The United Steelworkers of America will never subscribe to the idea that a man's wages are tied to the cost of living." [56] Thus, as 1951 drew to a close, the steel industry and the union prepared to defend their respective points of view at the upcoming contract talks. The stage was set for the historic confrontation of 1952.

four

THE GOVERNMENT SEIZES THE STEEL MILLS

"Whether our workers are to get a raise, and how much it will be if they do, is a matter which probably cannot be determined by collective bargaining, and will apparently have to be decided finally in Washington." [1] With this statement in November of 1951, Benjamin Fairless, president of the U.S. Steel Company, presaged the course of the steel dispute of 1951–1952. Contracts between the United Steelworkers of America, CIO, and the major steel companies were due to terminate on December 31, 1951. If no agreement were reached by that date, the union, which had given the notice required by the Taft-Hartley Act, was free to strike. And Fairless had plainly indicated that no wage settlement would be forthcoming until the government made clear how much it would permit the steel companies to raise prices. [2]

The goals of the parties to the dispute were far apart. The union sought substantial improvements in wages and fringe benefits, as well as a union shop in the steel mills. As many of the basic terms of the contracts between the United Steelworkers of America and the steel companies had not been open for collective bargaining since 1947, disagreements had arisen on such points as holidays, vacations, shift differentials, premium payments for weekend work, the interpretation and application of contract terms regarding incentives, local working conditions, and managerial rights. [3] Fairless admitted that the steelworkers deserved a raise: "Undoubtedly the union is entitled under the existing Wage Stabilization formula, to ask for some increase in wages to cover increases in the cost of living since the present wage scale became effective." [4] But the steel companies, fearful that they would not receive adequate compensation from the price control authorities, preferred to retain the status quo. As that was an unlikely outcome, the industry decided to seek a government commitment for a price adjustment commensurate with whatever wage increase the Steelworkers obtained.

During this time, the Truman administration was not unaware of the views of the union and of the industry. The Council of Economic Advisers (CEA) had studied the steel situation and had reported to the President as well as to stabilization officials. The council urged firmness in dealing with the steel dispute, because steel was basic to the economy. A reasonable wage increase, even one as high as 15 cents an hour, could be absorbed by the steel companies without necessitating any upward adjustment in prices, the council thought. It therefore noted that there was ample room for the parties to negotiate a noninflationary contract. The agreement resulting from collective bargaining should be approved by the government as long as the general price line remained unchanged.[5] Roger Putnam, the economic stabilization administrator, acting on the CEA's advice, warned Fairless that he was "bargaining with his own money." [6]

The government's refusal to give the steel companies any assurance on a price increase was reflected in initial negotiations between the Steelworkers and the U.S. Steel Company, the pattern-setter for the industry. Nothing was accomplished. The Federal Mediation and Conciliation Service then offered to mediate the dispute, but could not persuade the steel industry to make any offer on wages or working conditions.[7] As the strike deadline drew near, President Truman decided to refer the steel dispute to the Wage Stabilization Board. The President had concluded that a strike in this critical defense industry could not be permitted:

It is of the utmost importance to prevent an interruption in the production of steel. Steel is a key material in our entire defense effort. Each day of steel production lost is a day lost forever in the achievement of our production schedules. Continuous production, of this industry is essential in order to meet urgent demands for steel—steel for weapons, for factories, for highways and hospitals and schools.

It is for this reason that I have certified this matter to the Wage Stabilization Board. This will provide the parties with a forum where their differences may be resolved, and a fair settlement reached, without resort to a costly shutdown. [8]

In an appeal to patriotism, Truman urged the steel companies and the union to continue production and to cooperate with the WSB while it considered the issues. "The national interest demands it," the President declared. [9] He ended his request with a clear reiteration of the government's policy regarding adjustments in steel prices: Once a wage agreement had been reached, the Office of Price Stabilization would determine whether the wage increase warranted any change in prices. "The law and regulations assure that the steel companies will get price increases if they are entitled to them. No other advance assurances are necessary," the President asserted. [10]

Both the union and the industry acquiesced in the President's wishes; they agreed to adhere to regular production schedules, and the union postponed the strike to allow the WSB to come to a decision. The postponement, the steel companies later claimed, was the result of a deal between Truman and Philip Murray: If Murray would postpone the strike, the union would not have to fear the invocation of the Taft-Hartley Act. Murray may have given the industry this idea when, in a speech to the Steelworkers on June 22, 1952, he stated:

I remember distinctly just about the day before Christmas, 1951, President Truman communicated with me and, in the national interest, he suggested that the original strike scheduled for January 1, be postponed. He at that time said that "if you willingly and voluntarily agree to a suspension of the strike, I believe you need have no fear of the courts' imposing upon your membership the so-called 'Taft-Hartley injunction process.' " So . . . your international union postponed the strike. [11]

According to a White House staff member, however, Truman gave Murray the choice of complying with the President's request or being subject to a Taft-Hartley injunction; nothing was said about the future use of such an injunction.[12] But Arthur Goldberg maintains that when Murray spoke to the President no threat of a Taft-Hartley injunction was made nor were any future commitments discussed. In fact, Goldberg states, Murray did not want to speak to Truman at all. Murray wanted a strike and was angry at the President for certifying the dispute to the WSB. Goldberg had to convince Murray that the Steelworkers should cooperate with the government during a time of war.[13]

On their side the steel companies continued to press for an advance guarantee of a price increase. In a telegram to Nathan Feinsinger, chairman of the WSB, Admiral Ben Moreell, chairman of the board of Jones & Laughlin Steel Corporation, said that he assumed that the inquiry by the OPS into the amounts of price increases essential to maintain the "health" of the steel industry would proceed concurrently with the inquiry by the WSB. He also assumed that by the time the industry was called upon to consider the recommendations of the board "definite information relating to price adjustments" would be available to the steel companies.[14] Thus, as the various government agencies began their attempts to resolve the steel dispute, there was no common understanding of how the critical issue of price rises would be treated.

The Wage Stabilization Board had had little prior experience in handling labor disputes; by the time the board issued its recommendations on steel, it had made decisions in only three other certified cases.[15] Its task was made even more difficult, by the very nature of the board's disputes function. Cases were certified to the WSB in the hope of preventing production stoppages in essential defense industries. The board was therefore under great pressure to formulate settlements that both labor and management would accept. At the same time, the board, charged with stabilization duties as well, had to keep its recommendations within the general limits it had set up. Producing a settlement that satisfied the parties and was in line with stabilization rules often appeared impossible; the board was obliged to sacrifice one or the other of these goals. Industry members of the board frequently indicated their disapproval when they thought public and labor members had leaned too far in the direction of a settlement that was acceptable, rather than one that adhered to stabilization guidelines.[16]

The unseasoned wage board had not yet evolved a standard procedure for dealing with the disputes certified to it. In the steel case, the board decided to appoint a special panel to hear the union and industry presentations. The members of the panel representing the public were Harry Shulman, professor of law at Yale University and an expert on industrial relations and labor law, and Ralph T. Seward, a well-known arbitrator with experience in the steel industry; representing industry were John C. Bane, Jr., an attorney who had served as counsel for several steel companies, and Admiral Earl W. Mills, president of the Foster Wheeler Corp., a steel fabricator; representing labor were Arnold Campo, international representative of the United Steelworkers of America, and Eli L. Oliver, economist and manager of the Labor Bureau of the Midwest.[17] The panel was to write a report defining the issues between the parties, but without making any recommendations. On the basis of this report, the WSB would try to produce a settlement.

The most important demands presented to the panel by the union were for wage increases amounting to 18½ cents an hour, major improvements in fringe benefits, and adoption of the union shop.[18] The steel industry countered the union's argument for higher wages by declaring that steelworkers were among the best-paid industrial workers in the country, that their pay and benefits had kept ahead of the rise in the cost of living, and that any further wage increase would be inflationary. Moreover, the companies flatly asserted that additional labor costs could not be paid out of what they termed the industry's moderate profits.[19] The question of the union shop, the industry thought, should be settled by collective bargaining, not by a government board.

While the special panel of the WSB heard the steel labor dispute, the Office of Price Stabilization considered the issue of prices. The Steelworkers' request for an 18½-cents-an-hour raise plus fringe benefits, although well within the WSB guidelines,[20] disturbed OPS economists.[21] According to OPS studies, the steel companies could absorb as much as a 40-cent wage rise without requiring any price relief.[22] But a wage award of 18½ cents to the union and no price increase for the industry would appear to be a very inequitable settlement and was clearly one that the steel companies would not accept without a fight.

Thus, while the OPS hoped that the Wage Stabilization Board would recommend a wage increase of less than 18½ cents, price officials devised a stratagem they thought would be agreeable to the steel industry. The OPS would announce that the steel companies were not entitled to any price rise to offset whatever wage award the WSB made. At the same time, however, the OPS would grant the industry the price relief due it under the Capehart amendment. Notwithstanding the fact that the amendment authorized compensation for increased costs between June 1950 and July 1951, the steel industry had not applied for its benefits until December 1951, and—because of the special circumstances and highly complicated nature of the steel industry—the OPS had not yet determined the exact amount of price relief the Capehart amendment would yield.[23] When it completed its calculations, the OPS found that the steel companies would get a price rise of between $2 and $3 per ton.[24]

The White House staff, led by Steelman and Murphy, tried to follow developments at the WSB and the OPS and held informal meetings with members of those agencies. At first, the White House had hopes that the OPS strategy would be successful. But as time passed and reports leaked from the WSB that the wage award would be about 15 cents and that the steel companies were talking about a corresponding price rise of from $6 to $9 per ton,[25] the staff realized that their hopes of satisfying the companies with a $3 increase were just "wishful thinking." [26] Until the wage board announced its recommendations, however, price policy was in the hands of the OPS and the White House could do nothing. As President Truman told reporters, "As soon as they [the WSB] are ready to report to me, they will"; [27] then the White House would have the concrete information needed to evolve a realistic solution to the steel dispute.

Although the White House was reluctant to discuss the steel case before the wage board had made its decision, the media were not. Newspapers and periodicals frequently mentioned a "strike over prices." *Business Week* predicted that the steel industry would fruitlessly bargain with the OPS to raise the $3 Capehart figure. The industry would reject the WSB's recommendations, whatever they were; the union would go out on strike; and the lack of steel would force the OPS to change its mind. "No matter what it says on their sandwich boards, Phil Murray's pickets will be on strike to get steel prices raised," *Business Week* asserted. [28]

Furthermore, the steel industry initiated a propaganda campaign designed to capture public opinion by emphasizing the patriotism of the steel companies in advocating that neither wages nor prices be increased. [29] Typical of this effort was a statement issued in the name of "Steel Companies in the Wage Case":

Steel companies offered to give up a price increase the Administration admits they were entitled to if the Union would give up its demand for a wage increase. The Union rejected this.

Such a hold-the-line position was fair to employees and the companies *and* it PROTECTED THE PUBLIC AGAINST A NEW WAGE-PRICE SPIRAL. [30]

So many charges were made against the WSB, the union and the administration, that, as the Senate Committee on Labor and Public Welfare subsequently observed, the "outpouring of propaganda and scare advertisements before, during, and after the Wage Stabilization Board's deliberations was not calculated to create an atmosphere in which the union and management could come to a settlement on their own." The committee labeled the propaganda "patently untrue" and underlined the part that propaganda had played in turning a labor-management dispute into a major social struggle. "The processes of collective bargaining are difficult enough without the accompaniment of a hysterical chorus egging one of the parties on to battle," the committee concluded. [31]

In this tense climate the special steel panel of the WSB continued to hear the case. The issues involved were so numerous and the testimony so detailed that the panel could not complete its report until March 13, 1952.[32] Thus the Wage Stabilization Board had to postpone its decision date several times, and the union cooperated by extending its strike date. The new deadline was April 8, 1952.

Finally, on March 20, 1952, the WSB recommendations, along with the dissenting opinion of the industry members, were made public. They called for an 18-month contract with a 12½-cents-an-hour wage increase retroactive to January 1, 1952, and additional 2½-cent increases effective June 30, 1952, and January 1, 1953. Also included in the suggested settlement were various fringe benefits and adoption of the union shop.[33] The total cost to the steel companies per employee hour was calculated by adding up the money value of all the recommendations; 26 cents was the figure most popularly quoted, although estimates varied from a low of 18 cents to a high of 30 cents. Clarence Randall, president of Inland Steel Company, said the recommendations would increase employment costs by 30 cents per employee hour and total costs by 60 cents per employee hour.[34] According to Gardner Ackley, an OPS economist, Randall's calculation of total costs rested on rather tenuous economic reasoning: "[The steel companies] advance the absurd claim that for every extra dollar they pay in wages they will have to pay another extra dollar for other things they buy, even though these other things are mostly under ceilings, which we have no present intention to raise." [35] On the basis of the WSB's proposals and the total costs theory, steel industry executives raised their price relief claims to $12 a ton.[36]

The industry members of the WSB issued a minority report stating that the recommendations "in terms of stabilization are in serious disregard of the public interest." [37] The minority report asserted that the board's recommendations on money issues breached the stabilization guidelines, that it was unfair to put the weight of the United States government behind the Steelworkers' drive for a union shop and that the board had deliberately sought to suggest a settlement that would be agreeable to the union while making no effort to ensure its acceptance by the companies.[38]

Reaction to the wage board's announcement varied. Many

people thought the wage award was excessive and inflationary. Cyrus Ching, director of the Federal Mediation and Conciliation Service, later wrote that the suggested settlement provided the "coup de grace" to an already ineffective wage stabilization policy. He believed the WSB recommendation was much more generous than the union could have achieved through collective bargaining.[39] Other critics of the board maintained that the exorbitant wage increase made an upward price adjustment imperative and would therefore trigger a new round of inflation.[40] They accused the WSB of neglecting its stabilization duties in its search for a solution which the union would find satisfactory.[41] The *New York Times*, for example, noted that the WSB's public members had become just another mediation agency haggling with the parties and had failed to "keep an eye single to the stabilization function." The belief that any settlement that either the labor or the industry members could be induced to agree to would be a proper solution meant that whatever "solution" two of the three groups on the board could agree to would constitute a "wage stabilization policy."[42]

Republicans in Congress echoed this criticism. Representative Thomas Werdel of California charged the public members of the WSB with capitulating to the "demands of the labor demagogues in the steel industry." He was especially incensed by the fact that the board had recommended "a greater increase than has ever been granted by collective bargaining or by Government recommendation at any time in the past, war or peace."[43]

Some Democrats supported their Republican colleagues in their attack on the WSB. Representative Wingate Lucas of Texas reminded the House that he had predicted that the Wage Board would wreck the stabilization program if it were allowed to continue to handle disputes.[44] On March 28, 1952, the House Rules Committee, spurred on by the issuance of the WSB's recommendations, reported favorably on Representative Leo Allen's pending resolution to investigate the board.[45] Since the previous summer's debate over the disputes function of the WSB, little interest had been shown in how the board performed this task until Representative Allen, a Republican from Illinois, introduced his resolution on February 22, 1952.[46] At the time there was speculation that the steel industry had persuaded Allen to sponsor such a resolution.[47] The Democratic leadership of the House postponed floor action on the resolution until April 24, when it was passed by a vote of 255 to 88.[48]

The WSB's recommendations were not without defenders, however. In the Senate, a subcommittee of the Committee on Labor and Public Welfare held hearings on the proposed steel settlement.[49] Nathan Feinsinger, the only witness, argued that the wage board's decision had not breached the stabilization regulations, and he defended the WSB's role in resolving disputes. When the staff of the subcommittee issued its report on the recommendations, it upheld Feinsinger's contentions.[50] Journals as disparate as *Fortune* and the *New Republic*, *U.S. News and World Report* and the *American Federationist*, also justified the recommendations. They pointed out that, although the suggested wage increase appeared high, the steelworkers, who had not received a raise since December 1950, were only keeping abreast of the rising cost of living. Moreover, the steel settlement was not likely to set a pattern for other industries. In some industries, like automobile, rubber, and electrical manufacturing, long-term contracts were in effect and cost-of-living increases had already been granted. In other industries, such as coal, oil, and shipbuilding, where contract negotiations would begin soon, workers simply expected to get a raise commensurate with higher living costs. Furthermore, the fringe benefits that the wage board had endorsed for the steelworkers, and which added greatly to the cost of the settlement, were benefits that most other industries had already granted to their workers.[51]

When the wage board's recommendations were made public, Truman and most of his staff were in Key West, Florida, the President's favorite vacation spot. Their initial reaction, based only on the press version of the proposed settlement, was that the WSB had been too generous to the steelworkers.[52]

One week later, however, a close analysis of the WSB recommendations led the White House staff and the Council of Economic Advisers to conclude that the settlement would not adversely affect the stabilization program.[53] According to stabilization regulations, the White House staff found, the WSB could have approved an immediate wage increase of from 9 to 16 cents for a one-year contract. Instead, by advocating an 18-month contract in which total wage costs would begin at 12½ cents and rise above 16 cents only in the last six months, the WSB's suggested wage award represented a substantial contribution to stabilization policy. In addition, the board had not suggested the inclusion of an escalator clause in the contract, thus binding the Steelworkers to a set wage regardless of any rise in the cost of living. Moreover, the staff analysis pointed out, an 18-month contract would assure continued production, a most important factor in the administration's mobilization program.[54]

The only recommendations of the wage board that at first seemed excessive to the White House staff were those on fringe benefits. But further examination led the staff to conclude that the board's proposals were justified, because they would merely enable the steelworkers to achieve parity with workers in other industries.[55]

The staff report also called attention to some of the gains made by management in the proposed contract. Issues involving local working conditions and incentive systems were resolved in the industry's favor. The union's request for a guaranteed annual wage, along with other lesser demands, was returned to the parties for bargaining. In the area of "managerial prerogative," therefore, the steel companies fared well under the WSB's recommendations, the staff maintained.[56] In conclusion, the analysis asserted:

> The furore of the past few days should not be allowed to obscure two fundamental facts: (a) the wage recommendations are stabilizing and *not* by themselves a threat to the wage stabilization program, (b) the problem in the steel case has been, and is, the refusal of the steel industry to accept the Industry Earnings Standard of OPS. A different recommendation from the WSB would not have altered the position of the steel industry, or the dilemma of the Administration, one whit. The steel industry is on

strike against the present standards of price control and it will not settle its labor dispute until it wins substantial price concessions from the Administration.[57]

But this represented the views of the White House staff one week after the WSB recommendations were reported. During the critical days immediately following announcement of the wage board's decision—before this study was made—Truman and his staff continued to think that the recommendations were on the generous side.[58]

At that time, Charles Wilson, director of defense mobilization and nominally the head of the stabilization and mobilization programs, began to take charge of efforts to settle the steel dispute. During the wage board's deliberations, Wilson had informed the White House staff and his stabilization colleagues of his great concern about the steel case. He disapproved of stabilization officials meeting with White House staff members, and he frequently threatened to take action on the price front on his own authority.[59] On March 21, the day after the wage board's proposals were announced, he met with executives of the steel industry. The following day he consulted Roger Putnam, administrator of the Economic Stabilization Agency, and Ellis Arnall, director of price stabilization; and on Sunday, March 23,[60] Wilson flew to Key West to talk with the President.

Wilson and the President conferred immediately, and met again the next day. When Wilson left Key West for Washington late on March 24, he believed he knew what the President wanted him to do. Murphy and Stowe, with the President at Key West, reported to Harold Enarson, who had remained in Washington, that they thought Wilson had left the President with permission to settle the dispute for approximately a $4- to $5-a-ton increase.[61] Wilson's plan was to urge the union and the steel companies to begin bargaining again and to recommend a compromise settlement: a wage-fringe package smaller than the WSB-recommended 26 cents and a price increase not much beyond what the Capehart amendment allowed. It is apparent from Wilson's subsequent testimony before the House Committee on Education and Labor that he had no idea that the union would consider the WSB recommendations the least it would settle for.[62] To reporters' questions on what had been decided at Key West, Wilson repeatedly answered, "No comment." Almost as an afterthought, however, Wilson remarked to a reporter, "There is no question in my mind but that, if the wage increases contemplated under the WSB's recommendations are put into effect, it would be a serious threat in our year-old effort to stabilize the economy." [63]

The effect of Wilson's offhand statement was dramatic. Arthur Goldberg worked with Philip Murray on a statement declaring that Wilson had wrecked the stabilization program as well as the WSB itself; that Murray would attend no meetings with Wilson; that Wilson's comment would be responsible for precipitating a strike; that no honorable person would serve on the WSB and no self-respecting union could any longer bring its case before the board. At the behest of Nathan Feinsinger, the statement was modified before its release to read that Wilson had *attempted* to destroy the wage stabilization program and that there was no point in any union bringing a case to the board if the board's recommendations could be repudiated by Wilson.[64] "Accordingly," Murray announced,

The United Steelworkers will attend no such meeting with Mr. Wilson. In accordance with our commitments to the President and the Wage Stabilization Board, we shall enter collective bargaining discussions with the steel companies in the next 48 hours in the strong hope that a fair and equitable agreement may be reached.[65]

Feinsinger was extremely disturbed by Wilson's remark about the unstabilizing effect of the wage board's recommendations. In fact, he refused to believe that Wilson had been quoted correctly. Before the board had announced its decision, Feinsinger and Fred Bullen had met with Wilson and Putnam and had explained to them what settlement the public members of the board would try to obtain.[66] At that time, Wilson had appeared to accept a 15-cent wage increase for a one-year contract [67]—a result obviously no better from a stabilization standpoint than the recommendation actually made by the WSB.

Despite the excitement that his statement had caused, Wilson went ahead with his plans to settle the dispute. The first collective bargaining session between the United Steelworkers of America and the U.S. Steel Corporation since the previous December had been held the morning of March 26, as Murray had promised. Soon after the meeting began, it was recessed by John A. Stephens, chief negotiator for U.S. Steel, on the ground that the company wished to study the WSB recommendations further. Later that day it was leaked to the press that Wilson had a solution; [68] presumably, he had offered the company a palatable price increase.

At this point, the administration had to consider the future of the whole wage stabilization apparatus, as well as the steel controversy. If the President wished the WSB to continue to exercise its stabilization and disputes functions, he could not repudiate the board's work in the steel case. Thus, the White House made clear that it supported the recommendations of the wage board and that the suggested settlement should be the basis of collective bargaining. Wilson, too, in an attempt to soften the effect of his previous remark, declared that the WSB was a "duly constituted Government agency" and that its proposals would be used as the foundation for an agreement that would be satisfactory to both the Steelworkers and the industry. [69]

On Thursday, March 27, Wilson met individually with Putnam, Arnall, and Feinsinger. Nothing was said publicly about what was discussed at these meetings, but it was rumored that Putnam and Arnall had refused to agree to a large price increase for the steel companies. These stabilization officials were not convinced that Wilson had the President's permission to make sizable price concessions to the industry. To obtain a definitive statement of the administration's price policy, Putnam and Arnall arranged to see some White House staff members on Friday morning and scheduled an appointment with the President that afternoon. Wilson accompanied them to the conference with Truman. [70]

Reports after the meeting indicated that Putnam and Arnall had told the President that the OPS price policy should not be breached and that Wilson had offered the steel companies too large a price increase. Wilson related his version of his attempts to settle the dispute, which attempts he believed were in accord with the instructions Truman had given him at Key West. The President then stated his position; he had suggested that Wilson discuss price adjustments with the steel industry but had not meant to imply total capitulation to the companies' demands.[71]

Wilson returned to his office after the conference with Truman, summoned a staff meeting, and announced that he had resigned because his integrity had been called into question. "He did not stress the substantive disagreement between him and the President; what he stressed to us was that he felt he had been called a liar in front of other people," a member of his staff recalled.[72] In his formal letter of resignation, Wilson did not hesitate to inform Truman that the President had humiliated him:

I advised you my purpose was to meet with both sides to the controversy and try to effect a compromise. You fully agreed. On my inquiring whether, if the amount of wage increase jointly agreed upon was higher than the companies would absorb, your request to prevent a strike contemplated the possible necessity of allowing the steel companies price increases related to whatever wage increases were agreed to between the parties, your answer was "yes." You next urged me to follow the plan I previously proposed—try to effect a compromise by both sides so that the breaching of our stabilization formulas would be held to the absolute minimum necessary in carrying out your "prevent a strike" order. . . .

At our meeting today, however, you changed the plan we agreed upon.[73]

Wilson's departure from the government was made public on March 30, when President Truman released his letter accepting Wilson's resignation. Truman intimated that he had indeed originally thought that the WSB's recommendations were inflationary but upon further study had discovered that they did not have an unstabilizing effect. As for prices, the President admitted that he had agreed that some price increases might be necessary, but stated that he had authorized Wilson to explore only the possibility of such adjustments with the steel companies; he was not to make any final commitments.[74]

Wilson's resignation further impaired the administration's already diminished prestige. Although a few journals thought Wilson's behavior reprehensible,[75] most editorial opinion agreed with Wilson's contention that the government's stabilization goals had been compromised by the wage board's proposals and that justice required an upward adjustment in prices for the steel companies.[76] Buoyed by the favorable editorial response to its case, the steel industry remained insistent in its demand for a large price increase as a prerequisite to any settlement. The union, as the strike deadline drew near, was still determined to obtain all that the WSB had recommended.

Until Truman could find a permanent replacement for Wilson, John R. Steelman was appointed acting director of the Office of Defense Mobilization. If a strike was to be prevented, Steelman would have to find the means to a settlement.[77] The first order of business was to get collective bargaining started once again. With Nathan Feinsinger acting as special mediator,[78] the companies and the union were scheduled to meet on March 31. The steel companies postponed the meeting day by day until finally, on April 3, the session took place. On that day, the industry made its first offer to the union since discussions had begun in November 1951: an increase in wages and fringe benefits totaling 14.4 cents an hour. This offer appeared to be contingent on a price increase equal to the amount allowed by the Capehart amendment.

Not unexpectedly, the union refused to consider the offer.[79] The *New York Times* reported that "Mr. Stephens [chief negotiator for U.S. Steel] made it plain that he had no hope that an agreement could be reached unless the Government changed its attitude toward price increases. He said prices were 'the big issue' and added that the companies could not come to any conclusion on higher wages until there was 'clarity' on Government approval for higher prices." [80]

Price discussions with the government were indeed under way. On April 1, Ellis Arnall told Benjamin Fairless, president of U.S. Steel, that the OPS intended to hold the increase down to what the Capehart amendment permitted, between $2 and $3 a ton—a long way from the $12 a ton Fairless was demanding.[81]

Other steel company representatives met with Arnall too and received the same message. But on April 3, after talking with Steelman, Arnall secretly offered the steel companies a price adjustment totaling $4.50 a ton: $2.75 allowed by the Capehart amendment plus $1.75.[82] Although the steel companies were unenthusiastic about this offer, collective bargaining sessions were continued in the hope that a last-minute settlement might result.

But, at the same time, the White House prepared to act in the event bargaining failed and a strike became unavoidable. Throughout the executive branch of the government, it was agreed that a production stoppage in the steel industry could not be permitted to occur. The situation in Korea was uncertain. Fighting was sporadic, but the administration could not discount the possibility that military action would intensify, especially if the Communists thought the American position to be seriously weakened by a strike in the steel industry. Consultation with his major advisers on the war bolstered Truman's determination to prevent a strike.

As Truman himself recalls, Robert Lovett, his Secretary of Defense and one of the most respected men in the administration,[83] told Truman that "any stoppage of steel production, for even a short time,[84] would increase the risk we had taken in the 'stretch-out' of the armament program.[85] He also pointed out that our entire combat technique in all three services depended on the fullest use of our industrial facilities. Stressing the situation in Korea, he said that 'we are holding the line with ammunition, and not with the lives of our troops.' Any curtailment of steel production, he warned, would endanger the lives of our fighting men." [86]

Gordon Dean, chairman of the Atomic Energy Commission, informed the President that atomic weapons projects would be delayed by lack of steel. From Henry H. Fowler, administrator of the National Production Authority, Truman learned that the inventory of several kinds of ammunition was already low. Secretary of State Dean Acheson advised the President that the United States had huge commitments to its allies under the Mutual Defense Assistance Program. If a steel strike prevented the United States from delivering the promised material, not only would faith in America's ability to lead the free world be undermined, Acheson advised, but its allies might be too weak to resist Soviet aggression. [87]

Truman was also concerned with the way in which a shortage of steel would affect the domestic defense-related economy. Not only would the steel industry and the Steelworkers union suffer losses, but, as Secretary of Commerce Charles Sawyer informed him, "a ten-day interruption of steel production would mean the loss of ninety-six thousand feet of bridge and fifteen hundred miles of highway." [88] Ship construction and airplane production would be seriously curtailed. Petroleum, gas, and electric power plants, coal mines, and coke ovens were all dependent on steel in one way or another. [89] A slowdown in production in these fields would lead to significant unemployment—an unhappy result in an election year. [90]

The President was unwilling, however, to scuttle the administration's entire stabilization program by acceding to the steel industry's price demand. The White House staff therefore discussed the advantages and disadvantages of other courses of action to prevent a strike: seeking an injunction under the Taft-Hartley Act, seizure of the steel mills under Section 18 of the Selective Service Act of 1948, seizure under the "inherent powers" of the President, and requesting seizure legislation from Congress. The White House staff apparently did not consider requisitioning the steel mills under Title II of the Defense Production Act, because its procedure was too lengthy and involved to meet the needs of the steel crisis. [91]

The Taft-Hartley alternative was not given much attention by the White House staff primarily because the President had indicated that its use was unjustified in the present circumstances of the steel dispute. [92] Truman and his staff believed that Taft-Hartley was inadequate and unfair. Steel production would not be assured, because the union would be free to strike while the President appointed a board of inquiry that would engage in fact finding and then report to the President.[93] If the President decided to ask for an injunction against the strike, it was uncertain whether a court would grant the injunction, in view of the fact that the union had delayed a strike for more than ninety days already. Moreover, a Taft-Hartley injunction, it was thought, would impose a hardship on the workers: they would be forced to work for wages they now considered inadequate—yet it was the steel companies who were refusing to settle. Equity, the President's advisers reasoned, demanded action that would put the burden to settle on the companies, not on the union.[94]

Seizure under Section 18 of the Selective Service Act of 1948 was a more inviting alternative. Section 18 authorized the President to place orders with any manufacturer that produced material needed by the armed forces or the Atomic Energy Commission as long as procurement of that material had been approved by Congress. If the company failed to fill the order in the time specified by the President he was empowered to take immediate possession of the property "and to operate it, through any Government agency, for the production of such articles or materials as may be required by the Government." Fair and just compensation had to be paid for any property taken under this section.[95]

A week before the strike deadline, Milton Kayle, a member of the White House staff, discussed with attorneys from the Justice Department the possible use of Section 18 seizure authority. The consensus of the meeting was that the President could seize the steel mills under Section 18.[96] There were some difficulties, however. On the legal side, the Justice Department representatives noted that Section 18 did not specifically authorize seizure when a company failed to fill government orders because a strike had stopped production. The legislative history of that section was also silent on the subject of labor disputes that interrupt production.[97] But the lawyers concluded that, since the basic purpose of Section 18 was to assure the continuance of essential production, the President probably would be justified in seizing the steel mills under Section 18 authority.[98] On the practical side, there were obstacles too: Section 18 described a very specific procedure to be followed before any industrial property could be taken. The government had to decide with which plants it wanted to place orders and the amounts and types of materials it needed. This was a hard decision to make, because the armed forces and the Atomic Energy Commission did not buy steel directly, but rather purchased end products that contained steel. Placing these individual orders, moreover, would be very time consuming. The Justice Department attorneys, in addition, were not sure whether seizure could lawfully be invoked under Section 18 before an actual work stoppage occurred. From the law it appeared that the government had to allow a specific period of time in which the orders could be filled. When this time had passed and the material had not been delivered, the President could then issue an executive order seizing the property.[99]

At this meeting, Kayle and the Justice Department lawyers also considered the alternative of reliance on the inherent powers of the executive, under Article II of the Constitution, to seize the steel industry. The relevant sections of Article II state: "The executive Power shall be vested in a President of the United States of America"; "The President shall be Commander in Chief of the Army and Navy of the United States"; "he shall take Care that the Laws be faithfully executed." This approach had definite advantages. The President would be free of any statutory restraints on the manner and timing of the seizure. Immediately upon issuance of an executive order, the government could take over the mills. And specific requirements of just compensation—such as that contained in the Selective Service Act of 1948—would not be brought into play.[100]

The inherent powers authority, however, had one overriding difficulty: its uncertain legal basis. Prior to World War II, the inherent powers of the President had been invoked three times to justify seizure of property during labor disputes.[101] But the courts had never ruled on the legality of these seizures. Furthermore, since that time Congress had enacted the national emergency disputes provisions of the Taft-Hartley Act and Section 18 of the Selective Service Act. It could thus be argued that the specific remedies prescribed by Congress were preemptive, precluding other types of Presidential action in this field. On balance, the Justice Department attorneys concluded that, if seizure action were to be taken, Section 18 authority would be the most desirable basis.[102]

Another meeting was held on April 3, to discuss the problems involved in using Section 18. This time, Holmes Baldridge, Assistant Attorney General in charge of the Claims Division (now the Civil Division), along with another Justice Department lawyer, Ellis Lyons, and Assistant Secretary of Defense Charles Coolidge and his staff, attended the meeting with Kayle. The Justice Department attorneys believed that there was a 50-50 chance that the courts would uphold a seizure of the steel mills based on Section 18. The Defense Department representatives, however, were adamant in their opposition to the employment of Section 18: the procedure was too complex and time consuming. Their views ultimately prevailed. After listening to the opinion of the defense officials, Baldridge and Lyons concluded that reliance upon Section 18 authority would, in the current situation, render the government vulnerable to attack regarding compliance with the statutory requirements. They had become convinced that seizure relying upon the inherent powers of the President was the "more desirable approach." It would be less vulnerable to attack in court, involve less complicated administrative operations, and take much less time. The action could be defended in court on the grounds that neither the Taft-Hartley nor the Section 18 authorities were feasible or equitable in the steel situation.[103]

A fourth alternative, a request for congressional legislation to authorize seizure of the steel mills, was quickly rejected by the White House staff. The legislative process was too slow for a seizure bill to be enacted before the strike deadline. Moreover, even if there were enough time, the likelihood of Congress passing a seizure law was slim. Congress was generally unsympathetic to Truman, and it was reasonable to anticipate that many congressmen would expect the Taft-Hartley Act to be invoked before the President resorted to seizure.

While the White House staff was considering these options,

Arthur Goldberg, on behalf of the Steelworkers, was preparing to submit to the administration an analysis of the President's authority to seize the steel industry. The union realized that Truman was determined to avert a strike. Given a choice between working under a Taft-Hartley injunction or working for the government under seizure, the union preferred the latter. Transfer of control of the plants to the government would hurt the industry more than the union and would give the union the opportunity to bargain with the government for higher wages.[104] Thus Goldberg set out to show the administration that it had ample power to seize the steel mills. He concluded that the President had statutory authority to seize based on either Title II of the Defense Production Act or Section 18 of the Selective Service Act of 1948. He also made a cogent argument in favor of the President's inherent power to seize the steel industry.[105] But Goldberg revealed no preference for any particular route to seizure; in his opinion, Truman had sufficient authority whichever path he chose.[106]

Goldberg's view, however, was in the minority.[107] Although, in the last few days before the strike deadline, everyone talked about seizure, few really expected Truman to risk it. The Democratic party faced a contest for the presidential nomination; relations with Congress were vexatious; Truman was trying to clean up corruption in the government. The steel industry, Congress, and some members of the cabinet thought Truman would have to invoke Taft-Hartley because other alternatives were too controversial.[108] This pervasive feeling that a Taft-Hartley injunction would be the next step in the steel labor dispute gave the steel companies less incentive to settle with the union and more reason to believe that they were in a secure bargaining position. This, perhaps, accounts for the industry's rejection of Arnall's secret offer of a $4.50-a-ton price increase.

Upon rejecting Arnall's offer, however, the steel companies secretly made a counterproposal of between $5.50 and $6.50 a ton.[109] Apparently, the industry decided that it would accept a price increase covering the direct employment costs to the companies that, according to their own figures,[110] resulted from the

Wage Stabilization Board's recommendations and forgo the additional $6 it claimed was needed to compensate for the indirect costs of the WSB's suggested settlement. Steelman, who had been mildly optimistic that Arnall's offer would settle the dispute, immediately denied the new request. He concluded that the steel companies were intransigent in their refusal to come to terms with the union until they received price relief which the administration regarded as excessive.[111]

Serious work was consequently begun on a speech that would contain the President's declaration of seizure. Truman and his staff realized that seizure of the steel industry would not end the labor dispute; only the union and the steel companies could do that. But seizure would offer a new context in which the President could pursue his goals—continued steel production and reasonable price control—and intensify his efforts to effect a settlement.[112]

At 10:30 P.M. on Tuesday, April 8, 1952, President Truman told the nation in a radio and television address that he was ordering Secretary of Commerce Charles Sawyer to take over operation of the steel mills for the government of the United States.[113] He also announced that he was asking the parties to the dispute to come to Washington the next morning to resume collective bargaining in order to achieve a settlement. The President declared that the country faced a serious emergency, and that its security depended upon the ability of the nation to keep up its defense production, because steel was a major component of most defense supplies. Truman argued that the Taft-Hartley procedure would not meet the needs of the country at that time, because it would necessitate at least a short interruption in production while the board of inquiry appointed by the President studied the steel situation. Only after the board reported to the President could the government apply for an injunction.[114]

If that had been the sole content of the speech, perhaps the steel industry's reactions to the President's seizure of the mills would not have been so bitter. But Truman and his staff felt compelled to reveal to the nation their view that the real culprits in this dispute were the steel companies:

These [stabilization] rules have been applied in this steel case. They have been applied to the union, and they have been applied to the companies. The union has accepted these rules. The companies have not accepted them. The companies insist that they must have price increases that are out of line with the stabilization rules. The companies have said that unless they can get those increases they will not settle with the union. The companies have said, in short, that unless they can have what they want, the steel industry will shut down. That is the plain, unvarnished fact of the matter.[115]

The President argued that, because of the favorable profit situation in the steel industry, the companies' price demands were shocking:

Steel industry profits are now running at the rate of about $2½ billion a year. The steel companies are now making a profit of about $19.50 on every ton of steel they produce. On top of that, they can get a price increase of close to $4 a ton under the Capehart amendment to the price control law. They don't need this, but we are going to have to give it to them, because the Capehart amendment requires it.

Now add this to the $19.50 a ton they are already making and you have profits of better than $22 a ton.

Now, what would the Wage Board's recommendations do to steel profits? To hear the steel companies talk, you would think the wage increase recommended by the Board would wipe out their profits altogether. Well, the fact of the matter is that if all the recommendations of the Wage Board were put into effect, they would cost the industry about $4 or $5 a ton.

In other words, if the steel companies absorbed every penny of the wage increase, they would still be making profits of $17 or $18 a ton on every ton of steel they made.

Now, a profit of $17 or $18 a ton for steel is extremely high. During 1947, 1948, and 1949, the 3 years before the Korean outbreak, steel profits averaged a little better than $11 a ton. The companies could absorb this wage increase entirely out of profits, and still be making higher profits than they made in the 3 prosperous years before Korea. . . .

And yet, in the face of these facts, the steel companies are now saying they ought to have a price increase of $12 a ton, giving them a profit of $26 or $27 a ton. That's about the most outrageous thing I ever heard of. They not only want to raise their prices to cover any wage increase; they want to double their money on the deal.[116]

Truman explained that yielding to the price requests of the steel companies would mean higher prices for products throughout the economy.[117] He concluded with a plea to the industry and the union to come to terms:

We must have steel. We have taken the measures that are required to keep the steel mills in operation. But these are temporary measures and they ought to be ended as soon as possible.

The way we want to get steel production—the only way to get it in the long run—is for management and labor to sit down and settle their dispute. . . .

On behalf of the whole country, I ask the steel companies and the steelworkers' union to compose their differences in the American spirit of fair play and obedience to the law of the land.[118]

The tone and content of the President's speech, however, could hardly be expected to produce an atmosphere conducive to agreement. His vitriolic attack on the steel companies' position left the industry more determined than ever to resist a government-imposed settlement.

five

THE COUNTRY REACTS

Had the Truman administration realized that the steel seizure would precipitate a constitutional crisis, it might have been more reluctant to take that action. The White House had hoped, by threatening to increase the wages of the steelworkers while the government was in possession of the mills, to exert pressure on the steel companies to settle with the union on terms that would not harm stabilization policy. Since the seizure ensured the continued production of steel, the industry would, it was thought, have no leverage to obtain the price adjustment it wanted.

The steel companies, however—aided by the press and Congress—managed wholly to change the focus of the seizure episode. Rather than providing a new context for collective bargaining, the seizure became the subject of a great constitutional debate. The administration, hoping to direct the country's attention to the industry's exorbitant price demands, instead had to defend itself against charges of dictatorship. The legal maneuvers of the steel companies brought the issue of executive power to the forefront; wage and price matters were submerged in the resultant clamor. The labor dispute, far from being resolved, played only a minor role in the ensuing constitutional drama.

President Truman had announced in his speech that the government would assume control of the steel mills at midnight, April 8, 1952. Executive Order 10340, "Directing the Secretary of Commerce to Take Possession of and Operate the Plants and Facilities of Certain Steel Companies," [1] had been signed by Truman earlier that day. Citing the indispensability of steel to the troops in Korea and to defense programs, the President asserted his authority, based on "the Constitution and laws of the United States, and as President of the United States and Commander in Chief of the armed forces," [2] to order the Secretary of Commerce to seize the steel companies involved in the labor dispute with the United Steelworkers of America. The executive order declared that the Secretary of Commerce would specify the terms and conditions of employment while the government ran the mills.

Thus, as of midnight April 8, Charles Sawyer, the sixty-five-year-old Secretary of Commerce, a consistent champion of business, became the reluctant director of eighty-six seized steel companies. [3] Upon assuming his duties, Secretary Sawyer, in a revealing statement, declared that he had "neither requested nor wanted this job, but when our men at the front are taking orders in the face of great danger those of us further back can do no less." Sawyer expressed his dislike of witnessing, let alone participating in, the seizure of property. "We are, however, facing a situation of great peril where continued production of steel is essential to our national welfare." He pleaded for cooperation from both industry and labor so that a solution might be reached and the continued production of the weapons needed for the country's defense not be interrupted. [4]

Sawyer resented not having been consulted about either the seizure decision or the content of the President's speech, of both of which he disapproved, but he believed that considerations of national defense required that he be the good soldier. [5] In a meeting with the President and John Steelman late in the afternoon of April 8, Sawyer had agreed to do the job provided that he had "a free hand" after he took over. Truman reassured him on that point, and Steelman explained that, although there had been some talk of implementing the WSB recommendations, no "real commitments" had been given to Philip Murray. [6] The choice of Sawyer to run the steel mills was an attempt to blunt industry criticism and to demonstrate to the public that the administration's only intention was to assure a continuous supply of steel. [7]

Immediately after Truman's speech, Sawyer sent telegrams to the presidents of the companies to be seized informing them that he was taking possession of their properties at midnight. He asked the company presidents to serve as operating managers of their firms for the United States and to continue their usual functions— except that they would now be working for the government. The steel executives were ordered to fly the flag of the United States over their plants, to post a notice that the government had taken possession, and to set up separate books for the period of government operation. Sawyer also wired Philip Murray advising him of the seizure and urging that the steelworkers remain at work while the mills were run by the United States.[8]

At a press conference, Sawyer made clear that he was just following President Truman's orders. The need for steel was so great, the Secretary emphasized, that to obey the President seemed the proper thing to do. Sawyer announced that he had no plans to raise the wages of the steelworkers; he was counting on a quick settlement between the industry and the union by way of collective bargaining.[9]

The Steelworkers and the industry held negotiating sessions in the days immediately following the seizure, and the administration clearly hoped that a settlement would be reached before it had to take action on the wage issue. In his first press conference after the seizure Truman was asked if he planned to put the wage board recommendations into effect. "The parties to that dispute are negotiating for a settlement," the President declared, "and I want that negotiation carried through as a collective bargaining affair. I will not interfere with it." [10] When these initial bargaining sessions proved unsuccessful, the union began lobbying for a wage increase.[11] In response, Sawyer issued a statement on April 15, 1952: "Inasmuch as the negotiations which had been going on between industry and labor have ended, I shall proceed, promptly but not precipitately, to consider the terms and conditions of employment as I was instructed to do in Paragraph 3 of the President's Executive Order." [12]

The Secretary of Commerce apparently had two purposes in

mind: by announcing that he would consider wage adjustments he both placated the union and threatened the industry. He hoped that the industry, because of its distaste for being saddled with a government-imposed settlement, would come to reasonable terms both with the union and with the government. In pursuit of this objective, on Friday, April 18, Sawyer proposed one more meeting between the steel companies and the union and reiterated his intention to investigate the question of wage increases and other employment matters. He promised to consult with the President before taking action.[13]

Sawyer talked separately with representatives of the union and of the industry, but no joint session was held because the companies wanted a definite statement from the administration on price allowances and that was not forthcoming on April 18. Sawyer and his staff therefore spent the weekend preparing a wage order, and on Sunday, April 20, Sawyer asserted on "Meet the Press" that there would be some wage increases.[14] But the Secretary of Commerce was obviously reluctant to be the government official who imposed the wage adjustment. In an effort to gain time and share the responsibility, Sawyer met with Roger Putnam, administrator of the Economic Stabilization Agency. The two men agreed that the ESA would draw up recommendations for changes in the terms and conditions of employment, which Sawyer would submit to the President for his approval. Plans were also made for the steel industry to be informed that it could apply for the price increases to which it was entitled under the Capehart amendment.[15]

At this juncture the preparations to issue a wage order became entwined with the legal proceedings initiated by the steel companies, and events in the courtroom effectively restrained any administration action with regard to wage increases.

The entanglement of the wage issue in the court actions greatly disappointed the United Steelworkers. Philip Murray, pleased by the seizure order, had immediately issued a statement promising union cooperation with the government, and he ordered the local unions to call off the strike.[16]

To the union leaders, the seizure and Truman's speech vindicated their decision to cooperate with the government and postpone the strike until the Wage Stabilization Board could make its recommendations. Although there was no firm understanding with the administration that the steelworkers' wages would be increased while the government was in control, there was at least a possibility that this would be done. Moreover, even if the administration took no action on wages, the union thought the industry might settle sooner if the threat of a change in employment terms hung over its head.[17] And if neither of these occurred, seizure was still an important symbol, for, unlike the Taft-Hartley injunction, it was directed at the steel companies as well as at the union. Indeed, the President's speech clearly indicated that the steel seizure was being used as a weapon against the industry, as the recalcitrant party.

To rank and file union members, Harry Truman became a hero. Since January 1, 1952, the Steelworkers had stayed on the job without a contract. They had grown increasingly impatient with the three-sided negotiations and a strike had seemed the only way to gain their objectives. For most of them, however, a labor stoppage would mean severe hardship. A spokesman for a local union had described the impending walkout as one which "no one wanted—few could afford."[18] By April 8, union members were thus prepared for the strike while hoping for a miracle that would avert it. The huge steel plants, many of them situated in the Monongahela valley, had begun a gradual shutdown in anticipation of the stoppage. By the time the President spoke on radio and television an unusual darkness had already spread over the steel country. Steelworkers, congregating in local bars to hear Truman's speech, were pessimistic. As the President began, one union member remarked, "I'd give a thousand dollars right now to see that man drop dead." When Truman had finished, the same steelworker declared, "If he doesn't run again, by God, I'll write him in."[19] Jubilant over the news that the strike was off, the rank and file were grateful for the President's public support and hopeful that the dispute would now be settled without a strike. As Murray

Kempton reported from Homestead, "They were still standing at the corners at midnight talking about Mr. Truman. . . . They had come right under the gun and the President had paid off their patience." [20]

But, as the weeks passed and no settlement was achieved, union members were increasingly disturbed. When the administration's option to increase wages was precluded by the court action, resentment began to grow among the steelworkers.

Not surprisingly, news of the seizure astonished and dismayed the steel industry. On April 8, the chief executives of the major steel companies had met in New York to discuss the upcoming strike. The possibility of seizure had been mentioned but was dismissed because "the consequences that would flow from such action seemed so monstrous that we persuaded ourselves that it just could not happen," Clarence Randall, head of Inland Steel, later wrote. This conclusion was corroborated by reports from people "who professed to be close to the administration." [21] Exhausted by the days of tense negotiating sessions as the strike deadline neared, the steel executives had decided to take the night off. At about eight o'clock the news circulated that President Truman would be on all radio and television stations later that night. Randall recalled his immediate alarm, realizing that seizure was imminent. In his memoirs he relates his feelings as he watched the broadcast:

There on the screen before me was [Truman's] tense face, speaking the firm, crisp sentences that burned into my consciousness like the voice of doom. Seizure it was, . . . with the same finality [as] a death sentence. . . . He made it clear that this was the end of the matter: there would be no further talk.

I felt physically ill. It seemed to me that all that I had learned of government from school days on, all that I had believed in with respect to the balance of powers in a republic, all the safeguards conceived by the founding fathers for the preservation of our democracy, had suddenly been swept away. One man had coldly announced that his will was supreme, as Caesar had done, and Mussolini and Hitler. [22]

In his comments to the press, Randall was only slightly more restrained: "that is government by decree and not by law. . . . This is the end of stabilization, the end of collective bargaining and, if the courts sustain it, which God forbid, the end of the road for free enterprise." [23] Admiral Moreell asserted that President Truman's speech was "the most shocking statement made by a public official in my recollection." [24] The steel industry resolved to fight the seizure, both in the arena of public opinion and in the courtroom, with all the resources at its command.

The extreme reaction of the industry was paralleled in the nation's press. Antiadministration newspapers were vicious in their denunciation of Truman's action, and even previous defenders of the President censured him for abusing the powers of his office. The *Nation* noted that Truman had "made two serious mistakes: he exaggerated the crisis and exaggerated even more the 'inherent powers' with which the Constitution has invested him. . . . Put plainly, we don't like the arbitrary exercise of executive authority even to force a just settlement of a labor dispute or to prevent a strike in steel." [25] Only one newspaper with a sizable circulation, the prolabor New York *Post*, supported the seizure.

The majority of unfavorable editorials characterized the President's action as dictatorial—the New York *Daily News* titled its editorial "Truman Does a Hitler" [26]—and argued that because Truman had neither constitutional nor statutory authority to take over the steel mills, he should have consulted Congress before acting. [27] A columnist for the *Wall Street Journal* likened the President's attitude to a mayor in a Pacific coast town "who enjoyed a brief period of limelight after the First World War by going in for strong-arm arbitrary action. Closing down the office of an organization he did not like, he boasted: 'We didn't use any law; we used nails.' " [28]

Another theme stressed by condemnatory editorials was that Truman had overstepped his constitutional powers, since the United States was not at war. Some editorialists simply ignored the existence of the Korean conflict. The *Saturday Evening Post* commented: "If a great industry can be seized by the Government in time of peace upon such tenuous authority as that cited by Mr. Truman, then it is difficult to see what group of citizens can safely consider itself immune from arbitrary compulsions." [29] Other editorials acknowledged that the United States was involved in the Korean hostilities but emphasized that American participation had not been ratified by Congress. [30]

Disapproving press comment charged, furthermore, that Truman had abused his powers, not to ensure a supply of steel for the national defense but to promote the interests of a politically important pressure group. [31] One steel industry organ called it "confiscation—with labor dictatorship." [32] *Time* magazine noted:

In seizing the steel mills and violently taking sides, in unnecessarily stretching the vast powers of the presidency, Truman had acted primarily as a politician, not as a President.

A tipoff on the Administration's motive came inadvertently last week from Price Stabilizer Ellis Arnall. "The steel situation," said Arnall, "is the stuff on which campaigns—political campaigns—are won and lost." Politician Harry Truman was obviously operating on the axiom of political arithmetic that there are more votes in Big Labor than in Big Steel. [33]

Repeatedly, editorials declared that the United States government was no longer a "neutral referee" in a labor-management dispute; the President had put the administration squarely on the side of labor. [34] These same editorialists warned labor that the "sword has two edges. Mr. Truman can use it today in behalf of labor. Some future president, with Mr. Truman's precedent to guide him, can use it against labor. If industry can be blackjacked into paying higher wages, labor also can be blackjacked into taking a cut." [35]

The New York *Post*, virtually alone in its defense of the administration, could only plead, "Instead of shouting 'usurpment' 'socialism' 'totalitarianism' it would be wiser for spokesmen of the industry to give their answer to the President's figures." [36]

The President's figures, however, also caused controversy. Truman's seizure speech was attacked as "filled with distortions, half-truths and omissions." [37] The President's discussion of steel industry profits before rather than after taxes became a major bone of contention. Truman had claimed that the steel companies were making a profit of $19.50 a ton. He did not mention that this was a before-tax figure and that net profit after taxes amounted to only

$6.69 a ton.[38] In addition, he had failed to indicate that net profits for the steel industry in 1951 had fallen 12 to 13 percent below the previous year's level.[39]

The *New Republic* supported the President's use of the profits-before-taxes figure, arguing that it was the accepted practice for total profits to serve as the basis of most computations made by the government, just as total wages rather than take-home pay were used where labor was concerned. It noted, too, that Ellis Arnall, director of OPS, had proved to a Senate committee that the steel industry was not entitled to a price increase whether profits were calculated before or after taxes. The magazine accused the press of uncritically adopting the figures publicized by industry propaganda.[40]

Despite the mass of antagonistic editorials, it is by no means clear that public opinion opposed the seizure of the steel mills. In fact, judging from the mail received by the White House, Congress, Secretary Sawyer, and the Justice Department, and from letters to the editors of various newspapers, the public was closely divided. The White House reported that its mail was "running half anf half." [41] On the floor of the House of Representatives, Donald O'Toole, a Democrat from New York, complained about the misleading impression being fostered that the people of the country were incensed over the President's action. "I believe that I am truly representative of the membership of this House and I have received but five letters protesting the President's move. Yet, I have received 41 letters from the mothers and fathers of boys who are in Korea and who demand that their sons shall have the steel, armament, and ammunition essential to their task." [42] A conservative Michigan Republican, Homer Ferguson, speaking from the floor of the Senate, also referred to the lack of antiseizure mail, but his complaint was of a different kind: "Because of illegal acts being pyramided one upon another . . . the people back home are being led to think that it is just and right. They are not disturbed about it. I do not know about the mail that is coming to other Senators, but I am getting very few letters with respect to this great principle and with respect to the violation of the Constitution." [43] Letters to other congressmen, however, were more equally distributed. [44] The bulk of Secretary Sawyer's mail, not unexpectedly, was antiseizure, and many telegrams urged Sawyer not to make any changes in the terms and conditions of employment while he was in charge of the mills. [45] The Department of Justice, however, received many letters supporting the seizure and advising the department how to defend it in court. [46] Though business groups protested against the seizure as an "executive usurpation of power" for "political purposes," [47] several city councils in the steel-producing areas of the country endorsed the President's action. [48]

Polls reinforce the conclusion that the public had not reacted wholly adversely to the steel seizure. Interviews conducted by the Public Opinion Index for Industry in fifty-three eastern and midwestern cities soon after the President took over the mills found that 51 percent approved Truman's action while 43 percent opposed it. The split was largely along party lines: Democrats favored the seizure, Republicans disapproved of it. An even larger majority, 55 percent, thought that the President had the power to seize the steel mills. [49] Another poll, taken in Minnesota by the Minneapolis *Sunday Tribune*, found similar results: more people approved of the seizure than disapproved. This division, too, was along party lines. Among Democrats, 74 percent favored the seizure and 10 percent opposed it; among independent voters 41 percent approved, and 37 percent disapproved; among Republicans, 66 percent objected to the seizure and 22 percent endorsed it. [50] The Gallup poll, which interviewed from April 27 through May 2, came up with different results: 35 percent endorsed the seizure, 43 percent condemned it. [51] The President's popularity, at a low level before the seizure, increased by three percentage points after the seizure and continued to rise through June 1952. [52]

The steel industry, however, had no intention of allowing public opinion to develop in the President's favor. Clarence Randall spoke for the entire industry when he answered the President on nationwide television and radio the night after the seizure speech. [53] His address was impassioned, hard-hitting and every bit as vitriolic as Truman's speech the night before. Randall argued that Truman had no power, statutory or constitutional, to seize the steel mills and charged that he had done so to discharge "a political debt to the CIO. Phil Murray now gives Harry S. Truman a receipt marked, 'paid in full.' " [54] Randall attacked the public members of the Wage Stabilization Board for being prolabor and for awarding the Steelworkers twice as much as they had ever received as a result of collective bargaining. Just two weeks before, Randall reminded his audience, David McDonald, secretary-treasurer of the Steelworkers, in a public discussion of the union's political influence in Washington, had admitted that the Steelworkers were in a favorable position in the current dispute because of a "rather friendly gentleman in the White House." [55]

Randall also accused the President of grossly distorting the facts regarding the profits of the steel industry. Focusing on Truman's use of before-taxes profit figures, Randall asserted that the industry could not pay higher wages, and build new plants to produce more steel with what revenues were left to it after paying taxes. [56] Acceptance of the recommendations of the Wage Stabilization Board, he claimed, would increase costs by $12 per ton. Randall concluded his address with a rousing plea to the American people to join the steel industry in its fight against the President.

The steel industry followed up Randall's speech with an expensive advertising campaign, sponsored by the "Steel Companies in the Wage Case." In newspapers across the country, the group placed full-page advertisements designed to obscure the price issue by emphasizing the constitutional and political aspects of the seizure. And these advertisements, together with Randall's address and the speeches and congressional testimony of other steel executives, were widely distributed. [57] The group also prepared a detailed background memorandum with a statistical appendix "for the information of editors and others interested in the serious issues involved." [58] In this way, the industry was able to publicize the figures it wanted used.

The steel companies' advertising extravaganza aroused the ire of many of the President's defenders. Philip Murray, appearing before the National Press Club, assailed the industry for spending taxpayers' money to finance the advertising campaign by writing it off as a business expense against taxes.[59] Charles Murphy, the President's special counsel, complained to Secretary Sawyer about the inflammatory statements made by the steel companies in their published pamphlets. "In view of the intemperate and dishonest attacks on the President which are contained in this and similar propaganda by the steel companies, I must say it is very difficult to counsel a policy of restraint," Murphy declared.[60] And in Congress, Senator Wayne Morse of Oregon (still a Republican in 1952), criticized newspapers for "editorial policies [which] seem to follow rather remarkably the contents of the large full-page and half-page advertisements in the newspapers paid for by the steel companies of this country in misinforming the people of the country in regard to the facts in the steel case." [61]

In an attempt to counter the steel industry's campaign, the Truman administration first tried to bolster its position in Congress. The morning after the seizure order, the President sent a special message to Congress giving the reasons why he thought his action was the best course in the circumstances of the steel dispute and stating his willingness to follow any further legislative mandate on the subject. Truman argued that the only alternatives to his seizure were a shutdown of the industry or a sizable steel price increase. The Taft-Hartley Act was not mentioned by name although Truman made an oblique, loaded reference to it:

It may be that the Congress will feel the Government should try to force the steel workers to continue to work for the steel companies for another long period, without a contract, even though the steel workers have already voluntarily remained at work without a contract for 100 days in an effort to reach an orderly settlement of their differences with management.[62]

Congress could endorse any of these options, the President asserted, or it might wish to write a bill "establishing specific terms and conditions" for government operation of the mills. Such legislation, while not "essential," would, in his opinion, be "very desirable." Truman indicated, however, that in the absence of congressional action he would continue to take the responsibility for keeping the steel mills functioning while he tried to settle the dispute.[63]

The President's message seemed designed to show respect for Congress' prerogative to pass legislation while at the same time discouraging any action. Congressional approval of a seizure bill appeared unlikely (the very reason Truman did not ask for one before he announced the seizure); thus the message was really aimed at forestalling negative measures. The best the administration could expect was no legislation at all and, given the political situation in April 1952, this was a distinct probability. As one news magazine noted, "Congress in an election year, with its members eager to get home, is not likely to substitute a law for the President's order. The most it will do is investigate." [64]

The administration's hopes for congressional inaction were realized. The immediate response of Senators and Representatives to the steel seizure was largely verbal and emotional. The House of Representatives, after a few flights into oratory, voted to recess from April 10 to April 22.[65] The Senate, on the other hand, indulged its penchant for rhetoric throughout the month of April. The ostensible purpose of the initial Senate debate was to decide which committee should deal with the President's message. Because it involved a labor dispute, Vice-President Alben Barkley ruled that the message should be referred to the Committee on Labor and Public Welfare, where there were at least a few supporters of the President. The Republicans, led by Senator Styles Bridges, thought it should be sent either to the Committee on Banking and Currency, because the stabilization program was at issue, or the Judiciary Committee, because the constitutional power of the President to seize an industry was at issue. Both committees were chaired by conservative Democrats who would probably condemn the seizure. Vice-President Barkley stuck to his ruling, and the Republicans eventually dropped their opposition.[66] During the debate, several Senators pointed out that Congress had previously considered giving President Truman seizure power and had voted against it; clearly, in their view, the President was disobeying the will of Congress and usurping power.[67]

A small minority of Senators, however, voiced approval of the seizure. Senator Hubert Humphrey declared that the Senate should feel a sense of guilt for not having legislated seizure powers earlier: "Congress was perfectly willing to let the bodies of young men be seized . . . [but seems] unwilling to give the President the power of seizure . . . so as to protect the bodies and hearts and lives and souls of those young men and their families by making sure that the arms for defense are ready and available." [68] Humphrey was joined by Wayne Morse, who, on April 9, 1952, had introduced a bill (S. 2999) to provide more effective means of "dealing with labor disputes in vital industries which affect the public interest." [69] The bill described the procedure the President should follow when a strike was imminent and included seizure as a possible remedy. Morse, in a speech introducing his bill, noted that throughout its history Congress had accused Presidents of acting arbitrarily in various instances but never had the foresight to pass needed legislation before a situation became critical. He remarked that "in the cloak rooms of the Senate" it was generally agreed that the Taft-Harley law's national emergency disputes provisions were inadequate, but that Congress had evaded its responsibility. Morse supported the President's seizure as necessary to protect the lives of soldiers in Korea.[70] Another independent Republican, William Langer of North Dakota, wrote to a constituent, "It is my honest belief that the President was not conniving with the unions at all, but did what he did in an emergency." [71] But this assumption was questioned by those who believed that if a real shortage of steel existed and the lives of American soldiers were at stake, the steelworkers' union would never go out on strike.[72]

Congressmen introduced many bills pertaining to the steel seizure; few received any attention before the courts made the immediate object moot.[73] Senator Styles Bridges, for example, submitted a resolution calling for a study of the steel seizure by the Judiciary Committee to determine if there were any constitutional or statutory authority for it. The resolution was referred to the Committee on the Judiciary, but Judge Pine ruled before the investigation could begin in earnest.[74]

Lengthy hearings on some measures did take place, however. The Allen resolution to investigate the Wage Stabilization Board, which finally came out of committee, was favorably acted upon by the House on April 24, and the House Committee on Education and Labor commenced its investigation.[75] In the Senate, the Committee on Labor and Public Welfare held hearings [76] on the President's April message to Congress, Senator Morse's emergency disputes bill (S. 2999), and another Morse proposal (S. 3016) to "provide an orderly procedure for the relinquishment of possession of the steel plants under conditions which will assure the continued production of the articles and materials required for the common defense." [77] Both the House and Senate Committees on Banking and Currency had been considering the renewal of the Defense Production Act of 1950 as amended in 1951, and their hearings were greatly influenced by the steel seizure.[78] Representative Howard W. Smith, the powerful and conservative Virginia Democrat, introduced a bill "to protect the national defense against the consequences of certain labor disputes" by allowing federal courts to appoint receivers for the assets· of the union and the companies involved, if the dispute was not settled within the Taft-Hartley eighty-day injunction period.[79] Smith designed his proposal as an amendment to the Universal Military Training and Service Act then under consideration by the House, so it was referred to the House Committee on Armed Services, which proceeded with hearings on it.[80]

The congressional hearings provided the administration, the union, and the steel industry with a forum in which to air their views. Many of the same witnesses appeared before several committees, and the events leading up to the steel seizure crisis were retold numerous times. The most frequently called witnesses included Nathan Feinsinger, chairman of the WSB; Ellis Arnall, director of price stabilization; Philip Murray, president of the Steelworkers; John A. Stephens, vice-president of U.S. Steel; Charles E. Wilson, former director of defense mobilization; and John C. Bane, Jr., industry member of the WSB special steel panel.

The testimony of administration officials concentrated on five major arguments: the requirements of national defense made the steel seizure necessary; the use of the Taft-Hartley Act procedure would not have provided uninterrupted production of steel; the Wage Stabilization Board recommendations were fair and in accord with stabilization guidelines; the steel industry could absorb the cost of the WSB package with a minor diminution of profits and, therefore, a steel price increase was unwarranted; the price increase issue was responsible for the impasse in the steel dispute. The question of the President's power to seize was avoided where possible.[81]

Representatives of organized labor supported the administration on all counts in their testimony. Philip Murray gave an especially forceful presentation in which he stressed that the WSB recommendations were not exorbitant, that it would have been unreasonable and inequitable to make the union work under a Taft-Hartley injunction after it had voluntarily postponed its strike for almost one hundred days, and that he was shocked at industry charges that he had made a deal with President Truman. "My own personal relations with the President of the United States have been of a friendly nature, but it is beyond my concept of this thing we call manhood and ethics for me as president of a great organization to engage myself in any kind of a dirty, filthy practice," Murray declared.[82]

Industry spokesmen rebutted administration testimony by insisting that the wage board's proposals were unstabilizing; that the steelworkers were already the best paid workers in the country; that the companies desperately needed a large price increase; that the seizure was unjustified because the President should have proceeded under the Taft-Hartley Act; and that the authority to deal with labor disputes should be taken away from the WSB.[83] Hence, through the medium of congressional testimony, the controversial facets of the steel situation received a thorough examination.

But Congress took no concrete action to resolve the impasse. The most likely explanation for inaction was the fear that Congress would then be held responsible if a steel strike occurred. Moreover, once the steel seizure became the subject of court proceedings, many members hoped that the courts would invalidate the seizure and make congressional legislation unnecessary.[84]

Two measures did pass the Senate, but never became law. An amendment to the Third Supplemental Appropriations Bill enacted by the Senate forbade the use of any funds made available by the bill for the purpose of enforcing Executive Order 10340.[85] But, even if it had become law, this provision would have had little practical effect because the government did not need funds to run the mills; the steel companies managed their mills as usual, keeping separate records for the period of government operation. Senator Tom Connally, in explaining his vote against this amendment, noted, "I fully share with you the fear of the consequences of the undue extension of the theory of inherent powers; but it is too fundamental a problem to be met by meaningless restrictions on appropriations bills." [86] Another amendment to an appropriations bill, also added by the Senate, was prospective only:

No part of any appropriation contained in this act, or any funds made available for expenditure by this act, shall be used for the purpose of acquiring, seizing, or operating any plant, facility, or other property, unless the acquisition, seizure, or operation of such plant, facility, or other property is authorized by act of Congress.[87]

But by the time of final congressional action on these bills, the steel plants had been returned to their owners, and the amendments were therefore omitted.[88]

The administration also advanced its arguments in forums other than the Congress. Truman and his chief public spokesman on the steel issue, Ellis Arnall, sought to transform the steel crisis from a particular labor dispute into a broader battle against "big business." In a radio broadcast on April 18, 1952, Arnall developed at length the position that the industry's requests for price increases reflected greed rather than necessity. He concluded with a rousing "This is your fight! This is the people's fight!" [89] Truman, in an address to the Americans for Democratic Action, noted that the steel companies, despite record profits, had told the government "to give them a big price increase, or else." [90] The

Republicans, Truman declared, had immediately rallied to the industry's cause by demanding investigations and threatening to wreck price controls. "And what is the purpose of all this?" the President asked. "The purpose is to preserve high profits for the steel companies and prevent wage increases for the steelworkers. That shows exactly where the Old Guard stands. It shows that their hearts lie with the corporations and not with the working people. It proves that the old Republican leopard hasn't changed a single spot." [91]

The President also tried to use his press conferences to explain his actions during the steel crisis. Rather than offering clarification or building support for his position, however, Truman's answers to reporters' questions frequently exacerbated his difficulties. The most damaging blunder was the President's response when asked if, in view of his belief that he could seize the steel mills under his inherent powers, he also could seize the nation's newspapers and radio stations. "Under similar circumstances the President of the United States has to act for whatever is for the best of the country. That's the answer to your question." [92] His remark, widely interpreted to mean that he would seize the press and the broadcasting industries if he thought it necessary, was given great play in the newspapers.

At the following week's press conference, Truman attempted to mitigate the bad publicity. The President began with a short statement:

> There has been a lot of hooey about the seizure of the press and the radio. As I told you last week, the President of the United States has very great inherent powers to meet great national emergencies. Until those emergencies arise a President cannot say specifically what he would do or would not do. I can say this, that the thought of seizing press and radio has never occurred to me. I have difficulty imagining the Government taking over and running those industries. [93]

Truman then launched into a defense of the steel seizure, emphasizing the need to meet "as great an emergency as we have ever faced." "I am not the first President that has been abused under the same circumstances," he asserted, "so I know how to take it." [94] When queried if there were any limitations at all to a President's power to act during emergencies, Truman advised the reporter to "read your history and find out." The President then mentioned actions by Presidents Jefferson, Tyler, Polk, Lincoln, Andrew Johnson, and Franklin Roosevelt for which there had been no specific authority and from which the nation had derived great benefit.[95] But the correspondents were not satisfied with that answer. May Craig of the Portland (Maine) *Press Herald* informed the President that she had been listening to comments on Capitol Hill to the effect that "one of the greatest protections for our liberty is that we live by written law, and they dread a departure into unwritten or inherent powers, fearing that precedent. Do you recognize the danger of that too?" "Well, of course I do, May," Truman replied. "But then when you meet an emergency in an emergency, you have to meet it." [96] That glib reply did not impress the reporters, and their reaction should have served as a signal that a stronger, better-documented argument defending the inherent powers of the President would have to be produced if the administration expected to win its case before the public and in court.

six

THE STEEL COMPANIES GO TO COURT

At the same time that the administration was actively defending itself against criticism in Congress and in the press, the Justice Department was defending the seizure in the courts. As had often happened in American history, it appeared that the courtroom would be the crucial forum. The President considered this a favorable circumstance. In view of the Supreme Court's generous construction of presidential power during wartime, Truman and the White House staff were confident that the courts would uphold the seizure.[1] The staff therefore left the case to the Department of Justice to deal with as it saw fit. The fate of the steel seizure was in the hands of lawyers and, according to the newspaper accounts of the time, the future of democracy in America hinged on the legal proceedings.

Twenty-seven minutes after the President had concluded his seizure speech, attorneys for the Republic Steel Corporation and the Youngstown Sheet & Tube Company were at the door of Federal District Judge Walter Bastian's home in Washington. They submitted a motion for a temporary restraining order (TRO) to enjoin the President's seizure and also asked for a permanent injunction. Judge Bastian, refusing to consider the motions without giving the government a chance to be heard, set a hearing for 11:30 the next morning.[2] To waiting reporters, John J. Wilson, the attorney for Youngstown, denounced the seizure as a "coercive effort by the Government to compel us to enter into a union contract according to the recommendations of the Wage Stabilization Board." "The Government has reached out and taken our property away from us," he shouted at the press.[3] Another attorney averred less emotionally that this was the beginning of the "biggest lawsuit in years."[4]

Promptly at 11:30 A.M. on Wednesday, April 9, the court convened to hear motions for a temporary restraining order.[5] Attorneys for Youngstown and Republic were joined by those for Bethlehem Steel Company. As the task of sitting on motions was rotated routinely among the judges of the district court, it was simply a matter of chance that Judge Alexander Holtzoff was presiding. Holtzoff, a self-assured judge with a reputation for feistiness, was a former Justice Department official and a recognized authority on federal judicial procedure. He had been appointed to the bench by Truman in 1945.[6]

The arguments presented before Judge Holtzoff were a microcosm of those that would be offered at a future full hearing on the legality of the steel seizure. The major issues were set forth clearly, but lack of preparation time as well as the sense of urgency that surrounded the request for a TRO prevented an in-depth discussion of them. Basically, the steel company attorneys sought to show that the seizure was illegal and that the companies would be irreparably injured were it not immediately enjoined. They asserted that there was "no power in the President, and no power in Mr. Sawyer, to make the seizure."[7] Neither the Constitution nor any statute gave the President authority to take possession of the steel mills, they insisted. Youngstown, Republic, and Bethlehem rested their claims of irreparable injury on the probability that the government would conclude a contract with the United Steelworkers of America putting into effect the Wage Stabilization Board recommendations; that the government would have access to the companies' trade secrets; that normal customer relationships would be disrupted; and that property damage might result from inexperienced management.[8]

Judge Holtzoff's questions reflected two major difficulties with the arguments for the companies. First, irrespective of the legality of the seizure, Holtzoff believed it necessary to consider factors other than irreparable injury: "There is a question of balancing equities, when you apply for a temporary restraining order." John Wilson answered, "I will not dispute that with Your Honor at this time. I think, frankly, it is not a question of balancing the equities. I think the equities are 100 percent on our side." [9] If the public interest required the continued production of steel, Wilson pointed out, the President could invoke the Taft-Hartley Act, which contained the congressionally approved method for averting national emergency strikes and which would deprive no one of legal rights.[10] Thomas F. Patton, general counsel of Republic Steel, supported Wilson's contention that the government would not be hurt by the issuance of a TRO because each steel company had at least thirty days' inventory and could keep defense contractors supplied until the case was settled.[11]

Judge Holtzoff's second problem had not even been discussed by the steel companies' attorneys. As the judge described it,

> These actions are nominally directed against the Secretary of Commerce. . . . But the Secretary of Commerce is acting pursuant to a directive of the President, a specific directive, or a specific order of the President.
>
> Aren't you indirectly seeking a restraining order against the President though not nominally so? And, if so, does the Court have the power to issue an injunction against the President of the United States? I do not know of any case on record in which a Federal Court, or any other court, has issued an injunction against the President of the United States.[12]

Bruce Bromley, Bethlehem's attorney, insisted that the precedents indicated that the suit should not be treated as an action against the President. But under questioning from Holtzoff, Bromley also asserted that a district judge in any event had authority to enjoin the President.[13] Wilson admitted that he shared the judge's doubts on that proposition,[14] but agreed with Bromley that the motions were directed only at a subordinate of the President's. Wilson and Bromley pointed to a long line of cases where suits had been directed at inferior officers of the executive branch and had not been interpreted as suits against the superior; suits against cabinet members had not been considered suits against the President. Moreover, in the instant case, Secretary Sawyer in his orders to the steel industry had stated, "I, Charles Sawyer, deem it necessary. . . . I, therefore, take possession." This, according to Wilson, clearly made Sawyer the official responsible for the seizure, and it was he who should be restrained by the Court.[15]

Holmes Baldridge, Assistant Attorney General in charge of the Claims Division of the Department of Justice since December 1950, represented the Secretary of Commerce. A soft-spoken, conservative man, and a tireless worker, Baldridge had come up through the ranks of the Justice Department. He had spent most of his career in the Antitrust Division and had little experience in handling other types of civil cases.[16] His oral argument revealed that he was unfamiliar with, and ill prepared to discuss, the law concerning the pivotal issues in the steel seizure case.

Baldridge concentrated his argument on two points. First, he countered the irreparable injury allegations by emphasizing the limited nature of the seizure. Although the United States was the nominal owner of the companies, Baldridge demonstrated that all operations were left in the hands of the previous owners and that business would go on as it always had. Baldridge also discounted the possibility that a new contract would be signed between the government and the Steelworkers. Executive Order 10340, authorizing the seizure, specifically stated that collective bargaining between the industry and the union was to continue while the government was in possession of the mills. In addition, the history of previous seizures under the President's constitutional powers indicated the unlikelihood that the government would conclude an agreement with the union.[17]

Second, Baldridge argued that the request for a TRO was "untimely" because the steel companies had an "adequate remedy at law."[18] The government maintained that the seizure was a "legal taking under the inherent executive powers of the President"; thus, if the steel companies could prove that they had suffered damages as a result of the seizure, they were entitled to just compensation under the Fifth Amendment. Furthermore, if the seizure were found to be illegal, the steel companies could bring suit for damages under the Federal Tort Claims Act.[19]

Baldridge also argued briefly that the steel companies' contention that the President should have used the Taft-Hartley Act, instead of seizure, to block the strike was not a proper subject for the court's inquiry. Judge Holtzoff concurred, noting: "It is not for this Court to say which of several courses the President should have pursued. That is for the President. If he has legal power to pursue the course that the President has pursued, the mere fact that he had the choice of some other course is nothing for the Court to pass on." [20]

The government attempted to avoid the ultimate question of the President's power to seize the steel mills by asserting that there was no necessity to discuss it on a motion for a temporary restraining order. But Judge Holtzoff pressed Baldridge to argue the issue on the ground that a denial of the TRO would reflect the court's view that the seizure probably would be declared legal in the future court proceedings.[21]

Baldridge began his argument on this point with a recital of the provisions of Article II of the Constitution: the executive power shall be vested in a President of the United States; the President shall be Commander in Chief of the Army and the Navy; the President shall take care that the laws be faithfully executed. Baldridge maintained that these provisions in effect granted the President all executive power that was not specifically assigned to others elsewhere in the Constitution. Judge Holtzoff asked Baldridge what he meant by "executive powers." The judge thought it was the power to execute statutes. Baldridge answered that "among other things it is the power to protect the country in times of national emergency by whatever means seem appropriate to achieve the end." "Well, how far would you carry that?" Holtzoff wished to know. "Well, we don't think we have carried it too far in this particular instance, Your Honor. I don't know as I can discuss it." [22] But Baldridge did try to elucidate further the basis of the President's power to seize. The need for steel was so great, he declared, that lack of it would cause an emergency "sufficient to create the power to seize under these circumstances." Judge Holtzoff quickly refreshed Baldridge's memory by pointing out that Chief Justice Hughes, in an opinion many years earlier, had stated that emergencies do not create power but might supply the occasion for the use of powers already granted but not previously exercised.[23] "I think that whatever decision I reach, Mr. Baldridge," Holtzoff asserted, "I shall not adopt the view that there is anyone in this Government whose power is unlimited, as you seem to indicate." [24]

As Baldridge concluded his presentation, Bruce Bromley, the attorney for Bethlehem, tried to elicit from him a promise that, if the TRO were denied, the government would not sign a contract with the union. Baldridge said he had no authority to give any assurance on that point and had no knowledge of the administration's plans. He emphasized, in closing, the government's dire view of the impact of issuance of a TRO: It would be "an order, in effect, to strike." Baldridge explained to Holtzoff that a TRO would change the situation back to the status quo before the seizure and that the Steelworkers would then immediately go out on strike.[25] On this threatening note, Baldridge terminated his argument.

Barely ten minutes after the argument had concluded, Holtzoff reconvened the court to announce his decision. He sketched his cautious approach to the issuance of a TRO:

An application for a temporary restraining order involves the invocation of a drastic remedy which a court of equity ordinarily does not grant, unless a very strong showing is made for the necessity and the desirability of such action. The application is, of necessity, addressed to the discretion of the Court. It is not sufficient to show that the action sought to be enjoined is illegal. It is, in addition, essential to make a showing that the drastic remedy of an injunction is needed in order to protect the plaintiff's rights.

In arriving at its decision, the Court must arrive at a balance of equities, and consider not only the alleged legality or illegality of the action taken, but also other circumstances that will appeal to the discretion of the Court.[26]

Uppermost in Holtzoff's mind was the problem of granting an order that would nullify an action of the President of the United

States. He admitted that technically the TRO would run only against Secretary Sawyer, but he emphasized that its effect would be to prohibit the performance of the President's executive order and that he had grave doubts about the authority of a federal district court to enjoin an action of the President.[27] "The court," stated Holtzoff, "should not do by indirection what it could not do directly." Unless there was a "vital" reason, the court should not interfere with the President's attempt to forestall a national emergency. The possibility of irreparable injury in the immediate situation in which the steel companies found themselves was too remote to override this consideration, according to Judge Holtzoff. He indicated that if any of the fears of the steel companies materialized, they could again request a TRO and would have a stronger case. In Judge Holtzoff's opinion, the balance of equities was in the government's favor, and he therefore denied the motion for a temporary restraining order.[28]

The following day, April 10, 1952, four companies—Youngstown Sheet & Tube, Republic, Bethlehem, and Jones & Laughlin—sought to advance their cases for hearing on the merits.[29] Attorneys for the companies first appeared before District Judge Bastian, but he disqualified himself from hearing the cases because he owned thirty shares of stock in the Sharon Steel Corporation which, although not a party in the present actions, was one of the companies that had been seized. Holmes Baldridge, representing Sawyer, stated that he had no objection to Judge Bastian sitting on the case. The judge, however, declared that he would not preside because judges "like Caesar's wife, . . . have got to be above suspicion." [30] The hearing on the companies' motion was then transferred to Judge David A. Pine, who was acceptable to both the steel industry and the government despite his admission that his wife owned twenty to twenty-five shares of Bethlehem Steel Company stock.[31]

During his long tenure on the federal district court—he had been appointed by President Roosevelt in 1940—Pine had earned a solid reputation. He brought a competent knowledge of both private and public law to his job, a result of his previous employments. From 1913 to 1921, he had served in the Justice Department where he rose to the position of Special Assistant Attorney General.[32] For the next thirteen years, he had engaged in private practice in Washington, D.C. In 1934, Pine had been chosen chief assistant U.S. attorney for the District of Columbia and then, in 1938, was promoted to U.S. attorney.[33] Mild-mannered, "slightly stooped, with iron-gray hair and rimless spectacles," [34] Pine was known as a "methodical and hard-working," [35] "conservative" [36] judge with a very dry sense of humor.[37] Lawyers for both sides were pleased that he was sitting on the case.[38]

The attorneys for the companies asked Judge Pine to set the steel seizure case down for trial on the merits at the earliest possible date. "We feel," declared Edmund Jones representing Republic, "that it is of the utmost importance to the parties involved and to the country at large that this issue of the right of the President to direct the seizure of the property of these companies be tested out at the earliest possible moment." [39] But Holmes Baldridge refused to be rushed into a trial; he wanted all the time allowed him by the Rules to prepare the case thoroughly.[40] "I should have thought, Your Honor," Wilson sarcastically observed, "that the Government of the United States would have known what the law was on this subject before the Executive Order was issued; and they wouldn't have to make their research after the injunction suits were filed." [41] Since the government would not agree to advance the case for trial on the merits, Judge Pine advised the steel companies to file motions for a preliminary injunction, in response to which a hearing could be obtained much more quickly.[42] The April 10 proceeding was then adjourned.

At various times during the period April 10 to April 24, seven steel companies filed complaints for a permanent injunction against the seizure and a declaratory judgment that it was unlawful, accompanied by motions for a preliminary injunction that would enable them to obtain relief more quickly.[43] All the complaints were similar: the President was without statutory or constitutional authority to issue the seizure order; any actions taken under Executive Order 10340, therefore, were "invalid, unlawful and without effect." [44] Pending a final determination of the merits of the case, after which a permanent injunction could be granted, the steel companies sought a preliminary injunction restraining Secretary Sawyer from implementing Executive Order 10340 and from interfering with the plaintiffs' businesses in any manner whatsoever. On the ultimate merits, the steel companies asked Judge Pine to find the steel seizure was without authority, to declare Executive Order 10340 void, and permanently to enjoin the Secretary of Commerce from carrying out the seizure. [45]

The Justice Department, representing Sawyer in the cases, agreed to a consolidated hearing of all the plaintiffs' motions for a preliminary injunction on April 24, 1952. [46] The government's opposition to the motions was based chiefly on five propositions: (1) The threat of a strike in the steel industry, scheduled to begin on April 9, constituted a serious national emergency because the slightest interruption in the production of steel would adversely affect the "well-being and safety of the United States in a critical situation." [47] (2) In such circumstances, the President has inherent constitutional power to seize the steel companies in the manner described in Executive Order 10340. (3) The courts do not have authority to enjoin the President's action nor, if they had such power, could they enforce their decision against the President. (4) A preliminary injunction is never granted as a matter of right; the court must balance the interests of each of the parties to the suit and should not interfere before a full hearing on the merits unless the damage resulting from the denial of a preliminary injunction is shown to be irreparable and more harmful than the granting of an injunction would be to the opposing party. (5) The steel companies had not exhausted the legal remedies available to them and had not proved irreparable injury, and were therefore not entitled to the equitable relief requested. [48]

Both the Justice Department and the attorneys for the steel companies cooperated with Judge Pine's desire to handle the cases with dispatch. Baldridge and his associates filed their brief before April 24, thus giving Judge Pine and the steel industry lawyers a chance to study the government's case. The steel attorneys answered the Justice Department's points in their own briefs filed with the court on the morning of the hearing.

Since the government's brief was written first, Baldridge had the opportunity to present the steel seizure case in the posture most favorable to the administration. But he and his colleagues in the Justice Department made a critical error of judgment which may well have affected the final outcome of the case.

When a case is heard on a motion for preliminary injunction, ultimate questions of law as to the merits of the case need not be decided at that time. To be sure, the question of whether the plaintiff is likely to prevail on the merits is relevant. But it is only one factor; the court must also weigh the injuries that will be suffered by the parties and the public if preliminary relief is granted or denied.[49] In the steel seizure case, this required balancing the impact of a strike against the companies' claims of injury. It is most unusual for a court to determine the legal merits of a case on a motion for preliminary injunction, unless it is likely that no further facts or evidence will be produced at a later hearing on the merits. Baldridge's mistake was to invite the court to focus on the merits. In his brief, he emphasized too strongly the question of the President's constitutional power to seize private property, to the detriment of the equitable issue as to injury, where the administration's position stood on firmer legal ground.[50] The steel companies' attorneys, taking their cue from Baldridge, discussed the constitutional issue at length in their briefs. As the April 24 hearing got under way, it became evident that Judge Pine believed he should decide that issue before he considered whether or not to grant a preliminary injunction.[51]

On the morning of April 24, Judge Pine's courtroom was crowded with counsel for the opposing parties. Twenty-one attorneys, including members of the most prestigious New York and Washington law firms, appeared for the seven steel companies. Secretary of Commerce Sawyer was represented by Department of Justice attorneys Holmes Baldridge, who presented the government's case, and Marvin Taylor and Samuel Slade.

Theodore Kiendl, counsel for U.S. Steel and a member of

the New York firm of Davis, Polk, Wardwell, Sunderland, and Kiendl, delivered the major portion of the argument for the companies. From the order in which he dealt with the questions involved in the case, it was apparent that the attorneys for the steel industry had decided to emphasize the equity aspects of the case, rather than the ultimate issue of the President's power, in oral argument. This reflected the attorney's belief in the traditional practice that on a motion for preliminary injunction questions of equity assume prominence over the legal merits of the case. They had briefed the merits thoroughly only because the government had done so. Kiendl began his presentation by sounding the central theme that a preliminary injunction was necessary to restrain "what we consider to be the imminent threatened changes in the terms and conditions of employment of a steel employee." [52]

Before elaborating on the impending injury, however, Kiendl addressed himself to Judge Holtzoff's conclusion that the steel companies were not eligible for a temporary restraining order because they had an adequate remedy at law. Holtzoff had reasoned that, if the seizure was legal, the steel industry could sue for just compensation in the Court of Claims; if it was illegal, the companies could commence an action for damages against the United States under the Federal Tort Claims Act. [53] Kiendl asserted that if the steel seizure was illegal—as the industry contended—Supreme Court decisions showed that there was no possibility of the steel companies being compensated under the Fifth Amendment for the taking of their private property. [54] By the same token, Kiendl argued, the companies had no recourse under the Federal Tort Claims Act, for that statute establishes the United States' consent to be sued only where negligence on the part of an employee while acting within the scope of his official duties is alleged. The companies had not accused Secretary Sawyer of any negligence, Kiendl noted. In fact, if the industry's contention that the seizure was illegal and unconstitutional was correct, Sawyer was acting outside the scope of his official employment and the companies certainly would have no remedy under this act. Kiendl maintained further that the steel situation was specifically exempted from the Tort

Claims Act by a provision that excluded any claims based on the acts of employees of the government who exercised proper care in carrying out a statute or regulation, regardless of whether the statute or regulation was later found to be valid or invalid.[55]

Kiendl then turned to the question of irreparable injury. At great length he described the cost to the steel industry of putting the Wage Stabilization Board recommendations into effect—a cost not only in money but also in interference with, and possible destruction of, a collective bargaining relationship built up over many years. Kiendl insisted that the government intended to implement the WSB proposals immediately—Secretary Sawyer had said as much on nationwide television [56]—and that a preliminary injunction was the only means by which the steel industry could be saved from this irredeemable damage.[57]

Another issue that had to be considered by Judge Pine on the motions for preliminary injunction was whether the suits against Secretary Sawyer were in fact suits against the President and, if so, whether the court had jurisdiction to award an injunction. Kiendl argued that if an inferior officer's acts are neither within his statutory powers nor authorized by the President's constitutional powers, the officer can be sued as an individual; and if the relief sought requires no direct presidential action, the President is not an indispensable party to the suit.[58] The steel industry's case against Sawyer fell within this framework, Kiendl maintained.

Since the situation in which the steel companies found themselves met every requirement for equitable relief, Kiendl asserted, they need only convince the court that "there is a very serious basis for our contention—that this seizure is entirely unlawful and is wholly unconstitutional," [59] but he spent little time on this point. He denied the validity of the doctrine of inherent presidential powers beyond those expressly given in the Constitution. Commencing with a discussion of Magna Charta, Kiendl launched into a history of the struggle against the royal prerogative in England in order to show that "the Constitution was enacted with knowledge of what transpired in England, and, in spite of that, we are faced with the same attempt now to make operative that which was then condemned." [60] Kiendl shortly returned, however, to a discussion of those factors which he thought carried more weight in the determination of whether plaintiffs were entitled to equitable relief.

Kiendl attempted to refute Judge Holtzoff's conclusion, in his opinion denying a temporary restraining order, that on the injury issue the balance was in the government's favor. On the contrary, Kiendl declared, the balance was clearly on the side of the steel industry, for if the injunction were not issued the companies would suffer irreparable damage, whereas if it were granted the government would not be hurt. Administration claims that a strike would ensue should not be taken seriously, Kiendl pointed out, because under the decision in *United States* v. *United Mine Workers* [61] employees in a seized industry are employees of the government and a strike would be illegal. [62]

At this point, it suddenly occurred to Judge Pine that he might have misunderstood the nature of the relief desired by U.S. Steel. If the company was asking him to enjoin the seizure, the steelworkers would not be government employees after the injunction was issued. When Kiendl explained that U.S. Steel was requesting only a preliminary injunction restraining Secretary Sawyer from making any changes in the terms and conditions of employment until a final determination of the case, Pine was incredulous. He reminded Kiendl that the moving papers in U.S. Steel's case had asked that the seizure be enjoined. The judge inquired if the other companies were also only requesting preservation of the status quo, that is, that the seizure would continue but the government would be prohibited from changing the employment terms. Initially, only Bruce Bromley disagreed with Kiendl: "That is not all that Bethlehem Steel is asking for, Your Honor: We have filed a motion for a preliminary injunction and our position is 'the whole hog.' " [63] By further questions to the steel attorneys, Pine revealed both his sympathy for Bromley's position [64] and his willingness to listen to extended argument on the constitutional issues in the seizure case.

The industry lawyers were obviously surprised at Judge Pine's attitude. John Wilson tried to explain the companies' position to the judge: On the motion for a temporary injunction, the industry hoped only to raise a serious doubt about the legality of the seizure so that Pine could go on to consider the balance of equities in the case and give the companies the immediate relief they sought, an injunction to prevent the government from changing the terms and conditions of employment in the mills. The industry attorneys did not think the judge would have time at the present hearing to examine the ultimate question of the legality of the President's action, although they would be happy to convince Judge Pine of the illegality of the seizure. "If you should convince me of that, you wouldn't want me to perpetuate the illegality, would you?" Pine inquired. "I never look a gift horse in the face, Your Honor," Wilson replied. Pine announced that he had both the time and the patience to listen to argument on the merits of the case [65] and again asked the industry lawyers exactly what relief they were seeking. All but Kiendl now indicated that they wanted the seizure enjoined. When Kiendl stubbornly stuck to his original position, Pine asked the government if it would agree not to change the terms and conditions of employment in the mills, for if so he could then terminate "this hearing so far as Mr. Kiendl was concerned." [66] Baldridge, however, had no authority to speak for Secretary Sawyer.

Despite Judge Pine's repeated indications that he was more interested in argument on the merits of the case than on the issues in equity, the steel companies' attorneys apparently could not believe that the court would decide the merits on a motion for a preliminary injunction. Bruce Bromley, appearing for Bethlehem, and Luther Day, representing Republic Steel, obviously were reluctant to examine the constitutional issue in depth and merely took exception to the theory of presidential power advanced by the Justice Department in its brief. Bromley had planned to discuss the availability of the Taft-Hartley Act, but he was interrupted by Pine who wanted to explore the constitutional issues. When questioned on the inherent power of the President, Bromley referred Pine to U.S. Steel's brief which, Bromley said, offered the most complete treatment of the topic.[67] Luther Day told Pine that he would be glad to consider the merits but not "the ultimate merits, because you can't do that on this application for preliminary injunction." "Well, I have been doing it for two hours and a half," Pine replied sharply.[68] The argument of Charles Tuttle, counsel for the Armco Steel Corporation, was the first to confront squarely the issue of executive power. With a concise discussion of the Constitution, the Bill of Rights, and various Supreme Court cases bearing on presidential power, Tuttle showed that the government's interpretation of executive power was ill founded.

The government mistakenly asserted, Tuttle contended, that the fundamental conception of presidential power in our constitutional form of government was based on the theory of residual power: the President had the authority to perform any act for the good of the nation as long as it was not expressly forbidden by the Constitution or laws of the United States. That theory reversed the traditional interpretation of the American system of government, that it was one of delegated powers, Tuttle declared. Authority not delegated to the President, the Congress, or the judiciary, or not based on the necessary implications of the enumerated powers, was reserved to the states or to the people, not to the President. If this were not so, Tuttle observed, the Constitution would support a dictatorship.[69]

The Justice Department erred in its brief, Tuttle pointed out, when it used the case of *Myers* v. *United States* [70] as sanction for its theory of residual powers. The *Myers* case dealt with implied power, a concept very different from residual power. In *Myers* the central question was whether the President could unilaterally remove a postmaster who had been appointed with the consent of the Senate. The Supreme Court concluded that Article II of the Constitution granted the President the executive power of the government which the Court defined as "the general administrative control of those executing the laws," [71] and that the power to appoint and remove executive officers could be implied from that grant. But, Tuttle emphasized, nothing in *Myers* indicated that Article II granted the President the power to do "anything and everything" that he thought the public good required.[72]

The only restriction on the President's seizure power acknowledged in the Justice Department's brief, Tuttle indicated, was the Fifth Amendment's directive that just compensation be paid for the taking of private property. The government, he said, completely ignored the clause in the Fifth Amendment immediately before the just compensation mandate: no person shall "be deprived of life, liberty, or property, without due process of law." The government appeared to think, Tuttle argued, that the definition of "due process of law" was whatever the President thought was "in the public good." The Constitution lodges the power to care for the general welfare and the common defense in Congress: The Congress shall have power "to make all Laws which shall be necessary and proper for carrying into Execution the foregoing Powers, and all other Powers vested by this Constitution in the Government of the United States, or in any Department or Officer thereof." [73] It was Congress that made all the laws, Tuttle maintained, and the Fifth Amendment required that the taking of private property be authorized by law. [74] In concluding, Tuttle observed that the Supreme Court had repeatedly held that the judiciary has the power to "determine judicially whether, first, the power existed at all; and second, whether the circumstances under which it is exercised have been lawful." [75] Tuttle urged Judge Pine to find that the President was without power to seize the steel companies.

Before the remaining industry attorneys had a chance to criticize the government's theories, Judge Pine suggested that the government speak for itself. [76] Almost immediately, Baldridge got into trouble. He stated that "our position is that there is no power in the Courts to restrain the President and . . . Secretary Sawyer is the alter ego of the President and not subject to injunctive order of the Court." What would happen to you if the President ordered Secretary Sawyer to put you in jail right now and have you executed tomorrow? Judge Pine asked. Would the court be unable to protect you? Baldridge floundered for an answer until Judge Pine observed: "On the question of the deprivation of your rights you have the Fifth Amendment; that is what protects you." [77] Judge Pine wanted Baldridge to address himself to the proposition that the guarantees of the Fifth Amendment are applicable against the President as well as any other officer of the government. When Baldridge stalled for time, Pine told him to continue with the argument he had planned, but to be ready the following day to discuss that proposition.

Thrown off course by the judge's interruption, Baldridge was unable to present a coherent, persuasive argument even for what he considered his strong points. He urged Judge Pine to follow the well-known principle that constitutional questions should not be reached if the court could dispose of the case on any other basis. But Judge Pine thought that there was only one issue in the steel seizure case: did Secretary Sawyer have the constitutional power to seize the steel mills? Baldridge, who, like the steel industry lawyers, did not grasp that Pine was ready to decide the merits, replied that Pine did not have to consider that; he need only determine whether the steel companies had proved their claims of irreparable injury. Immediately Judge Pine demanded authoritative support for that principle: "I would like cases on that from you where there is a showing of invalidity of power where the Court must find that the equities when weighed in the balance favor no granting of relief." [78] Baldridge promised to submit the cases but he wanted a week in which to do it. Pine would not grant that request, however, unless the government agreed to maintain the status quo (i.e., preserve current wages and working conditions) until the case was decided. Baldridge responded that he did not have the authority to make such a commitment; thus his request for time was denied. [79]

Baldridge next argued that the equities when weighed in the balance clearly favored the government. At considerable length he described the affidavits filed by many government officials which sought to show that the nation was in the midst of a serious emergency and that national security would be jeopardized by any interruption in the production of steel. [80] Against this paramount interest of the government, Baldridge attempted to demonstrate that, in fact, the seizure would not cause irreparable harm to the steel industry. But Judge Pine cut Baldridge off in the middle of his exposition to ask him what power Sawyer had to seize the steel mills. As it was getting late in the day, Pine wanted some guidance from Baldridge so that he could think about the government's viewpoint overnight.[81]

Baldridge briefly stated that the President's power was based on "Sections 1, 2 and 3 of Article II of the Constitution, and whatever inherent, implied or residual powers may flow therefrom." [82] Baldridge refused to get involved in a quarrel over semantics; he thought the Justice Department's theory meant, in plain language, that in an emergency the executive had the responsibility to protect the national security and that any action the President took in that respect was legal. Judge Pine interpreted that statement to mean "If the emergency is great, it [the executive's power] is unlimited." [83] Baldridge observed that there were two limitations: the ballot box and impeachment. Could Baldridge cite any case that sustained such a broad conception of executive power? Judge Pine inquired. Baldridge mentioned a few cases but Pine pointed out that none was exactly applicable to the situation at hand. He told Baldridge to be prepared to discuss these questions the next day and adjourned the hearing.[84]

When the court convened the following morning, Baldridge gave every evidence of having spent a busy night. He filed with Judge Pine a supplemental memorandum on whether the court had to reach the constitutional issue before balancing the equities. The Assistant Attorney General then began to discuss his answers to the questions put to him by Pine the previous day. He attempted to show Judge Pine that in the hypothetical situation the judge had described (the Secretary of Commerce, under presidential order, had put Baldridge in jail and he was to be executed the next day), the court would have no power to enjoin the President from performance of the execution. Baldridge cited *Ex parte Merryman* [85] in support of this proposition.

The choice of this famous Civil War case was infelicitous. In *Merryman*, an assertion of power by the President, suspension of the writ of habeas corpus, had been invalidated by Chief Justice Taney. But the Chief Justice had admitted that, unless the President chose to follow the order of the court voluntarily, the judiciary lacked the physical power with which to enforce its order against the chief executive. Taney therefore decided not to issue an injunction but to file the proceedings in the case as well as his opinion with the Circuit Court of the United States for the District of Maryland and to send a copy under seal to President Lincoln. Baldridge maintained that the case was illustrative of the principle that the court had no power to enjoin the Chief Executive. Pine responded that he read the case differently. According to Pine, Chief Justice Taney, in *Ex parte Merryman*, indicated that the court had the constitutional power to enjoin the President; the Chief Justice merely acknowledged that since the prisoner was being held in an army fort, the court did not have the necessary physical power to compel compliance with its order in the event it was ignored.[86]

Judge Pine tried to obtain a better understanding of the Justice Department's theory of executive power by setting up another hypothetical situation for Baldridge to discuss: The President announced that the public interest in this time of a declared emergency required the seizure of Baldridge's home. An agent was ordered to seize it and dispossess Baldridge. In these circumstances, would the court have the power to restrain the agent from performance of the seizure, Pine wanted to know.

Baldridge declined to answer that question, preferring instead to discuss the court's power vis-à-vis the President in the actual circumstances surrounding the steel seizure. He asserted that there was a grave national emergency that "requires the exercise of rather unusual powers in these particular circumstances. I do not believe any President would exercise such unusual power unless, in his opinion, there was a grave and an extreme national emergency existing."[87] Judge Pine challenged this conception of presidential power. The government of the United States derives its power from the Constitution, Pine noted, and those powers are enumerated in and limited by that document. Baldridge agreed that was true, but only with respect to legislative powers. "Insofar as the Executive is concerned," Baldridge averred, "all executive power is vested in the President."

THE COURT: So, when the sovereign people adopted the Constitution, it enumerated the powers set up in the Constitution but limited the powers of the Congress and limited the powers of the judiciary, but it did not limit the powers of the Executive.

Is that what you say?

MR. BALDRIDGE: That is the way we read Article II of the Constitution.[88]

Baldridge, at the time of his argument, had no idea of the stir his statements would create.[89] He continued his presentation, noting the main points of the government's case and stressing what he considered to be the doctrine of *Mississippi* v. *Johnson*,[90] that the judiciary cannot enjoin the President. When Baldridge reviewed the government's contention that the steel companies had an adequate remedy at law and were therefore not entitled to equitable relief, Judge Pine asked him to discuss the cases cited by the steel companies to the contrary. Once again Baldridge indicated that he was not prepared. Since the steel companies' briefs had been filed only a short while before the hearing began, Baldridge had not had time to read them. He pointed out that in order to reply properly to the industry's case, he needed more time. Again Judge Pine indicated that he could give Baldridge additional time only if the government would agree to preserve the status quo in terms and conditions of employment. And again Baldridge answered that he could give the court no such assurance.[91] Hence Pine asserted:

I shall then have to act on the motions as expeditiously as possible, consistent with a complete, calm, and deliberate understanding of the case, and make my decision on the case. But I cannot assure you that that will be within a week. My impression is that it will be in much less time than a week, because I think the exigencies of the case required—that indeed justice requires—prompt action.[92]

Baldridge returned to the theme of the broad power given to the President by Article II of the Constitution and mentioned *In re Neagle*[93] and *In re Debs*,[94] where the Supreme Court had construed the executive power generously. Introducing what he called the "custom and usage approach," Baldridge recited the various instances in which previous Presidents had seized property and had taken other actions without statutory authority. The Assistant Attorney General then claimed that the courts had recognized that "the Executive, in appropriate circumstances, has inherent power in the nature of eminent domain and police power to seize, without statutory authority, and the Courts have been concerned not so much with whether the power existed but whether just compensation is required in view of the circumstances." [95] But when asked to name one case where the courts had assumed the existence of this inherent power, Baldridge could think only of *U.S.* v. *Pewee Coal Co.* [96] where the lower court had specifically stated that the legality of the seizure was not an issue. Baldridge contended that because the Supreme Court had upheld the award to the coal company of just compensation under the Fifth Amendment for its losses during government ownership of its properties, the Court had implicitly recognized the validity of the seizure. Judge Pine had a different interpretation of the case: the Court had treated the seizure as a fait accompli and had ordered the payment of damages to the company for the losses it had suffered. The Court, according to Judge Pine, had never considered the constitutional power of the President to seize private property. [97]

Baldridge, in concluding, presented a strong defense against the steel companies' argument that the President should have used the Taft-Hartley Act instead of seizing the steel mills. The Taft-Hartley Act was not mandatory, he noted. On the contrary, Baldridge asserted, the act, by use of the word "may" instead of "shall," left to the President's discretion whether or not to employ it. Moreover, Congress had passed two other acts, the Selective Service Act and the Defense Production Act, that contained provisions applicable to the circumstances of the steel dispute. Congress therefore had not intended the Taft-Hartley Act as an exclusive remedy, nor should its existence preclude the exercise of the President's residual or implied powers.

Baldridge then explained why the three pertinent statutes could not be used in the situation in which the administration found itself on the night of April 8. All three were administratively inadequate in the immediate circumstances of the steel dispute, Baldridge said, but he went into detail only about the Taft-Hartley

Act. The administration had taken no action before the strike deadline was imminent because it hoped that agreement would be reached through collective bargaining. Implementation of the Taft-Hartley Act had to be commenced at least several days before the scheduled walkout if an interruption in production was to be avoided. But experience had shown that as soon as Taft-Hartley proceedings were begun, collective bargaining collapsed. Hence the administration chose to allow negotiations to continue in the hope that a settlement would be concluded before the strike deadline. Had President Truman decided to use the Taft-Hartley Act on the night of April 8, at least several days of steel production would have been lost while a board of inquiry studied the dispute. In view of the military obligations of the United States, this interruption, as the many affidavits filed by the government demonstrated, would have been too costly.[98]

Furthermore, Baldridge observed, all the objectives of the Taft-Hartley Act had already been met by voluntary means. The steel dispute had been submitted to a thorough fact-finding procedure before the WSB, and the union had delayed its strike for ninety-nine days by April 8. Employment of the Taft-Hartley Act at that time would not have insured continued production, because a court might have thought that, after the union's voluntary postponement of the strike, the granting of an injunction was inequitable; if an injunction had been issued, the only possible gain would have been another eighty days in which to negotiate. But a Taft-Hartley injunction would, in fact, have hindered an agreement, for there would have been no incentive for a settlement: the Steelworkers would remain adamant in their demand for the full Wage Stabilization Board package; the steel industry could afford to wait, since the status quo was being preserved, for the government to acquiesce in its price requests.[99] Baldridge completed his argument with a plea to let the seizure stand so as to allow those government agencies that had been struggling with the steel dispute for months the opportunity to achieve a settlement.

In rebuttal, the attorneys for the steel companies concentrated their fire on the constitutional aspects of the steel seizure case.[100]

John Wilson challenged the government's theory of the President's inherent powers by discussing case law, an approach much appreciated by Judge Pine. Wilson noted that, in its brief, the Justice Department relied on the *Myers* opinion to prove that the Supreme Court recognized the inherent powers of the Chief Executive. However, Chief Justice Taft's language had been classified as dictum by the Supreme Court in *Humphrey's Executor* v. *United States* [101] and repudiated. Theodore Kiendl pointed out that Baldridge had not been able to find any case that specifically upheld the Justice Department's interpretation of executive power and the President's power to seize private property under that authority. Charles Tuttle observed that Baldridge's theories had been expressly disapproved by the Supreme Court in *House* v. *Mayes*; "The President is an executive having only enumerated powers," Tuttle reiterated. [102]

Baldridge, stung by the comments of the steel industry's attorneys, asked Judge Pine for two additional minutes to speak, and got into more trouble. The Assistant Attorney General stated that there were two cases, *United States* v. *Russell* [103] and *United States* v. *Pacific Railroad*, [104] dealing with the government's power to seize property without statutory authority and that these were discussed in his brief. But instead of explaining the cases he again urged Judge Pine not to decide the constitutional question on a motion for preliminary injunction. The emergency on the night of April 8 was great, Baldridge asserted. Resort to the legislature was too slow; employment of the Taft-Hartley Act was too slow. Both would have resulted in an interruption in steel production. Someone had to deal with the situation and Baldridge could not conceive of the national government as being without the power to act in a grave emergency. Judge Pine averred that that sounded like an argument for expediency. "Well, you might call it that, if you like," Baldridge responded. "But we say it is expediency backed by power." [105] On that unhappily phrased note, the Justice Department rested its case. [106]

Baldridge was unaware while he was arguing that he had made what one member of the White House staff called the "legal blunder of the century." [107] The Assistant Attorney General had touched the "ever-sensitive nerve of 'constitutionalism' " by espousing a theory "never dreamed of by anyone in the White House at any time." [108] Newspapers across the country carried headlines to the effect that the Justice Department asserted that the President's power was unlimited. The friendly New York *Post* declared, "President Truman can usually deal with his enemies, but who will protect him from his Justice Department? . . . Holmes Baldridge . . . sounded as if he were trying to hang the President." [109]

The reaction in Congress was equally severe. Senators who had defended the steel seizure took great pains to dissociate themselves from the view that the President's power was unlimited and to assure the Senate that President Truman did not hold this view either.[110] Senator Humphrey told the Senate that he had met with the President and that Truman had vigorously denounced the theory propounded by government counsel in the district court and had pointed out that the brief filed by the Justice Department made no such claim.[111]

The White House pondered how to counteract the misleading impression left by Baldridge's remarks. One alternative was to remove Baldridge from the case and seek to argue it again. But the staff decided against that. Instead, it was agreed that the President would answer in great detail one of the letters he had received concerning the steel seizure and that this reply would then be made public. By this device, the President's position would be made known to the country.[112]

The staff chose a letter from C. S. (Casey) Jones, which contained a number of questions to which the President could make a specific reply. The questions and answers were mainly concerned with the administration's handling of the steel dispute, not the events in the courtroom. But the letter gave Truman the excuse to address himself to Baldridge's remarks, and one of its concluding paragraphs noted:

I realized that the action I was taking in this case was very drastic, and I did it only as a matter of necessity to meet an extreme emergency.

In so doing, I believe that I was acting within the powers of the President under the Constitution—and indeed, that it was the duty of the President under the Constitution to act to preserve the safety of the Nation. The powers of the President are derived from the Constitution, and they are limited, of course, by the provisions of the Constitution, particularly those that protect the rights of individuals. The legal problems that arise from these facts are now being examined in the courts, as is proper, but I feel sure that the Constitution does not require me to endanger our national safety by letting all the steel mills shut down in this critical time.[113]

This vague reply, however, was incapable of repairing the damage done to the President's position by his counsel's argument in court.

Two days after the release of the letter, on Tuesday, April 29, 1952, Judge Pine handed down his opinion enjoining the seizure.[114] Fully aware of the importance of his ruling, Pine notified the clerk of the district court that his decision would be ready at 4:45 P.M. At the appointed hour the judge filed the fifteen-page document with the clerk of the court without convening court. The 500 copies of the opinion that had been mimeographed proved to be scarcely adequate for the crowd of lawyers, reporters, and others who had assembled in the clerk's office. "They had to file through a narrow gateway to pick up copies and the crush was so great for a few minutes that several cries for restraint went up from persons caught in the jam." [115]

Mincing no words, Pine ruled that the government's position was totally devoid of legal justification. His contempt for the Justice Department's argument permeated the brief opinion, which rejected the department's contentions, point by point. The opinion, however, was not a carefully constructed legal document. Rather, Judge Pine declared his faith in "immutable" principles of constitutional government, ignoring the more subtle legal problems presented by the case.

Pine began by affirming that the fundamental issue was "whether the seizure is or is not authorized by law"; therefore, that issue should be decided first.[116] The Constitution does not expressly grant the President the power to seize private property, Pine observed, nor does it contain a grant of power from which that authority might properly be implied. And no congressional statute gave the President seizure power.[117] The government relied, instead, on the theory that the President possessed a broad residuum of power, or inherent power, under the Constitution to take, in an emergency, whatever action he deemed necessary to protect the public interest. If the President has such power, Pine reasoned, it would have to be derived from Sections 1, 2, or 3 of Article II of the Constitution. Those sections, which provide that the executive power shall be vested in a President, that he shall be commander in chief of the army and the navy, and that he shall take care that the laws be faithfully executed, neither separately nor taken together, grant the President inherent power. "Their mere enumeration," according to Judge Pine, "shows the utter fallacy of defendant's claim." [118] It is in Congress that the Constitution places the duty "to provide for the common defense and general welfare" (Article I, Sec. 8).

In support of his rejection of the doctrine of inherent power, Judge Pine quoted a passage from *Our Chief Magistrate and His Powers* by former President William Howard Taft:

> The true view of the Executive functions is, as I conceive it, that the President can exercise no power which cannot be fairly and reasonably traced to some specific grant of power or justly implied and included within such express grant as proper and necessary to its exercise. Such specific grant must be either in the Federal Constitution or in an act of Congress passed in pursuance thereof. There is no undefined residuum of power.[119]

"I stand on that as a correct statement of the law," Pine declared.[120] The various cases put forth by the Justice Department to demonstrate judicial recognition of executive seizure power were dismissed by Pine as not being pertinent to the circumstances of the steel seizure. The department's argument that former Presidents had seized property in similar situations without statutory authority and that Truman's seizure of the steel mills was thus sanctioned by custom was not, in Judge Pine's opinion, credible. "Apparently, according to his [defendant's] theory, several repetitive, unchallenged, illegal acts sanctify those committed thereafter," Pine wrote. "I disagree." [121] In sum, Pine found an "utter and complete lack of authoritative support" for the President's position and concluded that the steel seizure was illegal.[122]

Although he believed that it was unnecessary in light of his disposition of the constitutional question, Pine also considered each of the Justice Department's arguments on the equity issues. He dismissed the assertion that the President cannot be enjoined, by noting that this was a suit, not against the President, but against Charles Sawyer.[123] To the claim that the balance of interest was in the government's favor, Pine replied that if a labor stoppage were to have the dire consequences described by the administration, he could not believe that the United Steelworkers of America would walk off the job. If a strike were called nonetheless, the government could employ the Taft-Hartley Act, and if that failed to bring a settlement, Congress could then be asked to enact the necessary legislation. Pine was unwilling to admit the inadequacy of these established procedures before they had been tried. Moreover, Pine thought that the contemplated strike, if it came, "with all its awful results, would be less injurious to the public than the injury which would flow from a timorous judicial recognition that there is some basis for this claim to unlimited and unrestrained Executive power, which would be implicit in a failure to grant the injunction." [124]

Pine also was convinced that the damages to the steel companies resulting from the seizure would be irreparable. He did not, however, "burden" his opinion with any facts to support his conviction.[125] To the Justice Department's contention that the steel companies had an adequate remedy at law, that is, a suit for compensation in the Court of Claims, and therefore were not entitled to equitable relief, Pine responded that monetary damages would not be sufficient for the injury incurred. More important, since the seizure had no authority in law, the companies would be unable to recover damages in the Court of Claims.[126]

Judge Pine directed one of his most scornful paragraphs to counsel for the United States Steel Company. Pine granted preliminary injunctions enjoining the seizure to all the plaintiffs except U.S. Steel. That company had verbally modified its original motion to a request for an injunction that would prevent the administration only from changing the terms and conditions of employment while it was in possession of the mills. Because of its "stultifying implications," [127] Judge Pine refused to issue such an injunction. He pointed out that granting U.S. Steel's motion for a limited injunction was tantamount to acknowledging that there might possibly be a valid basis for the seizure, a notion Pine did not entertain. He advised U.S. Steel that if it wished to withdraw its verbal motion and reinstate its original application for a preliminary injunction against the seizure, he would grant it leave to do so. [128]

Judge Pine's opinion created an immediate crisis for the administration. Without waiting for the injunction order to be signed, the United Steelworkers called a strike. The slow process of closing down the steel mills began virtually minutes after the news of Pine's ruling had been broadcast. The White House had to make decisions with utmost speed in order to avert the prolonged interruption in the production of steel that it had worked so hard to prevent.

seven

THE STEEL SEIZURE CASE IN THE COURT OF APPEALS

"Whatever the Supreme Court finally decides in this case, the universal judgment of informed mankind will sustain the opinion of Judge Pine," declared Senator Walter George.[1] The enthusiastic reception given to Pine's decision disheartened the White House. Overnight the previously little-known jurist became a hero. Editorials showered encomiums on him for upholding the traditional concept of constitutional government. Pine's enunciation of the so-called "immutable principles" of American government aroused a highly favorable response in the press and caused the seizure to be viewed as a major problem in constitutional government rather than as a measure taken to avert a strike in an emergency. Pine's decision and the laudatory press coverage of it apparently influenced public opinion, for the Gallup Poll taken after the announcement of the ruling showed less support for the seizure than had been evidenced in previous polls. This popular reaction, which theoretically should not have had any effect on the outcome of the steel seizure case as it traveled through the higher courts, as a practical matter became an important element in the legal decision-making process.

"It is hard to remember now," one journalist wrote several months after the Supreme Court ruled in the steel seizure case, "how much the country was startled—almost amazed—when

Pine's decision was first announced. It was like the Hans Christian Andersen fairy tale when the child, watching the royal procession, said 'The King has no clothes on.' . . . So it was when Pine declared that the President was not clothed with the authority so persuasively claimed for him. After a moment's incredulity, almost everyone took up the cry, 'The judge is right.' " [2]

Most of the editorials lauding the Pine opinion were preoccupied with the constitutional issue. The *New York Times* commented: "The important issues in the labor dispute in the steel industry . . . are only indirectly involved in this decision. Of far greater importance now and in time to come is the question of constitutional powers and relationships." [3] A Washington *Post* editorial labeled the opinion one "that will long be remembered in the annals of free government. . . . The court has grappled with the fundamentals of our constitutional system and come up with a fearless and, in our opinion, unanswerable conclusion, namely, that the President is not the whole Government and that he must act within the law." [4] According to the Wheeling *Intelligencer*, the decision was just the tonic needed by the nation:

It could very well be the reassertion by the Judiciary of its complete independence in its interpretations of various executive orders in light of Constitutional mandates. It could well reestablish our Congress to its rightful sphere of enacting specific laws for the general welfare of all the nation and not for any particular highly-organized pressure groups. It could very well mean the end to the Congressional abdication of power which has permitted the Executive branch to rule by fiat. [5]

Pine's opinion inspired the first serious questioning in the press of the extent of the emergency with regard to the need for steel. The judge's unusual declaration that he would prefer a strike "with all its awful results" to recognition of the President's claim to unlimited power apparently led some journalists to explore the possibility that the administration had been exaggerating the harmful effects an interruption in steel production would occasion. An editorial in the *Wall Street Journal* pointed out that the only evidence the country had on the scarcity of steel was President Truman's pronouncement that an emergency existed. [6] In an article in the same edition of the *Journal*, Joseph Evans accused the administration of "crying wolf." The actions of officials in charge of the defense mobilization program belied the White House argument that steel was in critically short supply, Evans wrote. He cited the steel companies' estimate that a seventy-day supply of steel was in the hands of all customers and the administration's decision to allow an additional year to complete the goals set previously for the defense buildup as proof that the country was in no immediate danger.[7] Administration spokesmen had not shown, Evans argued, that a work stoppage in the steel industry would have drastic effects on the fighting in Korea. The Pentagon calculated that it had more rounds of ammunition in reserve than had been used in the entire Korean war, Evans noted. From this information he concluded that a strike would be a setback for the defense program and for the economy, but it would be "supportable." [8]

Very little criticism of the Pine decision appeared in the press. One of the few journalists who attacked the ruling was Max Lerner, the New York *Post* columnist. Lerner took exception to the speed with which Pine had concluded the case: "A good deal of the blame must go to the incredibly inept handling of the government's case before Judge Pine. . . . But [the judge] ruled quickly, as though he feared that if he allowed the government to improve its argument, he might find himself without a thunderous decision, and the republic might not need him as saviour." [9] It would have been far better, Lerner observed, to rule on the constitutional question when the case was heard on the merits rather than on a motion for preliminary injunction. Had Pine waited until a hearing on the merits, the government would have been given a chance to answer the complaint fully, a record would have been established, and a judgment based on fact instead of rhetoric would have been the result. Judge Pine's statement that he had to rule immediately to prevent "irreparable damage" to the steel companies was not supported by any factual material presented in the Pine opinion, Lerner wrote.[10]

The strike called by the United Steelworkers of America as a direct consequence of Pine's decision did not detract from the editorial delight with the Pine opinion. Those editorialists who deigned to notice the strike after lavishing praise on Judge Pine condemned the union and called on the parties to negotiate a settlement immediately.[11] The St. Louis *Post-Dispatch* was one of a small minority of newspapers which warned against allowing Judge Pine's decision to distract attention from the more important business at hand, the settlement of the strike.[12] James Reston, too, recognized the importance of the strike in his *New York Times* column: "There is no time that can be described as ideal for a steel strike, but this particular walkout at this particular time is calculated to cause more trouble in the field of foreign policy than almost any other imaginable." [13]

The strike once again brought discussion of the Taft-Hartley Act to the forefront. In the press, only two alternatives to end the walkout were recognized: the administration could invoke Taft-Hartley or go to Congress for a new law.[14] In Congress, where Pine's decision had been received with relief and admiration,[15] sentiment also crystallized around these two options. The Senate Republican Policy Committee, led by Robert A. Taft, strongly favored forcing the President to use the Taft-Hartley Act; Joseph Martin, House Republican leader, concurred.[16] However, defenders of the seizure urged passage of a new law, because they believed the Taft-Hartley Act would be ineffective at the current stage of the steel dispute.[17]

It was the administration, however, that had to decide what step to take next. Events had moved more rapidly than expected and in a direction unanticipated by the President and his staff. Contingency plans had not been formulated, since the White House believed that Judge Pine would rule in favor of the President's seizure or, at most, enjoin a wage increase. According to a White House staff member, "we . . . never assumed for a moment that Judge Pine would . . . block seizure entirely." [18] Nor had the administration been prepared for the strike that was called immediately after the Pine ruling. Philip Murray, evidently infuriated by Judge Pine's opinion, issued the order for a work stoppage without consulting other union officials and without giving notice to the administration.[19] In fact, Secretary Sawyer had to call Murray to confirm the information he had received that the Steelworkers had walked out.[20]

The strike made the choice of action to be taken by the administration more difficult, while serving no purpose except as an outlet for the union members' bitterness and frustration.[21] The government could not allow the work stoppage to continue long enough to put economic pressure on the steel companies, for the administration had proclaimed many times since the seizure that even one day's shutdown of the steel furnaces would cause a significant loss of production, because it took many days more to heat up the furnaces again. Truman had no option but to stop the strike. Because the walkout had already begun, it was hard to resist employment of the Taft-Hartley Act. As one of the President's staff noted: "The public has been told that we could not use Taft-Hartley because there would be a strike before Taft-Hartley became effective. Now that we have a strike, the cry for Taft-Hartley will be louder and louder. Demands from the Hill for Taft-Hartley began last night and the tempo will undoubtedly increase." [22]

Within the administration, Secretary of Commerce Sawyer strongly advised Truman to use the Taft-Hartley Act. Although the President might be accused of turning against labor, Sawyer observed in a letter to Truman, the production of steel was more important and Taft-Hartley was the most expeditious way of halting the strike.[23] Sawyer noted that Congress would not be receptive to passing a new law unless the provisions of the Taft-Hartley Act were tried first. Once an injunction was obtained, it would give the parties an opportunity to negotiate again, Sawyer asserted, and if no progress was made in the eighty days "it will leave in the lap of Congress, without any excuse whatever, the sole responsibility for whatever chaos may result. You will then be in an impregnable position in stating to the Congress the need for some immediate legislation which will enable you to deal with this, and with similar situations." [24]

The Pine decision, however, created a new problem in using

Taft-Hartley that went unrecognized by Secretary Sawyer. At a meeting of the White House staff and some Justice Department attorneys, at which various courses of action to end the strike were discussed, it was suggested that the overriding difficulty with resort to the Taft-Hartley Act was that the seizure issue would then become moot and Judge Pine's opinion could not be appealed. Moreover, if the President invoked Taft-Hartley and the Pine ruling were allowed to stand, the administration would be subject to a very adverse public reaction. As one staff member described it, the public would think that "the President was wrong in the first instance in asserting that we could not afford a shutdown, wrong in not using Taft-Hartley, and wrong in using a legal theory which the courts would not and did not support." [25] The Taft-Hartley Act could not possibly be utilized in these circumstances, those present at the meeting concluded. [26]

Instead, the staff decided to seek a stay of Judge Pine's decision and to go directly to the Supreme Court for a definitive ruling. Justice Department attorneys would petition Judge Pine the following morning to issue a stay while the government appealed the steel seizure case to the highest court. If Judge Pine refused to grant a stay, the government could apply for relief to the United States Court of Appeals for the District of Columbia Circuit. If the stay were obtained, the steel plants would revert to government ownership and the steelworkers would be asked to return to their jobs, once again as employees of the United States. [27]

The application for a stay presented a ticklish problem because Solicitor General Perlman was convinced that the court would require the government to preserve the status quo in regard to wages and conditions of employment while a stay was in effect. If the government promised to do that, Philip Murray might be very reluctant to end the work stoppage and the government would have to go to court for an injunction to get the steelworkers back on the job. Perlman decided he would have to avoid making a positive commitment to maintain the status quo while somehow giving the court the impression that the government would not take any action on wages. [28]

At ten o'clock on the morning of April 30 counsel for the steel companies and the government again convened in Judge Pine's courtroom. Judge Pine signed the orders for a preliminary injunction,[29] received the government's appeal of his decision to the U.S. Court of Appeals, and then heard argument on the application for a stay pending appeal of the orders granting a preliminary injunction.[30] Baldridge briefly explained that a steel strike would be severely detrimental to the nation's defense efforts and that unless Pine stayed the injunction orders, the United Steelworkers of America had indicated that the walkout would not be called off. When the judge asked the Assistant Attorney General if he had any suggestions for conditions to be attached to the stay, Baldridge replied that he could think of no special terms that ought to be considered. Attorneys for the steel companies contended that Judge Pine, in deciding whether or not to issue a stay, should be guided by the same principles that had led him to grant a preliminary injunction. Circumstances had not changed, and if the stay were authorized the steel industry would be subject to the same irreparable injuries. Judge Pine agreed and summarily denied the application for a stay.[31]

The government immediately filed another application for a stay, in the United States Court of Appeals for the District of Columbia Circuit. Stating its intention to go directly to the Supreme Court to seek review of Judge Pine's decision, the government asked the court of appeals to stay the preliminary injunction orders until the Supreme Court disposed of the case. The application urged that a stay was necessary and justified because the preliminary injunction interfered with "sovereign functions of the United States," jeopardized the national security, resulted in a strike that terminated the possibility of collective bargaining between the parties, and was improperly issued because the court had no power, "under the circumstances of this case, to enjoin Presidential action." [32]

A hearing on the application for a stay began at 3:15 on the afternoon of April 30. The nine judges of the Court of Appeals for the District of Columbia Circuit, Chief Judge Harold M. Stephens and Associate Judges Henry Edgerton, Bennett Champ Clark, Wilbur K. Miller, E. Barrett Prettyman, James Proctor, David Bazelon, Charles Fahy, and George T. Washington, had decided to consider the case sitting *en banc*, a procedure reserved for very important cases only.[33] Philip Perlman, Holmes Baldridge, and Marvin Taylor represented the government,[34] and nineteen attorneys appeared for the steel industry.

For three hours counsel argued whether or not a stay should be issued and if so, for how long. The discussion was strange. Very little emphasis was given to the legal question of whether the President had the power to seize the steel mills and to the correctness of Judge Pine's decision.[35] The judges stressed practical matters: Would a stay assure the return of the steelworkers to the plants? How long would it take the government to file a petition for a writ of certiorari and how long did the government think it would take for the Supreme Court to act on it? Was it appropriate for the court of appeals to issue a stay that would be in effect until the Supreme Court completed consideration of the case? Questions like these frequently interrupted the oral arguments of counsel.

Baldridge, using the affidavits of government officials that had been filed in the district court, led off for the government with a strong presentation of the importance of the continued production of steel for the mobilization program of the United States. Although he could give no assurance that the union would end the strike if the steel plants were again under government control, Baldridge expressed the government's hope that the steelworkers would cooperate. The Assistant Attorney General explained that the government had decided to petition the court of appeals for a stay that would continue for the duration of the case because it seemed the most expeditious way of getting the union back to work and would minimize the number of interruptions in the production of steel. But the court of appeals judges seemed uncertain that they had jurisdiction to issue such a stay. Many of the judges believed that a stay lasting only until the Supreme Court had granted certiorari was all the court of appeals could allow. Baldridge agreed that a short stay would be satisfactory to the government, for it could then petition the Supreme Court for a stay at the same time as it filed for certiorari. He was not prepared to admit to any doubt, however, that the court of appeals could grant the longer stay requested by the government.

John J. Wilson, who had argued so successfully before Judge Pine, began the industry's presentation. Wilson maintained that the government had not proved there were sufficient grounds for issuance of a stay. The major point of government counsel, that a stay was necessary to terminate the strike, simply was not true. Wilson contended first that the government had no basis on which to assume that the steelworkers would return to their jobs once the government possessed the plants again. Technically, the union had already struck against the government, since the walkout had begun immediately after Pine filed his opinion but before the actual preliminary injunction orders had been signed. Second, the government could invoke the Taft-Hartley Act and the work stoppage would be ended just as quickly as it might be under a stay order. Moreover, Wilson insisted that the steel companies' ownership rights were equally as important as the government's desire to halt the strike. For the first time, the steel industry attorney was closely questioned as to what harm would be suffered by the companies if the government maintained possession of the mills but did not increase wages. Judge Edgerton suggested that all that the steel companies really needed was an order protecting them from irreparable injury, not one taking possession away from the government. Wilson's answer was evasive; he merely explained why Judge Pine chose to rule on the merits. He said nothing about what damage might have been done to the companies if Pine had granted a preliminary injunction enjoining only changes in the terms and conditions of employment.[36]

After several more industry attorneys had spoken, mainly urging the court of appeals not to let the existence of a strike affect its judgment and emphasizing the difficulties and expense involved in opening and closing the steel mills, Philip Perlman presented the government's rebuttal. Perlman had been appointed solicitor general in 1947, after a full and varied career both in private law practice and in a number of state offices under Democratic administrations in Maryland. By the summer of 1952 when he retired, Perlman, as solicitor general, had argued sixty-one cases before the Supreme Court and had won forty-nine of them.[37] Although he was not known as an unusually distinguished advocate, Perlman presented skillful, solidly grounded arguments.

Perlman's task in the court of appeals was to impress the judges with the government's need for a stay; he accomplished this dramatically. The solicitor general carefully pointed out that the steel companies would suffer no irreparable injury if the mills were returned to the government; the United States had agreed to assume all responsibility for any damage suffered by the industry because of the seizure. Turning to the merits, Perlman relied heavily on *United States* v. *Pewee Coal Co.*, arguing that the facts in that case were apposite to the circumstances of the steel seizure and, therefore, that the holding was directly controlling. President Franklin Roosevelt had seized without statutory authority the mines of the Pewee Coal Company. If the Supreme Court had believed that the seizure were illegal, Perlman maintained, the justices could not have ordered the United States to pay compensation for the losses the coal company had incurred while under government control. For when an official of the United States takes action that is beyond his statutory or constitutional authority, the Court usually will not hold the government liable for his act. Hence the Supreme Court had, in effect, approved President Roosevelt's action and, according to Perlman, Truman's seizure of the steel mills was, therefore, legal. But Judge Pine had ignored the precedents for Truman's action, Perlman contended, and had merely said that "six wrongs do not make one right." [38] Pine's opinion had invalidated every measure taken by a Chief Executive in time of national emergency, Perlman declared. Warming to his subject, the solicitor general described the current state of national and world crisis and suggested that the United States would be in a much weaker position if Judge Pine's order were allowed to stand. Gesticulating furiously, Perlman exclaimed with vehemence: "We are asking you to stay that order! Stay it—every minute counts.

We want the men back to the mills as soon as we can get them back." [39]

Perlman's impassioned plea appeared to be successful, for the judges immediately started questioning him on the type of stay that could be appropriately issued by the court of appeals. Sensing that the court was receptive to the government's request, counsel for the steel industry began asking for assurances that no changes would be made in the terms or conditions of employment while the stay was in effect. But Perlman made no commitment; he simply reiterated the government's offer to compensate the companies for any losses resulting from the seizure.[40] At this point, Chief Judge Stephens declared that he would like to consult with his colleagues, and he recessed the hearing until the judges were ready to announce their decision.

Forty minutes later the court reconvened and Stephens reported that the judges were in a division with reference to the decision to issue a stay. Judge Edgerton, representing the five-man majority,[41] read the order of the court. The preliminary injunctions granted by Judge Pine were stayed until 4:30 P.M. on Friday, May 2, that is, for only forty-eight hours. If petitions for writs of certiorari had been filed in the Supreme Court by that time, the stay would continue until the Supreme Court acted on the petitions. In the event that the request for certiorari were denied, the stay would remain in effect until the further order of the court of appeals. The four minority judges indicated that, in their opinion, the government had not made any showing that would justify granting its application for a stay.[42]

According to a newspaper account, the attorneys for the steel industry were shocked by the decision.[43] After the order had been announced, they detained Judges Stephens and Edgerton and begged that the order be changed to include prohibition of a government-imposed wage increase. The judges refused to do this. The lawyers then asked for such a commitment from Solicitor General Perlman; he reportedly laughed and left the courtroom. Counsel for the steel companies therefore decided to make a formal application to the court of appeals to attach a condition to its stay order that would require maintenance of the status quo unless the companies agreed to any proposed changes or unless a collective bargaining settlement were achieved. The attorneys noted in their application that a wage increase had been imminent when Judge Pine issued the preliminary injunctions. Hence they requested immediate oral argument on their motion to alter the stay order.[44]

On the morning of May 1, the court of appeals, receptive to the plea for a speedy hearing, convened at 10:27, barely an hour after the application had been filed. Chief Judge Stephens, apparently concerned about the long hours of argument the previous day, allotted a total of forty-five minutes to each side. He also cautioned counsel to exercise restraint in the manner in which they delivered their presentations; loud voices and forceful gestures were unnecessary, he advised.

The same arguments that had been used during the hearing on the stay order were restated in discussion of the motion to attach a condition to that order. Howard Westwood, counsel for U.S. Steel, and Edmund Jones, attorney for Republic, bore the burden of the argument for the industry. Westwood asserted that the Federal Rules of Civil Procedure required that the status quo be preserved when the court, in its discretion, granted a stay pending appeal of a previous judgment. If the government imposed a wage increase, not only would it cost the steel companies money but their bargaining position would be destroyed, Westwood declared.[45] Jones stressed his belief that the government was ready to implement changes in the terms and conditions of employment immediately and pointed to the fact that the steelworkers had not returned to work after the stay had been issued. Philip Murray had indicated that they would wait until after the President's press conference on May 1, before taking any action. Clearly the union expected Truman to announce a wage increase, Jones observed.[46]

When Perlman began to speak for the government, he tried to inform the court, without saying it outright, that the government had no plans to institute a wage increase before the petition for certiorari was filed. The solicitor general told the judges that he believed that the President would make no announcement at his meeting with the press.[47] The fears of the industry were premature, Perlman stated. As Acting Attorney General he would have to pass on the legality of an order changing the terms and conditions of employment in the steel industry, and since he had not even seen such an order he could hardly have approved it.[48] If the steel companies had not moved to have a condition attached to the stay, the men would have been back in the mills already.

The judges of the court of appeals, however, were not satisfied with Perlman's hints; they wanted a more definite statement on what steps the government would take before the Supreme Court had jurisdiction over the case. By repeated questions they indicated that they thought that the status quo should be maintained. Perlman attempted to explain that he did not want to say anything that would discourage the steelworkers from returning to their jobs but that he did believe that the government had no intention to implement a wage increase. Under the pressure of further questioning, however, Perlman could preserve this equivocal posture no longer. He advised the court of appeals flatly that the government would take no action before the filing of the petition for certiorari.[49]

After a noon recess, the court of appeals reconvened at 1:30 to deliver its decision. Once again Chief Judge Stephens announced that the court was divided. Judge Edgerton spoke for the same majority as had ruled on the stay order and declared that the application to modify that order had been denied. He noted that some of the judges might wish to file statements of their reasons.[50]

On the following day, the judges in the majority filed a joint unsigned opinion that must have heartened the administration. Although the opinion merely explained the reasoning that led the judges to issue a stay of Judge Pine's order, it emphasized that in the minds of these judges a serious doubt had been raised as to the correctness of the district court's decision.[51] The majority relied on three Supreme Court cases: *United States* v. *Russell,*[52] *United States* v. *Pewee Coal Co.,*[53] and *Yakus* v. *United States.*[54] In *Russell,* the court noted, the Supreme Court had recognized that in emergency situations private property could be seized; as long as the emergency was wholly proven, the officer taking the property could not be considered a trespasser and the government was liable for full compensation to the owner.[55] Regarding the granting of the preliminary injunction requested by the steel industry, the majority stated that the companies had an adequate remedy at law by a suit in the Court of Claims, citing the decision in *Pewee Coal Co.*, where the United States was ordered to pay just compensation for a taking under circumstances very similar to the steel seizure.[56] Moreover, the majority judges observed, the Supreme Court had ruled in *Yakus* that a court may withhold preliminary relief if an injunction would adversely affect a public interest that could not be compensated by an injunction bond, even though the deferment of relief might be onerous for the plaintiff.[57] In view of these considerations and the fact that defense officials had filed affidavits stressing the tremendous importance of the uninterrupted production of steel to the nation's security, the majority of the court of appeals agreed that a stay of Judge Pine's preliminary injunction order was warranted.[58]

Once the court of appeals had issued the stay order with no conditions attached, the administration had two tasks before it: to get the steelworkers back in the mills—a responsibility of the White House; and to file the petition for certiorari before the next day's deadline, 4:30 P.M., May 2, 1952, set by the court of appeals—the task of the Justice Department. Toward the first objective, the President, on the evening of May 1, asked Philip Murray to terminate the strike and to meet with the presidents of six steel companies at the White House on Saturday morning, May 3.[59] Murray heeded the President's request, and the strike was called off. The Department of Justice had little difficulty in fulfilling the second goal. In the petition for certiorari the department simply had to restate the arguments it had been making in all the papers filed since the steel seizure litigation had begun. By midmorning of May 2 the government's petition was ready, and it was taken to the Supreme Court immediately.

Much to the Justice Department's surprise, it found that the steel industry had already filed a petition for certiorari at 9 o'clock that morning.[60] It is most unusual, although it is permitted,[61] for the party who has won in the court below to request that a writ of certiorari be granted. Bruce Bromley, who, along with two young lawyers, John Pickering and Stanley Temko, filed the petition for the steel companies, informed reporters that the industry believed that in the public interest it was important for the Supreme Court to review promptly the correctness of Judge Pine's decision.[62] Moreover, by filing a petition for certiorari, the industry had the right to open and close the oral argument in the Supreme Court. The companies' petition stated that although Pine's ruling was proper in every detail, it should be given authoritative confirmation at the earliest opportunity. The steel industry urged further that the case was ripe for argument in the Supreme Court, even though the constitutional questions were presented on a motion for preliminary injunction, because nothing would be gained by having a final hearing on the merits in the court below as the case had already been fully briefed and argued. Of equal importance, there remained the possibility that the companies would suffer irrevocable injury if the government unilaterally imposed a wage increase, for the solicitor general had promised to maintain the status quo only until he filed the petition for certiorari. The steel industry therefore asked the Supreme Court to take the case immediately and requested that, if the stay of the district court's order were continued, a course of action opposed by the companies, the Supreme Court attach a condition requiring preservation of the status quo.[63]

The government's petition for certiorari stated the questions that the steel case presented for decision by the Supreme Court: Did the President have the constitutional authority to seize the steel mills in order to avert a nationwide strike? Had the district court erred by ruling on the constitutional issues on a motion for preliminary injunction? Was the granting of injunctive relief proper in this case? The Justice Department urged the Supreme Court to grant a writ of certiorari and hear the case promptly because Judge Pine's ruling had created an uncertain situation in the steel plants; national security required a quick resolution of these issues. As the petition noted:

The decision of the district court would strike down the power of the President to take action necessary to maintain such [uninterrupted] production [of steel]. More broadly, it casts doubt on the powers of the President under the Constitution—powers exercised on numerous occasions by past Presidents—to take possession, in time of strife and siege, of property outside the actual scene of military operations whenever such a temporary emergency taking is necessary to protect vital national interests. The decision of the district court, if not reviewed forthwith, will stand as a rigid and dogmatic barrier not only to the efforts of the President to maintain continued production of steel, but to any kind of executive action which may become necessary to meet other and unpredictable emergencies which may hereafter crowd upon the United States. [64]

The government also asked the Supreme Court to issue a stay of Judge Pine's decision patterned on the court of appeals order, in other words, with no conditions attached. [65]

Each party filed a response to the other's petition for certiorari stating that there was no objection to the granting of a writ of certiorari as requested by the other party. Each party strongly objected, however, to the opposing side's views on a stay order. Perlman noted in his memorandum that the steelworkers had returned to the mills that morning but that "any change in the nature of the stay now in effect would probably result in a new crisis, with danger of still another interruption." [66] The steel companies, in their reply, observed: "A suggestion that this Court should not do equity because a powerful union might strike against this Court's action has implications hardly less grave than the basic Constitutional issue in Mr. Sawyer's seizure." [67] The union filed a brief, amicus curiae, in support of the government's application to the Court for a stay with no conditions attached. [68]

While the Justice Department and the steel industry were doing verbal battle in the Supreme Court papers, the White House and Secretary Sawyer prepared for the negotiations between Murray and the company presidents to be held on the morning of May 3. Sawyer, who desperately wanted a settlement to be achieved at this session, disagreed with other members of the administration as to what attitude the White House should take toward the collective bargaining. Immediately after the court of appeals hearing on May 1, John Steelman had called a meeting to discuss the steel situation. At this gathering, Philip Perlman had suggested to Sawyer that he impose changes in the terms and conditions of employment in the steel plants on Saturday morning. Sawyer had flatly refused to do this while the status of the case in the courts was so uncertain. The discussion had led nowhere, and no concrete course of action had been decided upon.

Later that day, Steelman had called Sawyer and told him of the planned Saturday morning session with the industry presidents. Sawyer, irate at again not being consulted, wrote a letter to the President the following morning in which he recommended what he thought would be the proper stance for the administration to take at the negotiations. The Secretary of Commerce criticized Steelman's pessimism about the May 3 session (Steelman had talked with Sawyer on the phone about the prospects for a settlement) and urged that the President inform the union and the industry, not that he hoped for a settlement but that he expected one; each party should be prepared to yield something so that collective bargaining would be successful. Sawyer also noted that the government should be ready to make "reasonable" price concessions. In Sawyer's opinion, it was unnecessary for the government to take a strong stand against steel price increases because within a year there would be so much steel that prices would be regulated by supply and demand and not by the government. This prediction rested upon the assumption that there would be no steel strike so steel could be produced at capacity.[69]

On Saturday, May 3, at 10 A.M., the President, with Sawyer at his side, opened the meeting between Murray and the steel company presidents with a short speech. Truman declared that he had two primary concerns: the national security and the economic health of the nation. Uninterrupted production of steel was a necessity if the national security were to be maintained, and steel wages and prices within sound stabilization guidelines were essential to the economic well-being of the country. These goals were so important for every citizen that the union and the industry had a special responsibility to see that they were achieved. The President asked Murray and the company presidents to subordinate their emotions and work out a settlement within the next few hours. If they were unsuccessful, Truman advised them, ESA Administrator Putnam had completed his recommendations for changes in wages and working conditions in the steel plants and the government would put them into effect on Monday or soon thereafter. The President averred that a government-imposed settlement was distasteful to him, as the seizure itself was; hence, he concluded: "I didn't send for you just to make a speech. I sent for you for action and, gentlemen, I want it." [70]

From all reports, serious bargaining began immediately after the President and Sawyer left. By midafternoon the negotiating session had narrowed to three men: Steelman, Fairless, and Murray. They apparently arrived at a satisfactory settlement, and all that remained to be done was for Fairless and Murray to check with the other members of their respective groups. The bargaining session recessed to allow them time to do so. During the recess, however, word of the Supreme Court's action on the steel case flashed across the White House news ticker, and when the bargaining session reconvened, the attitudes of the negotiators had changed dramatically. [71]

At its normal Saturday conference, the Supreme Court had voted, 7 to 2,[72] to grant a writ of certiorari and had assigned the steel case for argument on Monday, May 12. But, more important with respect to the negotiations proceeding simultaneously at the White House, the Court stayed the preliminary injunction order of the district court until final disposition of the case, with the stipulation that Sawyer "take no action to change any term or condition of employment while this stay is in effect unless such change is mutually agreed upon by the steel companies . . . and the bargaining representatives of the employees." [73]

Unquestionably, the Supreme Court's stay order radically affected the collective bargaining at the White House. Whereas the industry and the union had been on the verge of settling before the negotiating session recessed, when the parties returned Saturday night the companies' will to settle had disappeared. No longer under pressure to resolve the dispute, for the threat of a government-imposed wage increase had been removed, the industry saw no reason to make concessions in order to arrive at a settlement when it had nothing to lose by waiting for the Supreme Court to decide the case.[74] Although the companies went through the motions of additional sessions on Sunday, it was obvious to the administration and the union that further talks were useless. The meetings were adjourned.[75]

Clearly, no progress toward an agreement in the labor dispute could be made before the Supreme Court resolved the basic issue of the validity of the seizure. Thus the country once again focused its attention on a courtroom and all watched anxiously as the case was argued in the Supreme Court.

eight

THE SUPREME COURT HEARS THE STEEL SEIZURE CASE

There was little more than a week between the Supreme Court's grant of certiorari, late in the afternoon on Saturday, May 3, and the time set for oral argument, noon on Monday, May 12. For the White House this was a period of calm—at least in regard to the steel dispute; because of the stay issued by the Court, Truman and his staff could do little but hope, until after the Supreme Court decision. For the steel industry, too, the pressure was off, since it was content to preserve the status quo as long as possible. The union also was precluded from taking any action; but the steel-workers grew more restless and irritated as the period in which they were forced to work at 1951 wage rates lengthened.

For the lawyers involved in the steel seizure case, however, the eight days were packed with feverish activity: Both the brief and the oral argument had to be prepared during this time. At the Justice Department, attorneys worked for five days from 9 A.M. until well past midnight and produced a 175-page brief which clearly, if not concisely, set out the government's case.[1]

THE GOVERNMENT BRIEF

The crucial questions, as framed in the brief, were:

1. Whether, on the facts recited in Executive Order No. 10340 and established by the uncontroverted affidavits, the President had constitu-

tional authority to take possession of plaintiffs' steel mills in order to avert an imminent nationwide cessation of steel production.

2. Whether, in the circumstances of this case, the district court erred in reaching and deciding the constitutional issues on motions for preliminary injunctions.

3. Whether the district court erred in granting injunctive relief.[2]

In its argument, however, the brief treated the questions in reverse order. The government claimed, first, that the district court had issued a preliminary injunction contrary to every relevant equitable principle, and second, that in view of uncontroverted testimony on the pressing need for steel, the President had the power under the Constitution and laws to seize the steel mills to avert a strike.[3] The brief emphasized that neither of these contentions could be considered abstractly; it was important to have a full understanding of the nature of the emergency in which the President acted and the tremendous damage to the public interest which would result if his action were not sustained. Thus the government presented a detailed account of the military and foreign affairs crisis that existed in the spring of 1952 and underlined the relationship between steel and the defense effort.[4]

When it turned to the specific points of law at issue, the government argued that Judge Pine, by reaching the constitutional question, had ignored the well-settled judicial tradition that if a case can be disposed of on nonconstitutional grounds, the court will not decide the constitutional question. Adherence to this tradition was particularly important in this case, the brief urged, since the constitutional issue involved required an interpretation of the separation-of-powers doctrine—one of the most basic and sensitive tenets of the nation's constitutional faith. The government therefore urged the Supreme Court to consider the merits of the case only if it were wholly satisfied that the equity requirements for granting a preliminary injunction had been fulfilled. Those requirements had not been met in this case, the government argued.[5]

Judge Pine had erred in finding an equitable basis for an injunction on two counts, the brief stated: The steel companies had an adequate remedy at law for any damage incurred because of the seizure and the industry had failed to show any irreparable injury that would override the harm the public interest would incur if the injunction were upheld. On the first point, the government presented an elaborate response to the companies' contention that a suit would not lie in the Court of Claims because the seizure was unauthorized. In the Selective Service Act of 1948 and the Defense Production Act of 1950,[6] Congress had given the President the power to take private property if necessary for the national defense. Truman, however, could not use the procedures described in these acts; on the night of April 8 the government had to take over the steel mills without delay, and neither of the statutory procedures provided for immediate possession. But the fact that Truman had not used the congressionally prescribed method for taking the property was no bar to the companies' obtaining compensation in the Court of Claims, the government argued, citing *Hurley* v. *Kincaid.*[7]

In *Hurley* v. *Kincaid,* the Supreme Court had been presented with a situation apposite to the steel seizure case. Congress had authorized the government to acquire land necessary for flood control work along the Mississippi River. The statutes required that, before the United States could take possession of the property, it had to file a notice of condemnation in court and deposit an amount of money adequate for just compensation to the property owner. Kincaid sought an injunction against Secretary of War Hurley because certain flood control projects would subject Kincaid's land to flooding and the government had neglected to file a condemnation petition or to deposit any money. He wanted the court to enjoin any further construction until the government had complied with all the statutory requirements.

The Supreme Court had held that Kincaid was not entitled to injunctive relief. Since Congress had authorized the acquisition of private property for the purpose for which the United States had taken Kincaid's land, Kincaid could seek compensation in the Court of Claims. The fact that the government had not followed the statutory procedures was not justification for the granting of an injunction. Kincaid would receive the same compensation whether the government filed a condemnation petition before taking his land or whether he instituted an action in the Court of Claims. Justice Brandeis, speaking for the Court, had noted:

> For even if the defendants are acting illegally, under the Act, in threatening to proceed without first acquiring flowage rights over the complainant's lands, the illegality . . . is confined to the failure to compensate him for the taking and affords no basis for an injunction if such compensation may be procured in an action at law.[8]

Similarly, in the circumstances of the steel seizure, the Defense Production Act authorized the taking of private property for defense purposes. The failure to follow the condemnation procedure stipulated by that act was not an obstacle to the steel companies seeking compensation in the Court of Claims, the government maintained.[9]

In addition to overlooking the presence of an adequate remedy at law, the steel companies had not shown that they would be subject to any irreparable injury which would exceed the damage to the public interest if the injunction were sustained, the brief continued. The case of *Hurley* v. *Kincaid* was a helpful precedent in this respect also, for in the Court's opinion Justice Brandeis had asserted:

> Even where the remedy at law is less clear and adequate, where less public interests are concerned and the issuance of an injunction may seriously embarrass the accomplishment of important governmental ends, a court of equity acts with caution and only upon clear showing that its intervention is necessary in order to prevent an irreparable injury.[10]

In declaring that the government's threat to change wages and working conditions would harm them irrevocably, the steel companies had neglected to point out that whatever damages might be suffered if the government actually did impose higher wages and new working conditions had to be weighed against the losses the industry would incur during a strike. By comparison, the increased wages might be a small price to pay for continued production, a benefit bestowed on the companies by seizure. Furthermore, it was obvious that the industry on its own would have to make some changes in wages and employment conditions if it expected the union to sign a new contract. It was wholly unfair, therefore, for the industry to measure contemplated damages against rates and conditions prevailing in 1951, the Justice Department brief suggested. Moreover, when the companies calculated their alleged damages, they took no notice of the price increase which would be allowed by virtue of the Capehart amendment. It was difficult to perceive how imposition of additional labor costs whose effect, if any, on the companies' profits would merely be to reduce them to the maximum permissible level, could be said to cause the kind of irreparable injury that would "warrant a court of equity in interposing its hand to enjoin action taken by the President to meet a grave national emergency." [11]

Even if the steel industry's claims of irreparable injury were accepted at face value, the government brief declared, the damages did not outweigh the harm that would befall the public interest if the industry received equitable relief. A strike in the steel industry would affect "our national safety, our discharge of international commitments, and the lives of our soldiers." [12] Judge Pine's issuance of an injunction was therefore clearly improper, the government maintained. As the Supreme Court had stated in *Yakus* v. *United States*:

> The award of an interlocutory injunction by courts of equity has never been regarded as strictly a matter of right, even though irreparable injury may otherwise result to the plaintiff. . . . Even in suits in which only private interests are involved the award is a matter of sound judicial discretion, in the exercise of which the court balances the conveniences of the parties and possible injuries to them according as they may be affected by the granting or withholding of the injunction. . . .
>
> But where an injunction is asked which will adversely affect a public interest for whose impairment, even temporarily, an injunction bond cannot compensate, the court may in the public interest withhold relief until a final determination of the rights of the parties, though the postponement may be burdensome to the plaintiff. [13]

Judge Pine, in granting the injunction, had mistakenly assumed that no strike would occur. Moreover, Pine had further declared, "I believe that the contemplated strike, if it came, with all its awful results, would be less injurious to the public than the injury which would flow from a timorous judicial recognition that there is some basis for this claim to unlimited and unrestrained executive power, which would be implicit in a failure to grant the injunction." [14]

The government maintained that Judge Pine's procedure transposed the normal order of adjudication in equity cases; the equities must be balanced before the merits are determined. Where an action of the executive branch is sought to be enjoined, the possibility of severe injury to the public interest if an injunction is granted should lead the court to presume the constitutionality of the executive's act and to withhold equitable relief pending final consideration of the legal rights of the parties. [15]

When the government brief turned to the merits, it narrowly defined the issue presented as "whether seizure is a method available to the President, in the exercise of his constitutional powers, to avert a crisis of this type." [16] It was unnecessary, the brief asserted, to identify a specific clause of the Constitution as the one from which the President derived the power to seize the steel mills. Rather, it had been recognized throughout American history that the Executive's authority to act in any particular instance may arise from several of the grants enumerated in Article II, taken together. No doubt existed that the seizure of the steel mills represented a valid application of federal governmental power; Congress unquestionably could have authorized the seizure. The issue was therefore whether seizure was an appropriate presidential response to an urgent situation. According to the brief, on numerous similar occasions during previous administrations the executive had seized private property in time of war or national emergency, and both the legislative and judicial branches had recognized executive seizure as a legitimate means of meeting crisis situations. [17]

The brief recited at length the many occasions, beginning with the Revolutionary War, when seizure had been employed. By the end of the War of 1812, the government asserted, the power to take private property during wartime emergencies on the basis of executive authority alone had been substantiated by frequent use. These instances, however, were mostly seizures of property needed by military officers to feed and equip their troops or to prevent supplies from falling to the enemy; the circumstances, therefore, were not directly relevant to the steel seizure. The actions of Presidents Lincoln, Wilson, and Franklin D. Roosevelt were more to the point.

The government brief noted that Lincoln had been the first President specifically to order the taking of private property when he had seized the railroad and telegraph lines between Annapolis and Washington. Issuance of the Emancipation Proclamation of January 1, 1863, provided another example of executive seizure during the Lincoln administration. Both these measures were justified as militarily necessary and therefore were based on the constitutional powers of the President as commander in chief. The brief quoted Lincoln on the constitutionality of the proclamation:

I think the Constitution invests its Commander-in-Chief with the law of war in time of war. The most that can be said—if so much—is that slaves are property. Is there—has there ever been—any question that by the law of war, property, both of enemies and friends, may be taken when needed? And is it not needed whenever taking it will help us, or hurt the enemy? [18]

Turning to President Wilson, the brief observed that, exercising his constitutional powers, he had seized the Smith and Wesson Company for failure to comply with an award of the National War Labor Board and threatened a similar sanction against the Remington Arms Company.[19] Many other actions, aside from seizure, taken by Wilson before and during World War I, without statutory authority, were also mentioned.[20]

But it was Franklin Roosevelt's administration which supplied the most convincing precedents for executive seizure. Twelve times before the passage of the War Labor Disputes Act authorizing plant seizures, President Roosevelt had taken possession of companies threatened by strikes which would impede defense efforts. Three of the seizures had occurred before Pearl Harbor. On the occasion of the first seizure, at the North American Aviation plant, then Attorney General Robert Jackson had justified the action as within the "duty constitutionally and inherently resting upon the President to exert his civil and military, as well as his moral, authority to keep the defense efforts of the United States a going concern." [21]

Having demonstrated that seizure was an accepted presidential response in certain circumstances, the brief embarked on a discussion of the nonseizure actions of other chief executives based on the inherent powers doctrine. Starting with President Hayes, the brief noted the various instances in which the President had sent troops to help maintain order during labor disturbances. The dispatching of troops by President Cleveland to Chicago at the time of the Pullman strike offered a most valuable precedent, the government argued: there the Supreme Court had upheld the President's claim that his duty to execute faithfully the laws of the United States and to protect its property required him to intervene, despite the lack of express authorization, and to seek an injunction against the strike. [22] Theodore Roosevelt, in accordance with his stewardship theory of the Presidency, had been prepared to seize the Pennsylvania coal mines during a strike in 1902. [23] Roosevelt, and subsequently Taft, had withdrawn land for forest and coal reserves from private entry, in the face of statutes which authorized only the withdrawal of land in which mineral deposits had been located. The Supreme Court had sustained this exercise of executive power also. [24]

The government also invoked the legislative history of statutes authorizing presidential seizure. The brief pointed out that when Congress had enacted a law giving the President the power to seize private property in emergency situations, congressmen during debate would state that the bill confirmed already existing executive power.

Congress' action in 1862 illustrated such an instance. After Lincoln had seized the railroad and telegraph lines between Washington and Annapolis, Congress had considered legislation which authorized the President to take possession of any railroad or telegraph line in the United States and which provided penalties for interference with government operation. Senator Wade, the bill's

sponsor in the Senate, had asserted during a discussion that the
bill conferred no additional power upon the government, "beyond
what they possess now. It attempts to regulate the power which
they undoubtedly have; for they may seize upon private property
anywhere, and subject it to the public use by virtue of the Constitution." [25] The bill's sponsor in the House, Representative Blair,
had similarly declared that the legislation did not confer on the
Secretary of War "any new or any dangerous powers. The Government has now all the powers conferred by this bill; and the simple
object of the bill is to regulate, limit, and restrain the exercise of
those powers." [26] And many opponents of the bill had based their
negative votes on the ground that the legislation was superfluous
and might be interpreted as a limitation on the President's powers.

Similar debates had occurred when Congress had passed seizure laws during World War I and World War II, the brief continued. Approval of these laws appeared to reflect the sentiment
that although the President already possessed the power to seize, it
was best to remove any doubt.[27] The brief reproduced many statements by congressmen, both those opposed to and those in favor of
the War Labor Disputes Act, which assumed the legality of President Roosevelt's seizure of the North American Aviation plant.
The brief also noted that Congress was not unaware of the inherent powers theory which President Roosevelt had used as ground
for many of the actions not specifically authorized by statute. In
1939, Attorney General Murphy, replying to a request from the
Senate for his opinion on the emergency powers of the President,
had stated:

You are aware, of course, that the Executive has powers not enumerated in the statutes—powers derived not from statutory grants but
from the Constitution. It is universally recognized that the constitutional
duties of the Executive carry with them the constitutional powers necessary for their proper performance. These constitutional powers have never
been specifically defined, and in fact cannot be, since their extent and
limitations are largely dependent upon conditions and circumstances.[28]

Apart from judicial precedent, the government maintained
that the frequent use of the seizure power by Presidents offered sufficient ground, by itself, to sustain Truman's seizure of the steel

mills. The brief quoted Justice Holmes' famous remark, that "a page of history is worth a volume of logic," [29] and cited *Inland Waterways Corp.* v. *Young* for the proposition that the validity of a constitutional power may be established by usage.[30] The most important support for this theory came from the Supreme Court's opinion in the *Midwest Oil* case:

Both officers, law-makers and citizens naturally adjust themselves to any long-continuing action of the Executive Department—on the presumption that unauthorized acts would not have been allowed to be so often repeated as to crystallize into a regular practice. That presumption is not reasoning in a circle but the basis of a wise and quieting rule that in determining the meaning of a statute or the existence of a power, weight shall be given to the usage itself—even when the validity of the practice is the subject of investigation.[31]

Turning to explicit judicial acknowledgement of the President's power to seize private property during a national emergency, the brief noted that controversy had been not so much over the existence of the power as whether just compensation was required as a result of a particular seizure. The relevant cases, the government maintained, upheld the authority of the executive, despite the lack of enabling legislation, to seize property in order to avoid a crisis. Court decisions had recognized two types of seizure, one based on the police power and the other on the power of eminent domain. Just compensation was required only in the latter category of seizures. According to the case law, seizures of the eminent domain kind may be undertaken regardless of the physical relation of the property to the field of battle and may result from a lesser degree of necessity than police power takings, because the owner will receive just compensation. The steel seizure was an eminent domain taking of this type, the brief asserted.[32]

Two cases, *United States* v. *Russell* and *Alexander* v. *United States*, the brief observed, were sufficient to sustain the government's position on the constitutional questions in the steel seizure litigation. In *Russell*, the Supreme Court had stated:

In cases of extreme necessity in time of war or of immediate and impending public danger . . . private property may be impressed into the public service, or may be seized and appropriated to the public use, or may even be destroyed without the consent of the owner. . . . Where such an extraordinary and unforeseen emergency occurs in the public service in time of war no doubt is entertained that the power of the government is ample to supply for the moment the public wants in that way to the extent of the immediate public exigency, but the public danger must be immediate, imminent, and impending, and the emergency in the public service must be extreme and imperative, and such as will not admit of delay or a resort to any other source of supply, and the circumstances must be such as imperatively require the exercise of that extreme power in respect to the particular property so impressed, appropriated, or destroyed.[33]

It was clear that Truman's seizure had met all these requirements for an acute emergency, the government contended, and once the emergency was established, *Russell* directly recognized the executive power to seize private property without statutory authorization.[34]

In *Alexander* v. *United States*, a Court of Claims case, the government had, without statutory authority, taken possession of a farm to use as a training camp, after the Spanish-American War had been concluded but before the peace treaty had actually been signed. The government had paid rent for the use of the land to the tenant, who had been dispossessed, but the property owner sued for compensation for the permanent injuries done to the land while the government was occupying it. The Court of Claims had noted in its decision that although the farm was more than one thousand miles from the closest approach of an enemy, the seizure was proper. However, the owner had to be compensated.

There was a military necessity that some land in that vicinity should be taken. There is always a necessity when property is taken, and it implies no wrong on the part of the Government that it does take property without the consent of the owner. Underlying the exercise of the right is grant of power upon the expressed condition that compensation be made.[35]

The government's brief also drew support, less directly, from two other cases, *United States* v. *Pewee Coal Co., Inc.* and *Dakota Coal Co.* v. *Fraser.*[36] In *Pewee*, the government had seized the coal mines, without congressional authorization, in order to avert a nationwide mine strike during World War II. The validity of the seizure had not been at issue in the ensuing suit because neither of the parties had raised the point. Instead, the central question had involved a determination of whether or not the seizure was an eminent domain taking within the meaning of the Fifth Amendment and thus, whether or not the owner should be compensated for losses suffered during government operation of the mines. The government had contended that the seizure was not in the nature of an eminent domain taking but that the government was merely acting as temporary custodian of the mines; hence, the property owner was not entitled to just compensation. The Supreme Court, unconvinced by the government's argument, had ruled unanimously that there had been a Fifth Amendment taking of Pewee's property which required compensation for losses suffered. The Court had been divided only on the issue of the measure of just compensation. According to the government brief in the steel seizure case, "Although the Court did not expressly so state, it is implicit in the decision in that case that there had been a valid exercise of executive power in the nature of eminent domain or requisition." [37]

Dakota Coal Co. v. *Fraser* had presented the district court in North Dakota with a situation practically identical with the one which had confronted Judge Pine except that the executive who had seized private property was a governor. A strike had been called in the lignite coal mines of North Dakota for November 1, 1919. After the strike had been in effect for several days, a severe snowstorm occurred and temperatures fell to extremely low levels. Lignite coal was desperately needed for fuel in the entire western half of the state. In view of this crisis, the governor ordered the coal mine owners to resume production. When the mines remained closed, the governor called out the militia and seized the mines. The mine owners then brought suit for an injunction. In denying equitable relief Judge Amidon stated:

I am asked to issue a writ of injunction which will necessarily say that the acts of the Governor have been illegal and unconstitutional. If I do that, I am not simply dealing with his acts; I am defining the powers of the Chief Executive of an American commonwealth to meet a crisis which threatens loss of life. I am not willing to strip the Governor of his power to protect society. I do not believe it comports with good order, with wise government, with a sane and ordered life, to thus limit the agencies of the state to protect the rights of the public as against the exaggerated assertions of private rights.[38]

With the Amidon opinion as a precedent, there was obviously no basis for Judge Pine's assertion that there existed an "utter and complete lack of authoritative support for defendant's position," the government brief maintained.[39]

To conclude its discussion of the constitutional question, the brief argued that the President's seizure of the steel mills was an action taken in accord with his duty to execute the laws faithfully. A vast array of statutes and treaties required the President to ensure the nation's security by seeing that the defense programs authorized by Congress were carried out properly. If a steel strike occurred, it would be impossible for the President to meet these statutory and treaty obligations. The seizure was thus a necessary and proper presidential response to fulfill a constitutional duty. Under the "aggregate of powers" theory, the Chief Executive was not limited by the express terms of statutes but derived power to act in a given situation from statutory law and constitutional grants taken in concert. Justice Jackson, then Attorney General, had clearly enunciated this theory, the government brief pointed out, on the occasion of President Roosevelt's seizure of the North American Aviation Company:

> The Constitution lays upon the President the duty "to take care that the laws be faithfully executed." Among the laws which he is required to find means to execute are those which direct him to equip an enlarged army, to provide for a strengthened navy, to protect Government property, to protect those who are engaged in carrying out the business of the Government, and to carry out the provisions of the Lend-Lease Act. For the faithful execution of such laws the President has back of him not only each general law-enforcement power conferred by the various acts of Congress but the aggregate of all such laws plus that wide discretion as to method vested in him by the Constitution for the purpose of executing the laws.

The Constitution also places on the President the responsibility and vests in him the powers of Commander in Chief of the Army and of the Navy. These weapons for the protection of the continued existence of the Nation are placed in his sole command and the implication is clear that he should not allow them to become paralyzed by failure to obtain supplies for which Congress has appropriated the money and which it has directed the President to obtain.[40]

Moreover, the brief continued, the Supreme Court had explicitly recognized the aggregate of powers theory in two cases, *In re Neagle*[41] and *In re Debs*.[42]

The final section of the government's brief dealt, defensively, with the Labor Management Relations (Taft-Hartley) Act of 1947. The existence of this statute did not preclude presidential seizure of the steel mills, the brief contended. The Taft-Hartley Act was not meant to be a mandatory or exclusive means of dealing with national emergency labor disputes. The act employed the words "the President may," rather than "the President shall," perform a particular action. Furthermore, the Defense Production Act of 1950 stipulated a different procedure for labor disputes that might interrupt defense production and affect the stabilization program. Critics could disagree with the wisdom of the President's use of the Defense Production Act procedure over the Taft-Hartley Act, but they could not question the legality of his choice. Moreover, by April 8, 1952, the objectives, although not the specific procedures, of the Taft-Hartley Act had been fulfilled. Imposition of the act at that time would not have brought a settlement closer but merely would have put off the strike again; and it was doubtful whether a court of equity would have granted the government a Taft-Hartley injunction after the union had voluntarily postponed the strike for ninety-nine days. Thus, the government argued, the President had reasonably concluded on April 8 that seizure was the most efficacious means of ensuring the continued production of steel and providing a favorable context for the settlement of the dispute.

The fact that Congress, during the debate on the Taft-Hartley Act, had specifically rejected a seizure provision should not be interpreted as a prohibition against seizure, the government maintained. As Senator Taft had stated at the time of the debate:

We did not feel that we should put into the law, *as a part of the collective-bargaining machinery*, an ultimate resort to compulsory arbitration, or to seizure, or to any other action. We feel that it would interfere with the whole process of collective bargaining. *If such a remedy is available as a routine remedy*, there will always be pressure to resort to it by whichever party thinks it will receive better treatment through such a process than it would receive in collective bargaining, and it will back out of collective bargaining. It will not make a bona-fide attempt to settle if it thinks it will receive a better deal under the final arbitration which may be provided. [Emphasis added by the brief.] [43]

In other words, the Justice Department argued, Congress had voted against seizure as a device to be used routinely, not against seizure as a remedy in a particular dispute. Similarly, because the Taft-Hartley Act provided that in the event a dispute was not settled while the injunction was in force the President shall make a report to Congress and submit recommendations for action, the act did not preclude the President, on the basis of his constitutional powers, from employing seizure to prevent a disaster while Congress considered appropriate action. "Nothing," the brief concluded, "in the language or history of the LMRA purports to restrict Presidential power stemming from sources outside that Act." [44]

THE INDUSTRY BRIEF

The attention paid to the Taft-Hartley Act by the government brief was not surprising or misplaced, because that statute formed the basis for the central contention of the steel industry's brief: that the President did not possess inherent power under the Constitution to seize private property in an emergency when Congress had supplied him with a different remedy for exactly that kind of emergency. [45] The Taft-Hartley Act, the companies argued, dealt with precisely the situation which had arisen in the steel dispute and clearly was intended to allow the parties to arrive at a settlement through collective bargaining and, if that were unsuccessful, to permit Congress to decide what action should next be taken. The evidence was overwhelming that Congress had deliberately determined to reserve the question of subsequent emergency action to itself. If the dispute was not settled during the eighty-day injunction period, the Senate report on the Taft-Hartley Act had stated, "the bill provides for the President's laying the matter before Congress for whatever legislation seems necessary to preserve the health and safety of the Nation in the crisis." [46] And, during the debate on incorporating a seizure provision into the act, the industry brief pointed out, Senator Taft had asserted:

We have felt that perhaps in the case of a general strike, or in the case of other serious strikes, after the termination of every possible effort to resolve the dispute, the remedy might be an emergency act by Congress for that particular purpose [seizure].

I have had in mind drafting such a bill, giving power to seize the plants, and other necessary facilities, to seize the union, their money, and their treasury, and requisition trucks and other equipment; in fact, to do everything that the British did in their general strike of 1926. But while such a bill might be prepared, I should be unwilling to place such a law on the books until we actually face such an emergency, and Congress applied the remedy for the particular emergency only. Eighty days will provide plenty of time within which to consider the possibility of what should be done; and we believe very strongly that there should not be anything in this law which prohibits finally the right to strike. [47]

The government's position that the Taft-Hartley Act could not have been employed on April 8 without causing an interruption in steel production should not be credited, the industry's brief urged. The union had given ninety-six hours notice of its intention to strike, which was sufficient time to invoke the act and obtain an injunction. [48]

The industry brief claimed that, although the Taft-Hartley Act might not be mandatory, the President could not simply disregard it and use another emergency procedure sanctioned neither by the Constitution nor by Congress. Nowhere in the Constitution was the Chief Executive given the power to take unspecified action to ensure the public welfare in an emergency. To the contrary, the brief declared, the Constitution categorically placed in Congress the express power to make all laws "necessary and proper for carrying into execution the foregoing Powers, and all other Powers vested by this Constitution in the Government of the United States, or in any Department or Officer thereof."

This included legislation to deal with crises, the brief insisted; hence the courts had recognized that the President retained no inherent powers.[49] As early as 1804, the brief noted, the Supreme Court, in *Little* v. *Barreme*,[50] had struck down an attempt by the President to usurp the power of Congress. During the undeclared naval war between France and the United States, Congress had enacted a law which provided that no American ship should be allowed to go to any French port and authorized the President to instruct commanders of United States armed vessels to seize any American ship bound to a French port. In written instructions sent to the commanders, President Adams had ordered them to seize all American ships—not only those heading for but also those coming from French ports. Following his instructions, Captain Little had seized a vessel sailing from a French port. The Supreme Court had unanimously upheld an order which returned the seized ship to its owner and found Captain Little liable for damages resulting from the seizure. Congress, the Supreme Court pointed out, had stipulated by its legislation the exact circumstances in which seizure could take place. Although the President's expansion of the seizure power to vessels coming from French ports would have made the legislation more effective, the Court concluded that the Chief Executive had no power, either as commander in chief or under the faithful execution clause, to institute his own procedure when Congress had clearly prescribed a different one.[51] The invalidity of Truman's seizure followed *a fortiori* from *Little* v. *Barreme*, the industry brief suggested. There the Court had struck down executive action which was distinctly compatible with the general intent of the congressional legislation. The case thus was a particularly compelling precedent, the brief argued, since the executive action at issue in *Youngstown* directly contradicted the congressional policy promulgated in the Taft-Hartley Act.[52]

Finally, the industry brief maintained that the President's position as commander in chief provided no constitutional basis for the seizure. Although the President on his own authority could take or destroy private property in the most serious military circumstances, this power did not include the function of eminent domain. A long line of cases, the companies' brief observed, had established that the power of eminent domain could be exercised only pursuant to an act of Congress. The brief quoted the holding in *Hooe* v. *United States:* "The taking of private property by an officer of the United States for public use, without being authorized, expressly or by necessary implication, to do so by some act of Congress, is not the act of the Government." [53] The clearest acknowledgment that the exercise of eminent domain power required congressional authorization, however, came from President Truman when he recommended seizing the coal mines in 1950. While a Taft-Hartley injunction was in effect, Truman sent a message to Congress urging it to enact legislation which would authorize the government to take possession of and operate the mines. If the President had inherent power to seize private property, as government counsel claimed, why did Truman think it was necessary to request congressional authorization in 1950, the industry brief wondered. [54] Such inherent power did not exist, the companies asserted; therefore, Secretary Sawyer's seizure of the steel mills was an arbitrary act executed only to enable the administration to impose a wage increase. And the brief warned that if arbitrary executive action to force a wage increase "is lawful today, then arbitrary executive action to force a wage decrease, or longer hours, or anything else, will be equally lawful tomorrow; and the constitutional rights of all citizens . . . will be gravely endangered." [55]

THE ORAL ARGUMENT

Although the brief is considered the most important representation of a litigant's position in a Supreme Court case, the oral argument gives counsel a further opportunity to impress his client's viewpoint upon the Justices and to respond to the Justices' particular concerns about the questions raised. The value placed on oral argument is reflected in the long-standing practice under which a Justice who misses the argument does not participate in the decision of the case. Moreover, the Court, at its weekly conferences, votes on the cases on which it has just heard argument.[56] In the steel seizure case it was obvious that the Supreme Court attached great weight to the oral argument, for it allotted five hours—two and one-half for each side—for the presentations. The Steelworkers and the railroad unions were given additional time to speak as friends of the court. This was a most extraordinary amount of time to be devoted to one case; the normal assignment was sixty minutes to a side, and frequently it was cut to thirty.[57] Clearly, the Supreme Court wished to treat the issue exhaustively. For the lawyers involved, this occasion furnished a rare challenge.

The attorneys for the steel companies, sobered by their unsatisfactory experience with a divided argument in the district court, wisely concluded that one attorney should present the entire argument for the industry. And the only person satisfactory to everyone was the eminent John W. Davis.[58] The unsuccessful presidential candidate of the Democratic party in 1924, Davis had been solicitor general of the United States during the Wilson administration and, in his many years of private practice, had achieved a reputation as a brilliant advocate before the Supreme Court. While the steel industry attorneys were writing the brief, Davis concentrated on preparing the oral argument, only occasionally asking his colleagues for their advice.[59]

Promptly at noon, on Monday, May 12, the Supreme Court convened to hear the steel seizure argument. All nine Justices—the five Roosevelt appointees: Hugo Black, Stanley Reed, Felix Frankfurter, William O. Douglas, and Robert H. Jackson; and the four chosen by Truman: Chief Justice Fred M. Vinson, Harold Burton, Sherman Minton, and Tom C. Clark—were in attendance. None had found reason to disqualify himself from participating. Although the Justices acted as if this case were no different from any other they might consider, the atmosphere in the courtroom indicated this was not so. By the time the Justices entered the courtroom, the more than three hundred seats for the public had been filled, mainly by congressmen and lawyers. In front of the section for the public were seats where counsel who would not be involved in the oral argument might sit. Over forty lawyers crowded into the steel industry's side to the right of the Justices. The government side had many fewer attorneys. After the Court had disposed of the routine motions of admitting new lawyers and announcing several decisions, two hundred more people were allowed into the courtroom to stand and listen to the argument. Hundreds more waited outside. Seated at the tables for counsel who would participate or assist in the oral presentations were, in addition to John W. Davis, Bruce Bromley and Theodore Kiendl, representing the steel industry, and Solicitor General Perlman and Holmes Baldridge for Secretary Sawyer. At approximately 12:25 P.M. the case was called.[60]

John W. Davis rose to begin his 138th argument before the Supreme Court.[61] White-haired, formally dressed, and standing erect, the seventy-nine-year-old Davis presented a distinguished figure at the lectern.[62] Glancing occasionally at notes in a large, loose-leaf notebook, Davis spoke smoothly, using few dramatic gestures. His oral argument was a masterful performance, the product of confidence gained from experience.[63]

After reviewing the steel labor dispute and the events preceding and following the seizure, Davis began a discussion of the legislative history of seizure provisions since 1916. He noted that the government had conceded that Secretary Sawyer and the President had acted under no statutory authority but, Davis pointed out, a brief examination of seizure legislation would reveal what Congress thought of the question of arbitrary seizure. From his summary, Davis drew three conclusions:

The first is that Congress fully understands the nature and scope of the power to seize private property for Government purposes;
Second: It recognizes the right of the Executive to exercise this power is drawn from legislative sources, and
Third: That the power should be granted, if at all, . . . for specific reasons and for limited purposes and with appropriate safeguards.[64]

Congress had deliberately not included a seizure provision in the Taft-Hartley Act, the current statutory embodiment of congressional policy for dealing with national emergency labor disputes. Although Davis could not claim that the provisions of the Taft-

Hartley Act were mandatory, he did believe that the fact of their existence made presidential creation of a new weapon for the same purpose an act of usurpation.[65] Moreover, the reasons given by the President for not invoking the Taft-Hartley Act were spurious, he stated.

Davis then turned to an analysis of the President's constitutional powers and their relation to congressional powers. He enumerated the many areas in which the Chief Executive exercised powers with respect to Congress. The President, he observed, had a constitutional responsibility to report to Congress on the state of the Union; he might recommend legislation; he could even convene Congress for an extra session or fix the date of adjournment of a congressional session, if necessary. That was all he could do, however. The President "cannot impair and cannot restrain, in any way," Davis maintained, "the power of Congress, which, in spite of his recommendation and reports, is perfectly free to ignore any quasi majestic views he may have to the extent it sees fit." [66] This was the basis for the separation-of-powers doctrine, a central tenet of the American system of government, Davis stated. Judge Pine had been correct in considering the constitutional question first. All other grounds were minor, Davis contended. Quoting the exchange between Holmes Baldridge and Judge Pine on the unlimited power of the President, Davis pointed out that although the solicitor general's brief disclaimed that argument, the brief really said the same thing. It reasserted "kingly prerogatives."

Davis then came to grips with the position he had advanced in the *Midwest Oil* case.[67] He admitted that in his brief in that case he had presented some "rather broad grounds," but he emphasized that the Court had not gone as far as the government had asked. The Court had upheld President Taft's withdrawal of certain government lands opened for occupation by an old but still effective statute because the President was acting as agent for the government and there was enough evidence that, despite the law, Congress approved of Taft's action. Furthermore, on 252 previous occasions the government had withdrawn public lands and Congress had not protested. Finally, the Court had held that since only government property had been withdrawn, no question of private rights was involved. Davis reminded the Court that a strong dissenting opinion had been filed by three Justices who thought Taft's act was a usurpation of power which attempted to dispense with the law. Indeed, after the opinions were handed down, Davis related, Justice Day, one of the dissenting Justices, called a page and sent Davis a note: "And you, a Jeffersonian Democrat, have done this thing." [68] But, Davis asserted, he, as solicitor general, had had a duty to represent the government. [69]

In concluding his eighty-seven-minute oral argument, Davis exhorted the Court to meet the responsibility which it faced. The steel labor dispute was transitory, he stated. The structure of the government was not, however, and the decision on the extent of executive power was of tremendous importance to the future of the nation. He repeated, for the Court, Jefferson's words in the Kentucky Resolutions, which, in Davis' opinion, summed up all of American constitutional theory: "In questions of power let no more be said of confidence in man but bind him down from mischief by the chains of the Constitution." [70]

When Solicitor General Perlman rose to speak, there was a perceptible change in attitude among the Justices. Perlman told the Court:

Your honors have just listened to an eloquent argument, an argument that is designed to turn the minds of this Court away from the facts in this case, away from the reasons which prompted the President of the United States to take the action that he did.

Very little, if anything, was said to this Court about the condition in the world today, about the struggle in which this nation is engaged. And, your Honors, practically nothing at all was said about the necessity, the vital necessity, to keep the plants owned by the plaintiffs here in operation without interruption of any kind, and it is argued here that your Honors should practically ignore that situation and pass on some Constitutional interpretation of the powers that the President exercised. [71]

Perlman recounted the facts of the case and launched into a discussion of the alternative procedures open to the President to help resolve the steel labor dispute in December 1951. Almost immediately he was interrupted with questions from the bench. Did the President's choice of the Wage Stabilization Board route mean that he could not pursue the Taft-Hartley route at the same time, Justice Frankfurter wished to know. That was correct, Perlman replied.

The solicitor general explained the events leading up to the seizure and the conditions in the world which made the seizure necessary. He asserted that President Truman had given Congress several opportunities either to ratify his action or to pass legislation terminating the seizure and providing a different solution. Justice Jackson inquired whether Congress could terminate the seizure, if the President had independent constitutional power to undertake it in the first place. Perlman agreed that Congress might not have the authority to end the seizure, but observed that the question was academic, since Truman had informed Congress that he would abide by any action it took. Perlman interpreted the lack of any contrary legislation to mean that Congress was content to let the President's action stand. Justice Frankfurter took issue with Perlman's interpretation, noting that a wide range of Supreme Court decisions had held that nonaction by Congress was not to be regarded as acquiescence in executive acts.[72]

Perlman next reviewed the statutory and constitutional basis of the President's power to seize private property in critical situations. Justices Reed and Black asked if the government relied on any specific statute as giving the President the power to seize the mills. Perlman answered that the President derived the power first from the Constitution and then from a number of statutes, although he had followed none of the prescribed procedures. Perlman then went over the same territory that his brief had covered, arguing that the executive, legislative, and judicial branches had recognized throughout the nation's history that the President had the power to seize private property in emergencies.

Perlman referred to President Roosevelt's seizure of the North American Aviation plant before Pearl Harbor as an exercise of the same power President Truman was asserting. Justice Jackson questioned Perlman as to the propriety of using the North American

Aviation seizure as a precedent for Truman's action. According to Jackson, that seizure could be easily distinguished from the present one: The North American Aviation Company was under contract to the United States. The government owned the materials and goods and the company was seized only when it could not comply with obligatory orders. No contest ensued between the company and the government because the owners had cooperated with the seizure. All these circumstances made the North American Aviation seizure different enough from the steel mill seizure to destroy its use as a precedent. Jackson admitted that he had again investigated the North American Aviation episode because he was afraid that much of the steel seizure would be laid at his doorstep. Perlman responded that the government did "lay a lot of it at your door." Perhaps that was right, Justice Jackson confessed: "I claimed everything, of course, like every other Attorney General does." He added pointedly, "It was a custom that did not leave the Department of Justice when I did." [73]

After being badgered by many questions reflecting the Court's desire to hear more about the Taft-Hartley Act and how it affected the President's seizure authority, Perlman finally began dealing with that statute. The Taft-Hartley Act, Perlman declared, contained no definite procedure for settling a labor dispute. Taft-Hartley merely provided a cooling-off period in which it was hoped that the parties would negotiate an agreement. At once Justice Burton interrupted Perlman: "Then it is a deliberate choice by Congress of collective bargaining instead of seizing the property. . . . It is a recognition by Congress that there is to be a report of the ballot and a settlement without seizure." [74] But, Perlman replied, in many of the nine instances in which the Taft-Hartley Act had been invoked, strikes occurred both before the injunction was issued and after it was discharged. President Truman wished to avoid any interruption in steel production. He therefore chose the route offered by the Defense Production Act because it seemed to present more possibilities for settlement than the Taft-Hartley procedure. The Justices then started questioning Perlman about the previous occasions when the Taft-Hartley Act had been used. It was already 4:30 P.M. the hour of adjournment, however, and in the middle of a comment by Justice Frankfurter, the other Justices left the courtroom.[75]

When Perlman resumed arguing at noon the following day, less than twenty minutes of his allotted time remained. He set out to deal with the equitable issues presented by the steel seizure case and strongly emphasized that the steel companies had an adequate remedy at law. The case of *Hurley* v. *Kincaid* was particularly instructive in this respect, the solicitor general stated.[76] But he soon returned to the Taft-Hartley Act and the President's reasons for not employing it. Perlman noted that the public had benefited from the Wage Stabilization Board procedure, which had postponed a strike for many more than eighty days. A special panel had studied the steel dispute at length, in contrast to the usual hurried investigation by a Taft-Hartley Board of Inquiry, and had arrived at well-considered recommendations.

Perlman's time for argument was extended to permit him to respond to further questions of the Justices. After many more inquiries concerning the Taft-Hartley Act and the reasons for not following the procedures stipulated by the Defense Production Act, Perlman began to sum up. He declared that the steel seizure must not be treated as a normal exercise of presidential power but as an extraordinary act taken to "avert a national catastrophe." The Court must recognize, Perlman asserted, that this was wartime. Instantly, the Justices attacked his statement. It might look like war, Justice Jackson said, but Congress "has specifically disclaimed this as war." "You cannot say," Justice Frankfurter interjected, "that you are not in a war on one hand and on the other, say that the President is exercising war powers, when he is not." [77] At the conclusion of this unhappy interchange, the Chief Justice told Perlman his time had expired.[78]

Davis, who had almost an hour of his time left for rebuttal, spoke very briefly and tellingly. He averred that he agreed wholly with the solicitor general's admonition to the Court that it must remember the steel seizure was not a normal case. "We, in this industry, are not living under normal conditions. We are faced with abnormal acts and . . . with a situation that, because of the extension of executive power—lacking legislative authority—we are deprived of our property and of our rights as citizens of the nation. The situation that is thus dictated is without parallel in American history." [79] To emphasize the industry's contention that the President had transgressed the bounds of his constitutional authority, Davis quoted from Chief Justice Marshall's opinion in *Little* v. *Barreme*, to the effect that when Congress prescribed the manner in which a law shall be executed the President must adhere to that prescription. It was the judiciary's role, Davis concluded, to enforce the separation of powers between Congress and the President. [80]

With the close of oral argument, the final opportunity for the parties to the seizure case to influence the judgment of the Court had passed. The Justices could retire to the privacy of their chambers to make their decisions. Events outside the courtroom occurred at their usual pace, however, and some of them were directly relevant to the steel seizure case. How they affected the process of judgment it is impossible to know, but that they formed some part of the reservoir of information with which the Justices worked we may be sure.

Most important was discussion in the press and Congress of the extent of the emergency should steel production be interrupted. Many observers questioned whether a steel strike would in fact have disastrous effects. The newspapers contained contradictory reports about the steel supply situation. On the one hand, the news that the government had liberalized restrictions on the use of steel for nonmilitary construction beginning July 1, 1952, created some doubt as to shortages. The press also noted that steel capacity was rising each week. On the other hand, articles also pointed out that the supply of certain types of steel was tight and that the future picture for these particular items was not bright. [81] In Congress, the same divisions were apparent. When administration officials testifying before congressional committees stressed the importance of averting a shutdown in the steel industry, they were faced with hostile questioning. Ellis Arnall, appearing before the

Senate Committee on Banking and Currency, stated that he could say unequivocally that a steel stoppage would be a catastrophe. Arnall was drawn up short by Senator Maybank who declared testily:

Who gave all this information in the Department of Defense, when they can't use what they have now? . . .

I just don't understand. We passed a Production Act. At the end of each quarter scarce metals are turned back after having been allotted to defense-connected claimants. I just got through talking to Mr. Fleischmann, and they are still being turned back. There are no uses for the materials. I don't know what is the matter with them. . . .

They claim all these fantastic requirements, and they never use the quantities allotted to them.[82]

Senator Morse, on the other hand, told his fellow committee members during the hearings on national and emergency disputes that he was also a member of the Senate Armed Services Committee and that no evidence had been brought forth there to show that the military emergency was nonexistent. Hence, he declared, he would continue to believe that the President had acted in an "hour of crisis." [83]

Other events in Congress which may well have received some attention from the Justices included a Senate committee recommendation on the future of the Wage Stabilization Board, Senate action on the nomination of Judge James McGranery to be Attorney General and a House Judiciary subcommittee's attempt to deal with the inherent power question in legislation extending the President's emergency war powers until June 30, 1952. The Senate Committee on Banking and Currency voted to abolish the current tripartite WSB and to set up instead a board composed of all public members with no power to hear disputes—a clear indication of disapproval of the performance of the WSB in the steel case.[84] On May 20, the Senate confirmed the McGranery nomination after Judiciary Committee hearings during which he had been closely questioned on the President's power to seize private property. McGranery had responded that inherent powers should not be used before statutory authority was exhausted and that the

President did not have the power to seize except in great public ca-
tastrophes.[85] The House Judiciary subcommittee considering the
extension of the President's emergency war powers urged Congress
to declare that the Chief Executive did not have inherent power
under the Constitution to seize private property and to preclude
the President from changing wages, hours, prices, or working conditions in any plant, mine, or facility of which he had taken possession since April 1.[86]

Another proceeding which probably did not escape the
Court's notice was the Sixth Constitutional Convention held by
the United Steelworkers of America during and after the oral argument. Many stories appeared in the press concerning the attitudes
of the union members at this juncture in the steel dispute. Most of
the workers interviewed were proud of their patriotism and responsibility for delaying the strike, but many believed that they would
have been better off if the President had used the Taft-Hartley Act
originally, for they would then have been free to strike, and that
seemed to them to be the only way to convince the steel companies to sign a contract.[87] The steelworkers were cheered by appear-
ances by Vice-President Barkley and Secretary of Labor Tobin,
who enthusiastically supported the union's position in the dispute.[88] The convention passed a resolution declaring that the
union would no longer work under the 1951 contract. If the Supreme Court ruled adversely, the union intended to strike and to
go to court to prevent a Taft-Hartley injunction from being imposed.[89]

A news conference held by President Truman in the interim
between the oral argument and decision day can hardly have
helped his position in the Supreme Court. A reporter asked the
President what kind of seizure law he would like to have Congress
enact. Would he like to have the statutory power to seize in reserve or would he rather go to Congress in each emergency? Truman responded: "The President *has* the power, and they can't take
it away from him." [90] Questioned as to whom he was referring to
by "they," Truman identified Congress and the courts: "Nobody
can take it away from the President, because he is the Chief Exec-

utive of the Nation, and he has to be in a position to see that the welfare of the people is met." [91] Reporters followed up this exchange by asking, if the Court could not take away this power from the President, what issue was the Court currently considering. Truman answered that he could not comment on what was now before the Court.[92] The press was obviously mystified by Truman's position, for soon after the conference Joseph Short, the White House press secretary, issued an additional statement clarifying the President's remarks. Short related that he had talked with Truman immediately after the news conference and that the President had said his point was that "neither the Congress nor the courts could deny the inherent powers of the presidency without tearing up the Constitution. . . . The Supreme Court, in the pending steel case, might properly decide that the conditions existing did not justify the use by the President of his inherent powers, but . . . such a decision would not deny the existence of the inherent powers." Short emphasized that the President would not have seized the steel mills if he had not "believed that he was taking a legal step." The President "did not anticipate that the Court would decide otherwise than that he [had] acted properly." The illustration that the President had used in talking privately to Short—that the Court might decide that conditions did not justify the use of the President's inherent powers—was "an illustration, and nothing more." [93]

In the meantime, the nation waited impatiently for the Supreme Court's decision.

nine

AWAITING THE DECISION

President Truman anticipated that the Supreme Court would uphold his action, and his expectation was shared by many. Twenty years of strong executive leadership had brought a change in the popular conception of presidential power. The decisions of the Supreme Court after the "constitutional revolution of 1937" had enhanced the general impression that an active President would not be restrained by the judiciary. Given this setting, few believed that the Supreme Court of 1952, composed of five Roosevelt and four Truman appointees, would be the instrument to call a halt to the accretion of power by the executive branch.

A variety of evidence reflected the assumption that the seizure would be upheld. Journalists, basing their predictions on the widely held belief of lawyers, filled their columns with theories on why the Supreme Court would not decide the constitutional issue. Most cited the Court's traditional practice of avoiding constitutional questions when a case could be disposed of on other grounds. Congressional defenders and critics of Truman alike presumed that the Supreme Court would rule for the President. Senator Dirksen stated on the floor of the Senate:

So as I appraise the action which was taken with respect to the steel mills and fit it, like a piece of a jigsaw puzzle, into everything else that has happened in the past 3 or 4 years, this action comes as no surprise to me. Nor . . . does the action of a Federal court which refuses to delimit the power of the President or so cautiously passes upon it that for practical purposes the decision has little legal value.

Can we expect anything different? If we are in an emergent situation, I doubt whether any Federal court will stand up on the side of Congress. If the job is to be done, it must be done in that branch of Government which James Monroe characterized as the very core and center of the Government of the United States, namely, the Congress.[1]

Emanuel Celler, chairman of the House Judiciary Committee, observed that the Supreme Court was likely to hold that it had no power to pass on the President's action:

Frankly, indeed, never in history has any judge attempted successfully to restrain a President. He could not enforce his order against the wishes of the President. And as a matter of fact there are very healthy decisions of precedence against such a step, notably, the famous case of the State of Mississippi against President Johnson.[2]

After his eloquent argument before the Supreme Court pressing the rights of the steel companies, John W. Davis received a letter illustrative of the popular expectation. The writer, who supported the steel industry in the seizure case, expressed anxiety that the Roosevelt-Truman Court would not rule against the President.[3]

Reinforcing the impressions gained from actual experience of the previous two decades was the scholarly literature which both described and approved the concentration of power in the President. Among political scientists and legal scholars, World War I had awakened a strong interest in documenting the changes in the government power structure caused by periods of crisis.[4] The depression and World War II had intensified these developments, and by 1952 the rhetoric of politics encouraged Americans to believe that a state of crisis was ever-present. Edward S. Corwin and Clinton Rossiter were the most important among the scholars who chronicled and explained the emergence of a new concept of the modern Presidency.

Corwin, writing in 1934, defined executive power as "what is left of the original competence of government after the relatively specialized functions of legislation and adjudication have been subtracted therefrom."[5] Some of the attributes ascribed to executive power by Corwin were its concentration in a limited number of hands, its control of the physical forces of government, and its ability to be exercised instantly. It therefore was uniquely adapted for action in dangerous circumstances. In wartime the President became the natural leader and steadily accumulated more and more power in order to prosecute the war successfully. Corwin, in 1948, noted a threefold development in the war power: (1) the constitutional basis of the war power changed from the theory of delegated powers to the theory of inherent powers; (2) the President's authority under the commander-in-chief clause shifted from a mere power of military command to one employed in pursuance of many objectives; (3) the undefined legislative powers claimed by Congress in wartime as an outgrowth of its inherent powers could be delegated to the President as Congress saw fit.[6]

The most startling consequence discerned by Corwin, however, was the stretching of the very concept of war, so that constitutional barriers were relaxed not only during the hostilities themselves but also for the periods preparing for the war and recovering from it. Unavoidably, the constitutional practices followed for such an extended term molded the character of the modern peacetime Presidency as well, Corwin contended.[7] Presidential responsibility for industrial relations was only one area where the effects of wartime procedures were apparent.[8]

Clinton Rossiter's main contribution to recognition of the expanded Presidency as a fixed part of the American government consisted of demonstrating how the Supreme Court had treated this new element in the constitutional scheme. In *The Supreme Court and the Commander in Chief*, Rossiter asserted that the Court had been very careful in its pronouncements on the powers of the President as commander in chief: "It has been respectful, complimentary, on occasion properly awed, but it has never embarked on one of those expansive flights of dicta into which it has been so often tempted by other great constitutional questions."[9] The Court had been reluctant to set specific limits to this power because it seemed unsure of the propriety of its intervention when the war power had been improperly used. Its estimation of its authority in this area had changed from case to case depending on the Court's immediate feelings regarding the substitution of its judgment for that of the President, his military officers, or Congress.[10] When possible, the Court would try to find some basis for approving a questioned presidential order other than the authority of the commander-in-chief clause; and usually the Court would grasp at any evidence which could be construed as congressional approval. As Rossiter pointed out,

This preference for statutory over constitutional authority actually works to the President's advantage, for the merger of his military powers and those of Congress produces something known simply but grandly as "the war powers of the United States," under which just about any presidential wartime action can be brought within the limits of the Constitution. Both President and Congress have constitutional powers of their own in military and foreign affairs; when these powers are merged they are virtually irresistible, at least in the courts. This is one instance in which two plus two equals five.[11]

Furthermore, Rossiter noted, the President had been able to exercise his war powers in periods of an undeclared war. On these occasions, Congress had either expressly or by implication ratified acts of war undertaken by the President solely on his authority as commander in chief.[12]

Other scholars supported the Corwin-Rossiter view that the presidential power had been irrevocably enlarged and that the Supreme Court would do little to interfere with its development.[13] Pointing to the probability of prolonged international tension, C. Herman Pritchett expected presidential power to expand with the emergency and doubted that the Supreme Court would uphold any challenge to that power which came before it:

Even apart from wars and rumors of wars, the prognosis is for limited judicial review. The present Court has tended to emphasize the restricted competence of judges, and to stress the inherent limitations of a lawsuit as a vehicle for the formulation of broad public policies. It has attempted, though not always successfully, to keep in mind its own limitations as a policy-forming agency.[14]

Moreover, Pritchett continued, the Court had not promulgated any "doctrinaire notions of the separation of powers" when dealing with problems concerning the relations between the executive branch and Congress. The Court preferred to allow the extent of presidential powers to be decided by the political process rather than by the judiciary. "And that is as it should be," Pritchett concluded.[15]

The backgrounds and records of the individual Supreme Court Justices lent support to the expectation that the Court would uphold the steel seizure. Because all the Justices had been appointed either by Roosevelt or Truman and because each of them—with the exception of Frankfurter, who was in any event unofficial adviser to Roosevelt—had had government experience in Congress or the executive branch during the New and Fair Deals, it was widely thought that their views of presidential power matched those of the Presidents in whose administrations they had served. It was also assumed that their political experiences had endowed them with sympathy for organized labor and antipathy for big business. It hardly seemed possible that Big Steel could win a victory at the bar of the Supreme Court as it was composed in 1952.

Born in 1886 in Clay County, Alabama, the son of a storekeeper, Hugo Black learned from an early age the difficulties involved in the struggle for existence. He paid his own way through the University of Alabama Law School, which he attended without ever having gone to college. For most of the years from his graduation until he entered the United States Senate in 1927, Black engaged in law practice in Birmingham, where corporate clients were not his stock-in-trade. Rather, he earned his keep successfully—too much so, in the minds of some—representing working people injured on the job.

In the Senate, Black became an extremely skillful investigator. He was frequently accused, however, of holding a cavalier attitude toward the constitutional rights of the people he questioned. As the New Deal unfolded after Roosevelt's election and Black's reelection, he became its ardent supporter, sponsoring labor legislation, investigating the utility companies, and working for passage of the President's court reform bill. In August, 1937, Black became Roosevelt's first nominee to the Supreme Court.[16]

On the Court, Justice Black soon earned a reputation for "decisiveness, forthright thinking, strong attitudes and impatience with legalism." [17] In most cases, he approved government intervention to regulate business and labor's right to help itself with the support of favorable legislation.[18] Black also achieved recognition as a consistent champion of civil liberty claims. According to Charles Reich, Black approached constitutional questions from the premise that the original structure of the Constitution must be maintained. In his mind,

The ultimate test of a constitutional decision is whether it retains, rather than alters, the grand design. Of course no contemporary decision can be reached by looking to the past and stopping there, and Mr. Justice Black, like every other judge, has had to decide issues composed of contemporary facts and problems. But the question he asks is always a question that has reference to the standards of the Constitution, even though the answers must be given in contemporary terms. It might be said that, in looking at present day issues, Black uses the Framers' scales and seeks to approximate the Framers' balance.[19]

Although Black generally interpreted the words of the Bill of Rights literally to prevent government interference with personal liberties, he admitted that civil liberty claims had to give way before governmental needs during wartime. When the war powers of the government came in conflict with individual rights, the safety of the nation required the sacrifice of the latter. Thus Black authored the opinion of the Court in *Korematsu* v. *United States*,[20] which upheld, as a proper exercise of the war powers, a military order to evacuate Americans of Japanese birth from the West Coast.[21] It was not unreasonable, therefore, to expect that Black might sustain the President in the steel seizure case, given the hostilities in Korea, the international tensions, and the administration's claim that steel was desperately needed for the United States and its allies.

Stanley Reed, the second of Roosevelt's Supreme Court appointments, was considered another Justice likely to approve President Truman's seizure of the steel mills. Throughout his judicial career Reed had deferred to the executive and legislative branches as the best judges of their own power and policies. Reared in Kentucky, where he was born in 1884, Reed went east for his education. He obtained a bachelor's degree from Yale University and a law degree from Columbia. After a year in Paris studying at the Sorbonne, Reed returned to Kentucky, where he entered private practice and became involved in Democratic politics. As general counsel of the Burley Tobacco Growers' Cooperative Association, he acquired some of the knowledge and contacts which prompted the Hoover administration to appoint him, in 1929, counsel for the Federal Farm Board. In 1932, Reed was promoted to the position of general counsel of the Reconstruction Finance Corporation. Evidently effective in this post, Reed was kept there by President Roosevelt when he took office.

After successfully arguing the Gold Clause cases before the Supreme Court, Reed was chosen in 1935 to be solicitor general of the United States. In the next three years Reed defended some of the most important New Deal legislation, including the National Industrial Recovery Act, the Agricultural Adjustment Act, and the Tennessee Valley Authority Act. His defeats at the bar of the Supreme Court when legislation was struck down undoubtedly tempered his own performance as a judge. Reed became a consistent supporter of the national government's actions, to such an extent that the rights of the individual frequently suffered. According to one observer of the Court, "Reed had no social philosophy to promote. He was a legal craftsman who believed in giving the political branches of government the power and the responsibility for governing" [22]—a likely vote for Truman in the steel case.

Felix Frankfurter was less predictable but also seemed likely to uphold the President. His decisions often rested on procedural points rather than on the merits of the issue. Years spent studying, teaching, and writing about the law—he was a professor at Harvard Law School from 1919 to 1939—had made Frankfurter into a champion of judicial self-denial.[23] Having come in 1894 from Vienna to the United States at the age of twelve, Frankfurter grew to respect and dearly love his new home, and especially to appreciate the democratic processes of government. Although convinced of the importance of the Supreme Court's function of judicial review, Frankfurter constantly warned against an enlarged view of the Court's power. The Court was not "an overriding promoter of the general welfare." [24] Such a view implied a want of faith in democracy, Frankfurter declared. Judges must allow the political branches of government the greatest scope legally possible under the Constitution, and in construing the acts of those departments judges must not impose their personal opinions as to the wisdom of these measures. Frankfurter had been acutely upset by the performance of the Supreme Court during the New Deal, and he strove to avoid a repetition of such a disaster while he was on the Court. [25]

On the Court, Frankfurter, during World War II, had been a consistent supporter of the government's war powers. Concurring in *Korematsu* v. *United States*, he had emphasized that the Constitution gave Congress and the President sufficient powers to prosecute a war. "Therefore, the validity of action under the war power must be judged wholly in the context of war. That action is not to be stigmatized as lawless because like action in times of peace would be lawless." [26]

He noted, however, that the Court's finding that the executive order to evacuate the Japanese from the Pacific Coast was a valid exercise of the war power in no way constituted approval of the action. Congress and the President were responsible for their policies. [27] In *Duncan* v. *Kahanamoku*, Frankfurter disagreed with the Court's majority, which struck down the trial of civilians by military tribunals when Hawaii was under martial law during World War II. The majority emphasized that conditions in Hawaii did not necessitate the supplanting of courts by military tribunals. [28] Joining in a dissent by Justice Burton, Frankfurter thought that the Court could not make the judgment that Hawaii was not a battlefield at the time of the challenged trial. Only the President could make that decision. [29]

Tempering his very generous construction of the war powers,

however, were Frankfurter's views on the separation-of-powers doctrine. Although the doctrine was not to be blindly adhered to—the functions of the branches of government would be defined by experience as well as by the Constitution—it must be respected for the limits it prescribed for those branches. Writing in 1930, Frankfurter stated that the Supreme Court had not treated the separation of powers as a "technical legal doctrine." The Court had no interest in defining abstract lines of separation; instead it had recognized that modern government demands interaction among the branches. Still, the separation-of-powers maxim must not be wholly dismissed for, in Frankfurter's words, the doctrine "embodies cautions against tyranny in government through undue concentration of power." [30]

Unlike Frankfurter, William O. Douglas, the fourth Roosevelt appointee to the Court, rarely allowed qualms about the exercise of judicial power to interfere with his desire to meet the issues squarely. Douglas came to the Court with special expertise in the areas of business law. As a professor of law at Columbia and then at Yale, he taught and wrote on business regulation, corporations, and bankruptcy. In 1934, he left Yale to work for the Securities and Exchange Commission (SEC). He became a member of the SEC in 1936 and its chairman in 1937. Although in his early years in Washington he was thought to be unfriendly to business, by the time he went on the Court in 1939 his reputation had changed somewhat. Douglas was confirmed by the Senate 62 to 4, with some of the dissenters accusing him of being a "reactionary tool of Wall Street." [31]

Douglas' work on the Court proved how wrong they were. In his autobiography, Douglas states that he and Black were "one with Brandeis in his insight into the corporate world and its chicanery. We also stood with Brandeis in his passion to protect small and medium-sized companies." [32] By 1952, few thought that Douglas would be sympathetic to the claims of the steel industry.

There were other elements which would enter into the judgment, however. Douglas was a staunch civil libertarian, who might view the President's seizure of the steel mills as an unconstitutional invasion of the steel companies' rights. On the other hand, he also had a generous conception of the war powers of the United States. During World War II, Douglas had voted with the majority to uphold the government even in cases involving a substantial infringement of personal freedom.[33] Since the Korean War had not yet ended, Douglas might very well judge the seizure in that context.

If his previous statements on executive power were any guide, Robert H. Jackson appeared to be a sure vote for the President. The most brilliant legal advocate the New Deal had brought forth, Jackson had been solicitor general from 1938 to 1939, Attorney General from 1940 to 1941, and a friend and adviser to President Roosevelt for many years.[34] His Justice Department service occurred during a period not altogether unlike the years of the Korean conflict: there was no declared war, yet the United States was involved in warlike acts and in defense preparations, and Jackson was called upon to render opinions on a variety of issues concerning the exercise of executive power. Not only had he offered legal justification for Roosevelt's seizure of the North American Aviation plant, but he was the architect of the legal underpinnings of the "destroyers for bases" deal with Great Britain. Jackson's expansive construction of the President's authority permitted Roosevelt to consummate that transaction without specific congressional legislation.[35] Nor would it be proper to dismiss Jackson's opinions as Attorney General as merely the lawyer's advocacy of his client's cause. Jackson once described the Attorney General as having a dual position—being lawyer for the president as well as "laying down the law for the government as a judge might. I don't think he is quite as free to advocate an untenable position because it happens to be his client's position as he would if he were in private practice. He has a responsibility to others than the President. He is the legal officer of the United States." [36]

Once securely placed on the Court, however, Jackson lost, in the words of one observer, "the first fine careless rapture of the New Deal crusade, and became more conservative." [37] Jackson voted frequently the way Frankfurter did and, like Frankfurter, was rather unpredictable.[38] According to one of Jackson's law clerks, Justice Jackson had several basic beliefs but did not let these beliefs interfere with his consideration of each case on its individual merits.[39]

During World War II, Jackson's attitude toward the government's war powers did not reflect a consistent point of view. Jackson had joined a unanimous Court in *Hirabayashi* v. *United States* to uphold a military curfew order on the West Coast for citizens of Japanese descent.[40] But in *Korematsu* v. *United States*, Jackson dissented. He was unwilling to put the Court's imprimatur on a military action, clearly unconstitutional in peacetime, the necessity for which the Court had no real evidence. If the Court approved the action, it would create a mischievous precedent of much greater weight than the military order itself.[41] Again, in *Duncan* v. *Kahanamoku* Jackson voted against the government, but this time he was in the majority. Balancing the government attorney's expansive view of presidential power against the Justice's more skeptical attitude, no one could predict how Jackson would treat the emergency claimed by the government in the steel seizure case. If, however, Jackson adhered to what he called the first principle of constitutional adjudication, "Namely, that a court will not decide a constitutional question unless such a decision is absolutely necessary to dispose of the case," [42] he might never get to a consideration of the merits.

It was widely assumed that the four Truman appointees to the Court would support the President. Each was a personal friend of Truman's. Each had a record of upholding government action whatever the issue, and all often voted the same way. Chief Justice Fred M. Vinson, a Kentucky Democrat who in fourteen years as a congressman had a "100% pro labor record," [43] firmly believed that the popularly elected branches of the government should be permitted to rule and that the Court must restrain itself in judgment of their actions.[44] His views undoubtedly were formed by his experience. Before Vinson became Chief Justice, he had served in all three departments of the federal government. After leaving Congress in 1938, Vinson spent five years as a judge on the

United States Court of Appeals for the District of Columbia. During the war, however, he gave up his judgeship for a variety of posts in the executive branch: director of economic stabilization, federal loan administrator and director of war mobilization and reconversion. When Truman had been President for a few months, he chose Vinson to be his Secretary of the Treasury and turned to Vinson for many kinds of advice.[45] Truman moved Vinson to the Supreme Court, not because he had shown any special talent for judging in his years on the court of appeals, but because Truman hoped that Vinson, who in his years in Congress and executive positions had proved to be an excellent mediator, could conciliate the badly divided Supreme Court.[46] While serving on the Court, Vinson continued to be an adviser to the President.[47]

Harold Burton, a Republican, was nominated by President Truman in the fall of 1945 to replace Owen J. Roberts. His Republicanism hardly distinguished him from Truman's Democratic appointments. Born in Massachusetts and educated at Bowdoin College and Harvard Law School, Burton left the East to practice law in other sections of the country and finally settled down in Cleveland, Ohio. There he was elected mayor in 1935 and twice won reelection. Burton bucked the state administration of Governor John W. Bricker in order to bring federal funds into Cleveland for relief projects. After a strenuous campaign to defeat the leaders of both the Republican and Democratic parties, he was elected to the Senate in 1940. In the Senate, Burton voted many times for the social and economic policies of the New Deal; but he was independent in certain areas, including labor issues. As a result of his assignment to Senator Truman's War Investigating Committee, Burton and Truman became close friends. In 1945, Truman thought it would be appropriate to fill Justice Roberts' seat with a Republican. Otherwise, Harlan Fiske Stone would be the only Republican left on the Court. The new President also believed that his choice of a Republican would foster good relations with the opposition party. Burton appealed to Truman because he was not highly partisan in his behavior and had a flexible approach to important issues.[48] In fact, on the Court, Burton was to be called by John Frank, a student of the Court, "a foremost example of judicial temperament . . . a fair man who apparently does not make up his mind until he has considered every aspect of a problem." Consequently Burton, "though basically conservative in his instincts, is highly unpredictable," Frank concluded.[49]

In 1949, Truman made two appointments to the Supreme Court. The deaths of Frank Murphy and Wiley Rutledge within a few months of each other gave Truman this unusual opportunity. To replace Murphy, Truman chose his Attorney General, Tom C. Clark. A Texan, who came to the Department of Justice in 1937, Clark had made his way first to an assistant attorney generalship in the antitrust division and then to head of the criminal division. During World War II, Clark dealt with the fraud cases referred to the Justice Department by Truman's investigating committee and this led to the friendship between them. Truman thought so well of Clark's work as a government lawyer that he picked him to be Attorney General in 1945, when the President selected a cabinet more to his taste than the holdover Roosevelt cabinet. In his four years as Attorney General, Clark energetically employed the government's powers in the antitrust field, in civil rights litigation and to develop the administration's internal security program. Known as a very hard worker, Clark once candidly admitted to a reporter, "I have to work long hours because I'm not as smart as some other fellows." [50] Throughout his government career, Clark had been a firm believer in the inherent power of the President to act in emergencies and everyone anticipated that Clark would vote for the President in the steel seizure case.[51]

Sherman Minton of Indiana, the last of Truman's Supreme Court appointments, was cast in the same mold as Vinson, Burton, and Clark: all had had many years of government service; none was noted for his intellectual prowess. Elected to the Senate in 1934, where he was assigned a seat next to Truman, Minton became an enthusiastic New Dealer. When the Supreme Court struck down major New Deal legislation, Minton was openly critical and wholeheartedly supported Roosevelt's Court-packing plan in 1937. Roosevelt reportedly considered nominating Minton to the Supreme Court in 1937 but chose Black, an equally ardent New Dealer, although the President did not think he was Minton's equal as a lawyer.[52] When Minton was defeated for reelection in 1940, Roosevelt employed him in the White House as a presidential assistant, and in 1941 appointed him to the bench of the United States Court of Appeals for the Seventh Circuit. Minton's record as a judge and as a Senator led many to believe that he would join the Black faction on the Supreme Court. This alignment did not materialize, however, for Minton had taken to heart the nation's experience under the Court of the 1930s and become a practitioner of judicial restraint. Minton tended to construe liberally the powers of the elected branches while constricting the judicial role, and few thought he would question the President's power to seize the steel mills.[53]

When the steel seizure case reached the Supreme Court, the Justices could have disposed of it in a variety of ways that would have avoided striking down the President's seizure order and been in keeping with the Court's traditional approach to constitutional questions and its self-proclaimed attitude of judicial restraint. First, the Court could have voted to deny, rather than grant, certiorari and to allow the case to be heard in the court of appeals, as Frankfurter and Burton had advised. During the time gained while the court of appeals considered the case, the companies and the union might have agreed to settle the dispute, and the seizure would then have been rescinded.[54] Furthermore, if the court of appeals refused to issue an injunction, which seemed a likely result judging from the majority's memorandum on the granting of a stay, that in itself might have spurred the steel industry and the union to resume negotiations.

Once the Court had granted certiorari, however, the Justices still had a number of options. The Court might send the case back to the district court for a determination of the facts. Despite Judge Pine's statements to the contrary, it was difficult to find on the basis of the record that the plaintiffs had been irreparably injured. Counsel for United States Steel had taken this view when, during the hearing in district court, he had asked only for an injunction to maintain the status quo. He had in effect admitted that the seizure did not harm the steel companies; only a government-imposed wage increase would cause an irrevocable injury. A hearing on the facts would give the government an opportunity to refute the claims of damage by the other companies, and also would provide more time in which a settlement might be reached.[55]

The Court also could decline to decide the steel seizure case and vacate Judge Pine's decision on the ground that the issues presented a "political question." A case may be dismissed if the Court refuses to exercise jurisdiction because the subject matter of the case concerns an issue "usually dealt with by political as contrasted with judicial methods, and is at one with, or included in, matters unquestionably and unequivocally delegated to the executive and legislative departments." [56] Thus, cases in the area of the President's actions in foreign affairs have frequently been dismissed as involving a political question which the Court does not have competence to resolve, both because it lacks the information necessary for a reasoned decision and because an independent judgment by the Court might embarrass the nation in the international sphere.[57] If the Court identified the steel seizure as a measure necessary to uphold the United States' treaty obligations and commitments to the United Nations with regard to the Korean War, it might decline jurisdiction.

Other grounds existed for the Court's possible rejection of the steel seizure case as involving a political question. Alexander Bickel has suggested that the foundation of this doctrine is:

The Court's sense of lack of capacity, compounded in unequal parts of (a) the strangeness of the issue and its intractability to principled resolution; (b) the sheer momentousness of it, which tends to unbalance judicial judgment; (c) the anxiety, not so much that the judicial judgment will be ignored, as that perhaps it should but will not be; (d) finally ("in a mature democracy"), the inner vulnerability, the self-doubt of an institution which is electorally irresponsible and has no earth to draw strength from.[58]

Certainly, some of these criteria could be applied to the steel case. If the Court took jurisdiction and invalidated the seizure, it would precipitate a strike, as had occurred after Judge Pine's decision. The repercussions of a strike would be felt throughout the mobilization program and the economy of the United States, and in its foreign relations, and the resulting problems would not be susceptible of solution by the Court. Rather they would have to be dealt with by the elected branches of the government. Arguably, it would be more responsible for the Court to abjure judgment and allow Congress and the President to resolve the issues.[59] Moreover, Congress had the constitutional power to override Truman's seizure order, as the President had conceded. The Court might conclude, therefore, that Congress should deal with the issue, that is, that the steel seizure case raised a "political question." [60]

Further, the Justices could refrain from ruling on the merits by disposing of the case on equitable grounds, as the government had so forcefully advocated. The Court could refuse to grant an injunction on either of two grounds: The steel industry was not entitled to equitable relief because it had an adequate remedy at law, that is, it could file a suit for just compensation in the Court of Claims; or, an injunction could not be issued where the injury to the public interest resulting from such a grant would be greater than the harm to the plaintiffs if no relief were given. Both routes had creditable Supreme Court precedents on which to rely, as the government's argument demonstrated. On the question of whether the steel companies had an adequate remedy at law, *Hurley* v. *Kincaid* and *United States* v. *Pewee Coal Co., Inc.* indicated that the Court of Claims would accept jurisdiction in a suit for just compensation.[61] With regard to the second route, the ruling in *Yakus* v. *United States* indicated that an injunction should not be granted if it would "adversely affect a public interest for whose impairment, even temporarily, an injunction bond cannot compensate," even if the delay of relief until a final consideration of the rights of the parties would be burdensome to the plaintiff.[62] Justice Brandeis had enunciated the same principle in *Hurley* v. *Kincaid*.[63] These cases were pivotal in the court of appeals decision to grant a stay of Judge Pine's injunction that had terminated the steel seizure. Surely they presented sufficient grounds to enable the Supreme Court to adhere to its normal procedure on constitutional questions and thus to avoid embarrassing the President.

Finally, the Court could uphold the President on the merits. There was a wholly respectable basis on which to declare the seizure legal. The United States was involved in the Korean conflict and had, in addition, many obligations to its European allies which could not be met without the continued production of steel. Viewing the seizure as a military or foreign policy measure, the Court could cite legal precedents sustaining the President's exclusive discretion in these areas.[64] If the Court did not wish to ground its decision on the President's powers alone, it would not be difficult to identify the seizure as an act taken in pursuance of a mass of legislative directives, which included seizure in various circumstances, to ensure the progress of defense mobilization and stabilization. The Court had approved this construction of the "Take Care" clause in a number of cases, [65] and the Chief Executive had been operating under this theory for decades.[66] The Court also had recognized, in *United States* v. *Midwest Oil*, that the President gained power from "the silences of Congress" [67] as well as from specific laws. Congress' failure to object to previous presidential seizures in similar circumstances or to pass a law denying seizure power to the President could be cited as corroborative evidence that no constitutional bar existed to Truman's seizure of the steel mills.

Thus, there were many ways by which the Supreme Court could let the seizure stand. Given the recent history of the presidency and of the Supreme Court, few anticipated that the Justices would invalidate the President's action. The country was, therefore, unprepared for the Supreme Court's decision.

ten

THE SUPREME COURT
RULES

Since the steel seizure litigation had been expedited, everyone expected the Supreme Court to rule speedily. The Justices voted on the case at their conference on May 16, several days after conclusion of the oral argument. But more than two weeks passed before they rendered their decision, on Monday, June 2. As the multiplicity of opinions filed suggests, the Justices, although arriving quickly at the result, had difficulty in agreeing upon the basis for their conclusion.[1]

The only direct evidence of the Court's internal processes between the argument and the decision comes from Justice Burton's diary. On May 10, before hearing the oral argument, Burton had lunched with the Chief Justice and Justices Black, Reed, Clark, and Minton.[2] From the discussion which took place that day Burton supposed that he might be "largely alone in holding the President was without power to seize the steel plants in the face of the Taft-Hartley Act providing a different procedure."[3] But by May 15, after Burton had lunched again with many of the same Justices, he observed that there might well be a majority to sustain Judge Pine's injunction. Burton recorded, however, that he expected Vinson, Reed, Clark, and Minton to support the President.[4] When the Court's conference took place on Friday, May 16, Burton noted that the Justices had spent four hours examining the steel case "with a most encouraging result."[5]

The vote taken at the conference is not necessarily final, however. After drafting of the opinions is completed or any time during the process of writing them, the Justices may change their minds. How the majority opinion is written may thus be an important factor in keeping the requisite number of votes for that result. The Chief Justice, if he is in the majority, assigns the writing of the opinion for the Court to himself or one of his brethren; otherwise, the senior associate Justice in the majority chooses the opinion writer. Similarly, the Justice with seniority among the dissenters assigns the writing of the dissenting opinion. When an opinion has been drafted, it is sent to the printing plant in the basement of the Supreme Court, where proofs are produced under tight security. All the Justices then receive a copy of the draft opinion and each is free to write his comments on it and return it to the author. Sometimes extensive changes in the line of argument or the style of the opinion may be suggested. The writer can reject these recommendations but at the peril of losing a vote for his side. Frequently, though, a member of the majority who is not satisfied with the opinion of the Court will choose to write his own concurrence rather than to change his vote.[6]

In the steel seizure case, the process of accommodating the views of all the Justices in the majority proved impossible. Although an official opinion of the Court was written by Justice Black, every other Justice in the majority wrote a concurring opinion of his own, [7] and one—Justice Clark—refused to join in the Court's opinion, concurring only in the result. The *Youngstown* decision produced a total of seven opinions.

When the Court convened at noon on Monday, June 2, the Chief Justice took the unusual step of announcing that admissions of attorneys to practice before the Supreme Court would be deferred, to allow the Justices to give their opinions in the steel seizure case. Those in the courtroom listened in suspense as Vinson called on Justice Black to read the opinion of the Court. Black took less than fifteen minutes to deliver the opinion. The Court's ruling, however, did not become clear until he had finished. Point by point, Black rejected the arguments that had been advanced by the government and in conclusion declared: "The judgment of the District Court is Affirmed." [8]

Thus, the Supreme Court invalidated the seizure of the steel mills. The ultimate tally was 6 to 3 in favor of sustaining Judge Pine's injunction. The seven opinions consisted of Justice Black's opinion for the Court, concurrences by Frankfurter, Douglas, Jackson, Burton, and Clark, and Chief Justice Vinson's dissent, which was joined by Justices Reed and Minton. For two and one-half hours the Justices revealed their reasons for striking down or upholding the seizure. The oral opinions were less formal than the written ones and gave the Justices an opportunity to make certain comments which they did not include in their official texts. Although the Court's appointed lunch hour arrived, the Justices continued reading the opinions until all had been completed. At 2:35 P.M. the Court recessed. The judgment of the Supreme Court was now history.

Hardly had the Justices spoken when the President acted. He sent a letter to Secretary Sawyer with instructions to return the steel plants to their owners. Sawyer immediately informed the companies that the government had relinquished possession and, consequently, the United Steelworkers of America went out on strike. [9]

In view of the number of opinions filed, the opinion of the Court in *Youngstown* did not carry the same force as a typical opinion of the Court. Technically, Justice Black's opinion was the only one that would become a controlling precedent for the future. [10] Aware of the weight an opinion of the Court carries, Justice Frankfurter appended a comment at the conclusion of the Black opinion:

Although the considerations relevant to the legal enforcement of the principle of separation of powers seem to me more complicated and flexible than may appear from what Mr. Justice Black has written, I join his opinion because I thoroughly agree with the application of the principle to the circumstances of this case. Even though such differences in attitude toward this principle may be merely differences in emphasis and nuance, they can hardly be reflected by a single opinion for the Court. Individual expression of views in reaching a common result is therefore important.[11]

A close reading of the opinions leads to the conclusion, however, that the differences are much more than "in emphasis and nuance." Notwithstanding that five Justices united with Black in the opinion of the Court and that that opinion ostensibly represented their judgment, the individual concurrences reflect views quite different from those of Black on a number of important points.

According to Justice Black, the *Steel Seizure* case presented two crucial issues for examination by the Supreme Court: Was it proper to make a final determination of the constitutional validity of the President's executive order on a motion for preliminary injunction? And, if it was, did the President have the constitutional power to seize the steel mills? [12] The government had argued, Justice Black noted, that the district court should have denied equitable relief to the steel industry because the seizure did not cause it to suffer irreparable injury since there was an adequate remedy at law which would provide compensation for any possible damages. The denial of a preliminary injunction to the companies, the government maintained, would have been in keeping with the usual judicial practice of reserving judgment on constitutional questions until there were no other grounds on which to dispose of the case. Justice Black took exception to the government's contentions. He observed that previous Supreme Court decisions, including *Hooe* v. *United States*, made uncertain the right to recover damages for properties illegally seized by government officials. Furthermore, Black declared that government operation of the steel mills would probably lead to damages "of such nature as to be difficult, if not incapable, of measurement." [13] From this perspective, the district court had found no reason to delay a decision on the merits. "We agree with the District Court," Black stated, "and can see no reason why that question was not ripe for determination on the record presented." [14]

Having thus briefly disposed of the equitable issue, Black

turned to the merits: whether the President had lawfully seized the steel plants. Presidential power, Black wrote, must stem from either an act of Congress or the Constitution. Clearly, no act of Congress authorized Truman's seizure. The government had not met the requirements of the two statutes which permitted seizure under certain conditions. Additionally, Congress had specifically rejected seizure as a method of solving labor disputes when it voted against a seizure amendment during the debate on the Taft-Hartley Act.[15] If, therefore, Executive Order 10340 were to be upheld, the President's power had to come from the Constitution.

The government conceded, Black observed, that there was no express language in the Constitution which authorized the seizure. But the government maintained that the power could fairly be implied from the aggregate of powers granted to the President by the three principal sections of Article II: "The executive Power shall be vested in a President of the United States of America"; "The President shall be Commander in Chief of the Army and Navy of the United States"; and "he shall take Care that the Laws be faithfully executed." Justice Black rejected this argument. The steel seizure could not be sustained, he asserted, as an application of the President's military power as commander in chief. The Court was unconvinced by the government's analogy of the seizure to taking possession of private property by military officials engaged in combat, which had been upheld in numerous cases. "Even though 'theater of war' be an expanding concept," Black affirmed,

We cannot with faithfulness to our constitutional system hold that the Commander in Chief of the Armed Forces has the ultimate power as such to take possession of private property in order to keep labor disputes from stopping production. This is a job for the Nation's lawmakers, not for its military authorities.[16]

Nor could Executive Order 10340 be supported by the constitutional provisions granting the executive power to the President, Black declared. These presidential powers are distinct from the authority to make laws. The Constitution plainly assigns the lawmaking power to Congress, and the President is limited solely to recommending, vetoing, and executing the laws. The President's seizure order was obviously a piece of legislation: In form, it read like a statute; in content, it promulgated policies and authorized a government official to execute them. However, Black noted, only Congress had the power to prescribe such policies.[17] "The Founders of this Nation entrusted the lawmaking power to the Congress alone in both good and bad times," Black observed. "It would do no good to recall the historical events, the fears of power and the hopes for freedom that lay behind their choice. Such a review would but confirm our holding that this seizure order cannot stand."[18]

Hence, for Justice Black, the division between executive and legislative powers was fixed and exact. Under no circumstances, including, presumably, grave emergencies, could one branch exercise the powers assigned to the other.

Only Justice Douglas, among the majority Justices, shared Justice Black's view that the separation of powers doctrine was absolute and inflexible. Although he arrived at the same result— seizure was a legislative power which could not be exercised by the President without congressional authorization—the route Douglas took differed from the one traveled by Black, and this led him to write a concurring opinion. Douglas evinced a concern that illegal acts of the President might be allowed to stand because the executive could move more expeditiously than the legislature, an attractive prospect in time of crisis. The same standard must be used, however, whether or not a crisis existed, to determine if the President's seizure of the mills was legal. That standard, Justice Douglas stated, was "the allocation of powers under the Constitution."[19]

The Constitution, Douglas noted, placed all, not some, legislative power in Congress. Truman's seizure of the steel plants clearly was a legislative act. The seizure was "a taking in the constitutional sense." Even if the seizure was only temporary, Douglas observed, it was still a total condemnation of the property and the government must pay compensation under the Fifth Amendment. Congress was the only branch of government which could authorize the payment of compensation; therefore it was the only branch which could authorize seizure or make legal one that the President ordered. "That seems to me to be the necessary result of the condemnation provision in the Fifth Amendment," Justice Douglas reasoned. [20]

If the Court were to uphold Truman's seizure of the steel industry, Douglas declared, it would be rewriting Article II of the Constitution to simplify the task of dealing with the current crisis. Although this might seem like the right thing to do in the present circumstances, it certainly would not conform to the structure of government described in the Constitution. The framers of that document were afraid that the concentration of executive and legislative power in one branch would lead to tyranny; hence they assigned the exercise of each power to a different department. The American system of checks and balances exacts a price which might seem excessive to many people. Justice Douglas felt the price well worth paying. In conclusion, he warned:

Today a kindly President uses the seizure power to effect a wage increase and to keep the steel furnaces in production. Yet tomorrow another President might use the same power to prevent a wage increase, to curb trade-unionists, to regiment labor as oppressively as industry thinks it has been regimented by this seizure. [21]

The four remaining majority Justices rejected the absolute stance of Black and Douglas, for they regarded this treatment of the separation of powers doctrine as simplistic. Experience had shown that the functions of the three branches of government hardly fitted into the neat compartments Black and Douglas described.

Justice Frankfurter, especially sensitive to this problem, explained in his concurrence:

The content of the three authorities of government is not to be derived from an abstract analysis. The areas are partly interacting, not wholly disjointed. The Constitution is a framework for government. Therefore the way the framework has consistently operated fairly establishes that it has operated according to its true nature. Deeply embedded traditional ways of conducting government cannot supplant the Constitution or legislation, but they give meaning to the words of a text or supply them. It is an inadmissibly narrow conception of American constitutional law to confine it to the words of the Constitution and to disregard the gloss which life has written upon them. [22]

Nevertheless, Justice Frankfurter took issue with those people, in the not too distant past, who had ridiculed the system of checks and balances as old-fashioned, a bar to competent government. If anything, World War II should have made the citizens of the United States more aware of the wisdom of the framers of the Constitution, Frankfurter asserted. It was nonsense, he declared, "to see a dictator in a representative product of the sturdy democratic traditions of the Mississippi Valley." [23] To prevent dictatorship, however, it was still necessary to check unauthorized, albeit well-meaning, assertions of power.

Frankfurter carefully reconciled the Court's decision with his own philosophy of judicial restraint. The Constitution did not make the judiciary the "overseer of our government," and it was the duty of the Court to avoid where possible decisions based on broad constitutional interpretations. [24] One of the cardinal rules of this approach to constitutional adjudication, Frankfurter stated, was to refuse to pass on a constitutional issue "however narrowly it may be confined, if the case may, as a matter of intellectual honesty, be decided without even considering delicate problems of power under the Constitution." [25] Hence it was imperative in the *Steel Seizure* case to show that it could not be disposed of without a determination of the relation between the powers of the President and those of Congress. To do this, Frankfurter demonstrated that the granting of an injunction to the steel companies was entirely proper. He agreed with the Black opinion that damages flowing from the seizure would not be totally compensable by money. Further, Frankfurter, following Judge Pine, rejected the government's argument that an injunction should not be issued if an important public interest would be seriously harmed by it. At the same time he dissassociated himself from Pine's reasons for reaching this result:

One need not resort to a large epigrammatic generalization that the evils of industrial dislocation are to be preferred to allowing illegality to go unchecked. To deny inquiry into the President's power in a case like this, because of the damage to the public interest to be feared from upsetting its exercise by him, would in effect always preclude inquiry into challenged power, which presumably only avowed great public interest brings into action.[26]

Frankfurter concluded, with professed reluctance, that he could not escape an investigation into the powers of the President and of Congress in order to decide whether Executive Order 10340 was legal.

Frankfurter approached the merits of the case more narrowly than Black. He would make no sweeping statements on presidential power, the Justice declared. He would not consider what powers the Chief Executive might have had if Congress had not enacted statutes concerned with labor disputes and seizure. He would adjudicate only the problem at hand—the President's assertion of seizure power in the specific circumstances in which it had occurred.

Most important of the circumstances surrounding the steel seizure was the congressional legislation bearing on the subject. From the Taft-Hartley Act itself, and from the legislative history of its passage, two conclusions could be drawn, Frankfurter stated. First, Congress had thoughtfully dealt with the problem of national emergency strikes and had decided that when the procedures stipulated in the act failed to bring a resolution of a dispute, Congress would choose what, if any, further government action was to be taken. Second, Congress had considered the possibility of giving the President seizure authority in the Taft-Hartley Act, but had deliberately decided not to include this power. Congress thus had made a conscious choice: if seizure were necessary, it would provide the authority on an ad hoc basis. "Congress has expressed its will to withhold this power from the President as though it had said so in so many words," Justice Frankfurter averred.[27]

No statute enacted since 1947 had in any way changed Congress' intention to deny seizure power to the President, Frankfurter continued. Title V of the Defense Production Act, the section dealing with labor disputes, merely offered the President another means of achieving a voluntary settlement. The Wage Stabilization Board created by the President in Executive Order 10233, which had the authority to make recommendations in labor disputes referred to it, had no coercive powers. The WSB was restricted to mediation efforts similar to those contemplated by the Taft-Hartley Act. It still remained for Congress to choose what steps should be taken, after these voluntary procedures had been exhausted.[28] The seizure of the steel mills in total violation of Congress' will, upset the constitutional balance of power between the executive and legislative branches, Frankfurter asserted.

The highest obligation of the President, Frankfurter wrote, was to "take Care that the Laws be faithfully executed." Justice Holmes, in the *Myers* case, had described the perimeters of that authority definitively, at least for Justice Frankfurter. "The duty of the President to see that the laws be executed," Holmes had written, "is a duty that does not go beyond the laws or require him to achieve more than Congress sees fit to leave within his power." [29] Frankfurter was willing to acknowledge, however, that in certain instances, executive acts in the teeth of legislation might be permissible:

[A] systematic, unbroken, executive practice, long pursued to the knowledge of the Congress and never before questioned, engaged in by Presidents who have also sworn to uphold the Constitution, making as it were such exercise of power part of the structure of our government, may be treated as a gloss on "executive Power" vested in the President by §1 of Art. 2.[30]

Within this framework, Frankfurter distinguished the *Midwest Oil* case from the seizure case at hand. In *Midwest Oil* the executive action under consideration had been practiced previously on 252 occasions. In contrast, Frankfurter noted, the record of presidential seizure of private property when war had not been declared did not constitute an established executive practice. Only three times in the history of the nation—the three seizures by President Roosevelt before Pearl Harbor—had an executive taken measures comparable to the seizure of the steel mills. It was obvious, Frankfurter observed, that these three seizures "do not add up, either in number, scope, duration or contemporaneous legal justification, to the kind of executive construction of the Constitution revealed in the Midwest Oil Case." [31] Nor could the government claim that the President's interpretation of his seizure powers had been sanctioned by Congress. There was, therefore, no doubt in Justice Frankfurter's mind that in seizing the steel plants the President had exceeded his constitutional powers.

Frankfurter stated that it was difficult for him to make this decision, in view of the fact that President Truman had acted only for the good of the nation. But the structure of the American government at times foreclosed swift action in emergency situations, and it was important to the future of freedom in the United States to adhere to constitutional restrictions. To emphasize his point, Frankfurter quoted the words of Justice Brandeis:

> The doctrine of the separation of powers was adopted by the Convention of 1787, not to promote efficiency but to preclude the exercise of arbitrary power. The purpose was, not to avoid friction, but, by means of the inevitable friction incident to the distribution of the governmental powers among three departments, to save the people from autocracy. [32]

Justice Jackson's concurrence also revolved around the separation-of-powers doctrine but, perhaps because of his years of experience in the executive branch, his opinion concerned itself more with the actual practice of governing under the Constitution than with theoretical considerations. "While the Constitution diffuses power the better to secure liberty," Jackson observed, "it also contemplates that practice will integrate the dispersed powers into a workable government. It enjoins upon its branches separateness but interdependence, autonomy but reciprocity. Presidential powers are not fixed but fluctuate, depending upon their disjunction or conjunction with those of Congress." [33]

The authority of the Chief Executive was at its height when the measures he took were the direct consequence of an express or implied command of Congress, Jackson stated. In such a situation the President was relying on his own powers as well as those delegated to him by Congress. There was a second set of circumstances where, in the absence of any congressional action, the President might choose to act, but here his authority was less because it rested only on his independent power. According to Jackson a "zone of twilight" existed in which Congress and the President might have concurrent power or in which the apportionment of authority was ambiguous. If the President's power were questioned in this kind of situation, the legality of his act was "likely to depend on the imperatives of events and contemporary imponderables rather than on abstract theories of law." [34] The authority of the Chief Executive was weakest, Jackson asserted, when he acted in defiance of the express or implied intention of Congress. In this circumstance, the President's reliance was on his own constitutional powers minus whatever constitutional authority Congress might have over the matter. The Court could uphold the President in this event only by ruling that the Congress could not act upon the subject.

It was clear, Justice Jackson declared, that President Truman's seizure of the steel mills fell into the third category. It could not be considered in the first classification for, as the government admitted, the seizure was not authorized by Congress. Nor did it fit into the second grouping, because Congress had legislated in the field of seizure of private property when it enacted the Taft-Hartley Act, the Selective Service Act of 1948, and the Defense Production Act. The President had employed none of these. Thus Truman's action had to be placed in the third category and could be sustained "only by holding that seizure of such strike-bound industries is within his domain and beyond control by Congress." [35]

The President's power to seize, the government had stated in its brief, stemmed from three clauses in Article II. Justice Jackson, at some length, refuted each of the government's contentions. The government's claim—that "The executive Power shall be vested in a President of the United States of America" meant that all executive power inhering in the government was granted to the President—was rejected by Jackson as illogical. The Justice could not comprehend why the framers had enumerated, after the general grant, several particular powers, some of a trivial nature, if the government's interpretation were correct. "I cannot accept the view that this clause is a grant in bulk of all conceivable executive power," Jackson declared, "but regard it as an allocation to the Presidential office of the generic powers thereafter stated." [36]

The government's argument based on the commander-in-chief clause particularly disturbed Justice Jackson. The solicitor general had maintained that President Truman had sent troops to Korea by an exercise of his constitutional power, and that from that action the President derived the power to seize the steel mills to assure a supply of steel for the troops. This was a pernicious doctrine, Jackson asserted. When the Constitution designated the President Commander in Chief of the army and the navy, it did not make him Commander in Chief "of the country, its industries and its inhabitants. . . . While Congress cannot deprive the President of the command of the army and navy, only Congress can provide him an army or navy to command," Jackson continued. [37]

The duty of raising and supporting armies and providing and maintaining a navy was specifically assigned to Congress in the Constitution. The Third Amendment required that the acquisition of necessary military housing be authorized by law, and there were other limitations in the Constitution on the Commander in Chief's authority. By no stretch of the imagination could it be said that the President controlled use of the war power "as an instrument of domestic policy," Jackson declared.

Jackson disavowed any intent to circumscribe or contract the constitutional role of the President as Commander in Chief. Indeed, he believed that the command function of the Chief Executive should be given the greatest latitude when military force had to be employed against other nations for the security of the United States. "But, when it is turned inward, not because of rebellion but because of a lawful economic struggle between industry and labor, it should have no such indulgence," Jackson wrote. [38] The Court therefore could not uphold the government's claim of seizure power based on the commander-in-chief clause.

Jackson dismissed summarily the government's argument that the responsibility of the President to see that the laws were faithfully executed gave him the authority to seize. Any authority the President might have derived from that clause was canceled by the words of the Fifth Amendment: "No person shall be . . . deprived of life, liberty or property, without due process of law." [39]

Finally, Jackson dealt with the government's reliance on the inherent powers doctrine to support the seizure. From the rhetoric and practice of previous administrations, the government claimed, certain powers not expressly granted to the President in the Constitution have come to be accepted as necessary to the proper functioning of his office. One of these is the authority to act in critical situations. As persuasive as the inherent powers theory might seem, it would be unwise, Justice Jackson declared, for the Court to approve its use as grounds for presidential action during emergencies. The recent experience in Germany, he said, should be sufficient to convince anyone of the danger in embarking on such a course. Moreover, the government of the United States had already evolved a successful method of operating during emergencies: Congress can grant the President extraordinary powers in such urgent situations. The crisis facing President Truman on April 8, 1952, certainly was no worse than many with which previous Chief Executives were confronted, and Congress in the past had willingly supplied the necessary authority to deal with them. Hence Justice Jackson did not believe that the Court should sustain the President's possession of these powers without benefit of statute. "With all its defects, delays and inconveniences," Jackson concluded, "men have discovered no technique for long preserving free government except that the Executive be under the law, and that the law be made by parliamentary deliberations." [40]

In his short concurring opinion, Justice Burton emphasized what he considered to be the crucial factors leading to the ruling that the President's seizure order was illegal. The validity of the steel seizure depended on the constitutional division of power between Congress and the President, Burton wrote. Congress retained the power to act in emergencies, because the Constitution assigned it all legislative power. When Congress enacted the Taft-Hartley Act, it provided a procedure for just such an emergency as faced President Truman on April 8. Furthermore, it was clear from the legislative history of that act, Burton asserted, that

Congress reserved to itself the prerogative of authorizing seizure.[41] Burton rejected the government's contention that the President's use of the Wage Stabilization Board procedure fulfilled every objective of the Taft-Hartley Act and therefore must be treated as its equivalent. There was no need to assess the pertinence of the government's claim, Burton pointed out, for neither procedure contemplated seizure of private property as undertaken by President Truman.

Burton was cautious on the inherent power issue, however. The immediate circumstances, Burton felt, provided no occasion for the exercise of inherent power. But Burton did not totally repudiate the doctrine. He emphasized that the Court was not dealing with the issue of what the President's powers might be in "catastrophic situations." The pivotal fact in the steel seizure case was that Congress, in accord with its constitutional duty, had prescribed procedures to be followed in the current type of emergency. By issuing Executive Order 10340, Burton declared, the President had "invaded the jurisdiction of Congress," and "violated the essence of the principle of the separation of governmental powers." [42]

Justice Clark was the only member of the majority who did not concur in the opinion of the Court as well as its judgment. Clark disagreed with the Black opinion because it seemed to rule out the existence of any inherent power in the President to act in emergencies. According to Clark:

The Constitution does grant to the President extensive authority in times of grave and imperative national emergency. In fact, to my thinking, such a grant may well be necessary to the very existence of the Constitution itself. . . . In describing this authority I care not whether one calls it "residual," "inherent," "moral," "implied," "aggregate," "emergency," or otherwise. I am of the conviction that those who have had the gratifying experience of being the President's lawyer have used one or more of these adjectives only with the utmost of sincerity and the highest of purpose.[43]

But Clark found that he could not sustain the steel seizure under the inherent power of the President because Congress had provided several methods which the President could have employed in the steel crisis. The precedent set by *Little* v. *Barreme* was conclusive for Justice Clark: where Congress has legislated a specific procedure the President must follow it and not enlarge upon it in the relevant circumstances. The President could have chosen to use the Taft-Hartley Act injunction procedure, the provision for seizure in the Selective Service Act, or the condemnation method prescribed in the Defense Production Act. Given the existence of those three statutes, the President had exceeded his constitutional powers when he ordered seizure of the steel mills on his own authority.[44]

Justice Vinson, who was joined in his dissent by Justices Reed and Minton, issued a stirring defense of President Truman's action. Noting that there was some disagreement among the members of the majority, several thinking that the President was wholly without power to act in critical situations unless he had express statutory authorization and others believing only that the existence of certain statutes precluded the seizure in this instance, Vinson declared that the dissenters disagreed with both positions. The questions presented were of "transcending importance," Vinson observed, and the answers would affect the ability of the President and future Presidents to take action in moments of crisis; hence the minority Justices felt compelled to put their views on record.

With the world crisis, Congress' reaction to it, and the administration's efforts to carry out the congressional defense program as background, Vinson examined the emergency facing the President on April 8. Reviewing the affidavits submitted by government officials, Vinson asserted that there could be no doubt that a steel strike of any length would imperil the national security. If, therefore, the President had any authority at all under the Constitution to act in an emergency without specific congressional direction, the exercise of this power in this instance was entirely warranted.[45]

Vinson then turned to a discussion, much like the one in the government's brief, of previous examples of executive leadership, to show that many Presidents had acted for the good of the country regardless of whether Congress had provided prior authorization for the measures they took.[46] The holdings in the cases of *In re Neagle* and *In re Debs* demonstrated the Court's approval of the inherent powers doctrine, Vinson declared.[47] Moreover, the many instances of executive seizure of private property established the propriety of such action, the Chief Justice asserted.

The fact that temporary executive seizures of industrial plants to meet the emergency have not been directly tested in this Court furnishes not the slightest suggestion that such actions have been illegal. Rather, the fact that Congress and the courts have consistently recognized and given their support to such executive action indicates that such a power of seizure has been accepted throughout our history.[48]

Vinson made sure to include in the opinion Attorney General Jackson's justification, both moral and legal, of President Roosevelt's seizure of the North American Aviation plant.[49] In the oral delivery of his opinion, the Chief Justice commented that Justices Jackson and Clark had held very different views when they had been Attorneys General. To change one's mind appeared to be "evidence of strength," Vinson observed sarcastically.[50]

From the provisions of the Selective Service Act and the Defense Production Act authorizing seizure in certain circumstances, Vinson drew a different conclusion from the one favored by the majority Justices. "Where Congress authorizes seizure in instances not necessarily crucial to the defense program, it can hardly be said to have disclosed an intention to prohibit seizures where essential to the execution of that legislative program," [51] the Chief Justice stated. The President is charged with the duty of carrying out a mass of legislative enactments, he went on. Truman's seizure of the steel mills was an act taken to preserve the status quo and to prevent the destruction of the legislative defense programs until Congress could consider the crisis.[52] His two messages to Congress stated that he would abide by the wishes of that branch. This could hardly be construed as an act of a Chief Executive bent on tyranny, Vinson commented.[53]

The Chief Justice took the majority opinions to task for their paucity of authoritative precedents, for their frequent reliance on dissenting opinions, and especially for their disregard of the seriousness of the crisis facing the President on April 8. The Court had decided, Vinson declared, that

The broad executive power granted by Article II to an officer on duty 365 days a year cannot . . . be invoked to avert disaster. Instead, the President must confine himself to sending a message to Congress recommending action. Under this messenger-boy concept of the Office, the President cannot even act to preserve legislative programs from destruction so that Congress will have something left to act upon.[54]

The facts in this case offered no basis for charges of "arbitrary action, unlimited powers or dictatorial usurpation of congressional power." [55] The President had acted wholly in accordance with his responsibilities under the Constitution, the Chief Justice asserted. Hence, the dissenters would have reversed the district court's order.[56]

The Supreme Court's decision enormously pleased many people.[57] The immediate result of the Court's judgment—the strike by the United Steelworkers of America—was virtually ignored as the press, Congress, and steel industry executives congratulated the Court for saving the nation from dictatorship. Only the White House seemed worried about the practical consequences of the Court's ruling.

A *New York Times* editorial typified newspaper reaction, stating that the steel case had given rise to a Supreme Court decision far more important than the immediate production or nonproduction of steel. "We have, in the opinion delivered by Justice Black yesterday and sustained by five other justices, a redefinition of the powers of the President. Under this opinion the trend toward an indefinite expansion of the Chief Executive's authority is deliberately checked." [58]

The Chicago *Sun-Times*, generally regarded as a liberal organ, noted that the decision was a "stinging rebuke to President Truman" and offered its view that government by presidential fiat constituted a far greater danger than a steel strike.[59] Applauding

the Supreme Court for not dodging its responsibility, the Pittsburgh *Press* asserted, "The high court fulfilled its constitutional
duty to check a headstrong President gone rampant beyond the
limitations prescribed for him by both Constitution and law." [60]
The Los Angeles *Times* quoted a statement by California Governor
Earl Warren on the Court's decision:

> I consider this to be a red-letter day in the history of our country.
> The decision of the Supreme Court holding that everyone in the na-
> tion, including the President, is subject to the written provisions of law
> removes the challenge to the basic American principle that we have three
> equal and coordinate branches of government, all of which must live
> within the framework of the Constitution. [61]

Only the New York *Post* discussed the harmful effects of the strike
and pointed out that the Supreme's Court's decision resolved nothing with respect to the steel dispute. [62]

In Congress, the reaction of Senator Charles Tobey expressed
the feelings of a majority of Senators and Representatives. When
news of the Supreme Court's decision was read on the floor of the
Senate, Tobey shouted, "Hurrah. Thank God for the Supreme
Court." [63] Senator Harry Cain asserted, "The Supreme Court has
declared that our Constitution is not to be employed to serve the
purposes and whims of individual men. The Constitution is to
remain inviolate." [64] In the House, Republican Representative
Charles Vursell announced that the people of America "this
morning have unbounded praise for the Supreme Court." [65] Senator Lehman, however, reminded his fellow legislators that a steel
strike impended and exhorted his colleagues: "Congress must act,
without a moment's unnecessary delay, to pass essential legislation
to give the President the power he needs." [66]

Steel industry executives exulted upon hearing that the steel
seizure had been declared illegal. Charles M. White, president of
Republic Steel, commented that it was "the greatest Supreme
Court decision of my lifetime." A strike was better "than further
intrusion of Government into private rights," he concluded. [67]
Clarence Randall, president of Inland Steel, lauded the Supreme
Court's ruling: "This is a great day for America. The whole

country will take new hope for the future." As for the walkout, Randall remarked, "Phil Murray has struck the country again." [68]

According to reports in the *New York Times*, the steelworkers greeted the Supreme Court decision with relief, not bitterness. Seizure had been of no benefit to them. Many workers commented that they would have been better off had they gone out on strike immediately after their contracts had expired. Now, however, they were free to " 'slug it out' for the showdown." [69] But they were concerned that the President might seek a Taft-Hartley injunction.

The union vowed to contest the legality of such an injunction if one were granted. Arthur Goldberg issued a statement immediately after the Court's judgment had been made known, which pointed out that the Court had only declared the seizure unconstitutional. The Justices had not decided that the President must invoke the Taft-Hartley Act. Goldberg noted that the Union's contention that the President should not use a Taft-Hartley injunction in this situation had been "vindicated by the failure of any one of the justices of the Supreme Court to accept the companies' Taft-Hartley argument." [70]

Goldberg's interpretation of the opinions in the *Steel Seizure* case was literally correct; no Justice had stated in so many words that Truman must invoke Taft-Hartley. The consensus was, however, that where Congress had provided a procedure to be followed in national emergency strikes, the President could not ignore it in favor of a procedure of his own design. But there was nothing in the Justices' opinions that would preclude him from simply doing nothing if he thought that was the proper course.

To President Truman the Supreme Court decision came as a great blow. He had clearly expected a favorable judgment. Events of the past two decades had given Truman little reason to suspect that the Supreme Court would curtail presidential power. He therefore took the Court's ruling as a personal rebuke. Charles Sawyer recorded that the President was very emotional and declared that the Court evidently had not examined the matter carefully. [71]

The Supreme Court was well aware that its ruling had disturbed the President. Justice Black, concerned that the President not take the judgment personally, invited Truman and the Justices to a party at his home. According to Justice Douglas, the get-together somewhat soothed Truman's ruffled feelings.[72] Douglas reported that "Truman was gracious though a bit testy at the beginning of the evening. But after the bourbon and canapés were passed, he turned to Hugo [Black] and said, 'Hugo, I don't much care for your law but, by golly, this bourbon is good.' "[73] Truman's opinion on the law never changed, however: the Supreme Court decision was wrong. He later wrote in his *Memoirs:*

> I would, of course, never conceal the fact that the Supreme Court's decision, announced on June 2, was a deep disappointment to me. I think Chief Justice Vinson's dissenting opinion hit the nail right on the head, and I am sure that someday his view will come to be recognized as the correct one. . . .
>
> I am not a lawyer, and I leave the legal arguments to others. But as a layman, as an official of the government, and as a citizen, I have always found it difficult to understand how the Court could take the affidavits of men like Lovett, Chapman, and many others, all of whom testified in great detail to the grave dangers that a steel shutdown would bring to the nation—affidavits that were neither contradicted nor even contested by the companies—and ignore them entirely.
>
> I could not help but wonder what the decision might have been had there been on the Court a Holmes, a Hughes, a Brandeis, a Stone.[74]

Most lawyers and legal scholars did not agree with the President's assessment of the case. The *Youngstown* decision brought forth a large number of comments in the law reviews, most of which applauded the judgment of the Court.[75] The majority of writers viewed with approval the Court's reaffirmation of the separation-of-powers doctrine. But many found reason to criticize the opinions.

The Court was taken to task for the multiplicity of its opinions, which made it impossible to discuss the Court's holdings with any authority. Justice Black's opinion for the Court was clear enough. It was difficult, however, to reconcile that opinion with the concurring opinions of several of the majority Justices.[76] The opinion of the Court rested on the proposition that the President lacked power to seize the steel mills because neither Congress nor the Constitution expressly supplied the required authorization. Black's analysis appeared to rule out any exercise of inherent power; lawmaking was assigned solely to Congress in "good and bad times." Only Justice Douglas agreed with that rigid conclusion, however.

Justice Jackson's position on this question was ambiguous. Although Jackson explicitly rejected the concept of inherent power and emphasized that the legislature should make the laws, his description of a "zone of twilight" in which authority was not clearly apportioned and in which "the imperatives of events" would establish the criteria to judge the legality of a particular action, seemed to allow the President to exercise inherent power in certain instances.

Justices Frankfurter and Burton stated that they were not passing judgment on the inherent power issue because the *Steel Seizure* case did not compel them to do so. Their decision relied on the fact that Congress had enacted specific policies with which the President's seizure action conflicted.

Justice Clark, the only member of the majority who did not join in the opinion of the Court, indicated that he believed a President, in extraordinary circumstances, could act pursuant to inherent power. And the dissenting Justices agreed with Clark. Thus, it can reasonably be concluded that while the opinion of the Court declared illegal all unauthorized "legislative" acts of the President, seven Justices believed that future acts of the President in emergencies should be judged on an ad hoc basis.[77]

The Black opinion itself was the object of criticism. Corwin called the opinion an "arbitrary construct created out of hand for the purpose of disposing of this particular case." He found the opinion "altogether devoid of historical verification." [78] There was a long history of presidential action unauthorized by statute in fields later occupied by Congress, he asserted.[79] Others believed that Justice Black's description of the American government was marred by a too precise division of powers. They thought the

Frankfurter-Jackson approach more consistent with reality; the executive, the legislature, and the judiciary shared a variety of powers, and interaction among the branches was a common practice.[80] Paul Freund agreed and pointed out that Black's opinion, "in its rigid conception of the separation of powers," was "hardly conformable" with the positions taken by the other majority Justices in their concurring opinions. Freund regarded Black's opinion as a needless pronouncement on the President's inherent powers and noted that the Court's few sweeping judgments on executive power had always created problems:

Chief Justice Marshall in *Marbury v. Madison* proffered some rather restrictive notions of the President's removal power. Chief Justice Taft in *Myers v. United States* found it necessary to disavow those observations as dicta, drawing comfort from the fact that Chief Justice Marshall himself had pointed out the inconclusive nature of dicta. Then Mr. Justice Sutherland in *Humphrey's Executor v. United States* found it necessary to disavow certain dicta in the *Myers* case, drawing comfort from the fact that Chief Justice Taft had drawn comfort from the fact that Chief Justice Marshall had pointed out the inconclusive nature of dicta. And so perhaps *ad infinitum*.[81]

Freund predicted a similar future for the opinion of the Court in *Youngstown*. Had the Court found it unnecessary to make such broad statements on presidential power, it would not have been an "unmitigated loss," Freund concluded.[82]

Although some writers preferred the Frankfurter-Jackson approach to the constitutional questions in *Youngstown*, Justice Frankfurter's opinion also presented problems. His detailed analysis of the legislative history of the Taft-Hartley Act laid the groundwork for his conclusion that Congress not only had entered the field of industrial relations and had prescribed a procedure for dealing with national emergency strikes, but that that action constituted a positive prohibition against presidential seizure as "if it had been written into . . . the Labor Management Relations Act of 1947." [83] As proof that such a prohibition was intended, Frankfurter reported that when the Taft-Hartley Act was being considered, Congress had rejected an amendment which would have given the President the power to seize. Wayne Morse, who was respected as an authority on constitutional law, viewed such an interpretation derisively:

> This afternoon we had a novel statement of law in the negative. . . . Why . . . this afternoon we were told that the President of the United States violated a law when he issued his order the other night, because in the course of a debate on proposed legislation . . . an amendment was offered in an effort to give the President seizure power, but the amendment was defeated; and therefore it is said that the President of the United States is now violating the law. Imagine that. . . . That is a new doctrine. In other words, it is claimed that by rejecting an amendment, we pass a law; and thus it is claimed that because we rejected that amendment . . . the President of the United States is violating a law—a law not even on the statute books.[84]

Legal scholars have questioned such a use of legislative history.[85] And the Supreme Court itself, in *United States* v. *United Mine Workers*,[86] had rejected the finding of positive prohibitions merely from defeated amendments which would have established the power.

One of the issues considered in *United Mine Workers* was whether the United States government could move for an injunction against a strike in a seized industry. An argument offered in support of the proposition that the United States could not apply for such an injunction maintained that Congress, in enacting the War Labor Disputes Act, had effectively restricted such a resort to the courts by rejecting an amendment which would have given the courts the power to issue injunctions to restrain violations of the War Labor Disputes Act. But Chief Justice Vinson, writing for the Court, pointed out that only the Senate had defeated this amendment. It was not voted on in the House.[87] The Supreme Court was unwilling to read the Senate rejection of an amendment specifically giving the courts the authority to issue injunctions in these circumstances as a positive denial to the courts of this power. The Chief Justice observed: "It is not at all clear that the rejection of a proposal in this form should, in any event, be of determinative significance in the case at bar." [88]

The legislative history of the Taft-Hartley Act presented an apposite situation. This time it was only the House which had defeated an amendment giving the President seizure power.[89] The only evidence adduced by Justice Frankfurter to prove that the Senate had expressed the same intent were statements by the chairman of the Senate Committee on Labor and Public Welfare, Senator Taft, that the committee had considered and rejected a general grant of seizure powers.[90] By enacting the Taft-Hartley Act, did the Senate endorse its committee's decision to deny seizure power? Further, did passage of the act mean that the legislative majority agreed with the reasons given by Senator Taft for rejecting a grant of authority to seize? Senator Taft had stated that he did not want such a general seizure law on the books. Others may have thought the President already had seizure power in emergencies, as many Senators had asserted during the debate on the War Labor Disputes Act. It is impossible, without documentary proof, to ascribe the same motive to every legislator who votes for a bill. It therefore seems highly dubious to interpret passage of the Taft-Hartley Act, in view of its legislative history, as a positive prohibition against presidential seizure.

Moreover, in his opinion, Justice Frankfurter disingenuously declared: "No authority that has since been given to the President can by any fair process of statutory construction be deemed to withdraw the restriction or change the will of Congress as expressed by a body of enactments, culminating in the Labor Management Relations Act of 1947." [91] Frankfurter neglected to mention that in June 1949, while it was considering repeal or revision of the Taft-Hartley Act, the Senate passed an amendment sponsored by Senator Taft which gave the President the power to seize.[92] The House never voted on this amendment. Surely Frankfurter should have given the Senate's action in 1949 as much weight as he gave the House's defeat of a seizure amendment in 1947. In addition, Frankfurter never took notice of the power to seize given the President in the Selective Service Act of 1948 or the authority to requisition granted in the Defense Production Act. These acts did not give the President the specific power to seize industries as a means of dealing with labor disputes, but they did grant him seizure authority as a method of preventing bottlenecks from developing in national defense industries. A strike in the steel industry might very well have been considered such a bottleneck. Thus, it appears that Justice Frankfurter's conclusion that the actions of Congress amounted to a specific prohibition against seizure rests on shaky ground.[93] As a result, his contribution to a more realistic assessment of the separation of powers doctrine is marred by its dependence on a questionable congressional negative that precluded presidential action.

In view of the fact that both Justice Jackson and Justice Burton also relied on the presence of a positive prohibition of presidential seizure,[94] the most sensible approach to the issue presented in the *Steel Seizure* case would appear to be that of Justice Clark.[95] "Where Congress has laid down specific procedures to deal with the type of crisis confronting the President," Clark wrote, "he must follow those procedures in meeting the crisis." [96] Truman could have chosen any of three methods to avert a steel strike. He could not go beyond the procedures prescribed by Congress. As Corwin noted, "Only Justice Clark . . . guided by Marshall's opinion in the early case of *Little* v. *Barreme,* had the courage to draw the appropriate conclusion: Congress having entered the field, its ascertainable intention supplied the law of the case." [97]

Since the views expressed in the concurring opinions were in certain respects incompatible with the opinion of the Court, some legal and academic scholars observed that the Court majority was united only in the holding that President Truman's seizure of the steel mills was unlawful. The *act* of ruling against the President, they believed, constituted the primary significance of the *Steel Seizure* case. Carl Swisher commented: "Because of the diversity the only thing we can assert with confidence is that in some degree the enhancement of the power of the presidential office has been restrained." [98] Bernard Schwartz noted that American jurists had been reluctant to review actions of the Chief Executive and stated that this delicacy stemmed from a "perverted construction of the separation of powers." [99] In *Youngstown* the Supreme Court did

not expressly reject the maxim that the federal courts cannot enjoin the President, but in fact the Court exercised just such a
power.[100]

A few commentators, however, disagreed with the Court's
judgment. Glendon Schubert, a political scientist, vehemently
denounced the Court's ruling. Schubert believed that the central
question raised by the *Steel Seizure* case should have been: "Was
presidential seizure of the steel industry a 'means' reasonably related to a legitimate end?" Had the Court approached the case
from this perspective, it would have reached the opposite conclusion, he asserted.[101]

The *Harvard Law Review* also questioned the Court's ruling.
In its view, the constitutional questions presented in *Youngstown*
should not have been decided, because, apart from declaring the
seizure illegal, the majority was not united on the constitutional
grounds for such a decision. Justice Black had invalidated the
seizure because it lacked express authorization. His opinion denied
the existence of inherent power in all cases. But only Douglas, and
perhaps Jackson, approved that result. Frankfurter, Burton, and
Clark had based their conclusions on the presence of congressional
legislation relevant to the situation in which the President had
acted. Thus, the *Harvard Law Review* observed, the *Steel Seizure*
case might not preclude a future Court from holding that inherent
presidential power does exist.[102]

On the merits, the *Harvard Law Review* thought that the dissenters were correct in their view that the seizure was an act taken
to execute the laws of Congress, that is, the defense mobilization
and economic stability statutes. The intention of those laws was
clear, and it seemed sensible that the President should have discretion in the methods he wished to use in pursuance of that intention.[103] But Paul Kauper, in his article on the *Steel Seizure* case,
objected to this position taken by the dissent. The idea that the
President may use a variety of means to effect the programs and
policies of Congress, unless a particular method has been specifically proscribed by Congress, "states too broad a view of executive
power." Kauper gave some examples: In an emergency the Presi-

dent might wish to draft men for industry, to levy more taxes to support the legislative program, or to use heavier penalties against those who violate the laws. "But it would hardly be contended that presidential prerogative would extend to these areas of legislative authority." Kauper warned of the danger in this view: To say Congress can take positive action to disaffirm the President's action is not a sufficient safeguard. In many instances "presidential initiative opens the way for the executive *fait accompli* often leaving no choice for Congress except to follow the course prescribed by the President." [104]

Bernard Schwartz noted that the dissenters' interpretation would have been proper if there had been a declaration of war. Such a declaration would be treated as a statute and the President would have the duty to execute it faithfully, Schwartz stated. As long as one hundred fifty years ago, Justice Story had asserted that when a declaration of war is enacted, the President receives the authority to take any measures necessary for its execution. But barring a declaration, a presidential action must "find its source in a law which the action concerned is seeking to execute." [105]

Learned Hand, the distinguished jurist, took issue with the need for a declaration of war. He indicated in a letter to Justice Frankfurter that he would have upheld the seizure:

I should not have decided whether the President had the power to seize until I had said whether Congress had vetoed. I doubt that I should have held the Taft-Hartley Act stood in the way; I rather think I should have said that its implications do not extend to a condition of war. . . . I don't know enough about the Defense Production Act to have any opinion. If I had found no sufficient indication in these acts that Congress meant to confine the President's powers, I should have held that he had constitutional power to seize the industry and keep it in operation, bello flagrante, regardless of any "declaration of war." But the whole business only fortifies my belief that such jobs are not for judges, though I don't see that much harm will be done here, since Congress is left supreme.[106]

In view of the widespread expectation that the Supreme Court would declare the seizure legal, many wondered what had prompted the Court to strike it down. Why had the Court declined to take advantage of its option not to consider the case at all or, after it had granted certiorari, why had it chosen to make a decision on the merits? And why, once the Court had determined to meet the constitutional issue head on, did it rule against the President?

The answer to the first two questions lies in part in the atmosphere and circumstances in which the *Steel Seizure* case was presented to the Justices. The publicity attendant upon every event in the course of the steel dispute and litigation, and the issues raised, made Supreme Court consideration of the case imperative. If the Court was to retain respect as an institution, the politics of the situation required a ruling on the merits.[107] Judge Pine's decision, so widely acclaimed, made it virtually impossible for the Court to resist a determination of the constitutional issue. Had it tried to avoid judgment, the country might have thought that the Court was evading its responsibility.

Moreover, in retrospect it seems clear that the Court deliberately chose to make a judgment on the merits. As a political institution,[108] the third branch of government felt a need to assert itself. Alpheus Thomas Mason wrote: "With the banner of 'self-restraint' still flying, the Court made it clear that Presidential action is not immune to judicial review." [109] "Whether the ruling be rated good or bad, right or wrong, the outstanding fact—little noted at the time or since—was that the Court . . . appropriated the top power and the last word, as usual, for itself," Fred Rodell commented.[110]

The Court, however, had another interest beyond affirming its power of judicial review. The constitutional issues brought forth in the *Steel* case, it thought, were ripe for adjudication. The prospect of many more years of international tension may have encouraged the Justices to think that the time had come for a careful examination of the direction in which the United States government was headed. There are suggestions in several of the opinions that the Justices' views had been influenced by the events of World War II. The growth of the fascist governments in Europe and the resultant horrors were still fresh in the minds of the Justices. Robert Jackson especially seemed affected by what had occurred there. His experience at Nuremberg, according to Justice Frankfurter, had "a profound influence on [Jackson's] endeavor to understand the human situation. An essentially good-natured, an even innocently unsophisticated temperament, was there made to realize how ultimately fragile the forces of reason are and how precious the safeguards of law so painstakingly built up in the course of the centuries." [111]

The *Steel Seizure* case impressed the Justices again with the necessity for a firm foundation in law for all government action. The United States was involved in the Korean conflict because of a presidential commitment of troops, with no declaration of war by Congress. It was the President—not Congress—who had declared a national emergency after the hostilities had begun in Korea. Now the President was asserting the power to seize private property to avert a strike whose effects might harm United States forces in Korea. This trend of events undoubtedly aroused the Justices' concern. As Justice Jackson stated in his opinion: "No doctrine that the Court could promulgate would seem to me more sinister and alarming than that a President whose conduct of foreign affairs is so largely uncontrolled, and often even is unknown, can vastly enlarge his mastery over the internal affairs of the country by his own commitment of the Nation's armed forces to some foreign venture." [112] Given the opportunity to consider the propriety of the President's action, the Court believed it was its duty to analyze these recent developments and their influence on the practice of government under the Constitution. Hence, the Court determined to face squarely the constitutional issue in the *Steel Seizure* case.

The Court's resolution of the constitutional issue against the President surprised many people. In the past, the Court had upheld a number of presidential actions based on tenuous evidence of congressional approval or authorization, or even of congressional silence. [113] As Chief Justice Vinson showed, such tacit approval could have been found in the case of the steel seizure. What led the Court, in *Youngstown*, to ignore these precedents and to rule against the President?

The Court simply was not convinced that the crisis confronting the nation was sufficiently grave to justify the President's assertion of power. To be sure, Justices Black and Douglas stated that the President could never exercise legislative power; the seriousness of the emergency was immaterial. But the other four Justices in the majority indicated that in certain circumstances the President might take action based on his inherent power. They were unpersuaded, however, that the emergency claimed by the Truman administration warranted such an exercise. As John Roche noted, the dissenters had asserted that this was a *real* emergency, and had taken the traditional judicial attitude toward executive action in *real* emergencies. "The legal arguments between the two divisions of the Court were consequently of little significance; the vital disagreement was over premises." [114]

Chief Justice Vinson pointedly referred to "the complete disregard of the uncontroverted facts showing the gravity of the emergency" in the majority opinions.[115] The Court, however, did not disregard the crisis; it just did not believe that one existed.[116] The problems presented by the Korean conflict and the cold war were not yet comparable to those the nation had faced in the Civil War and the two world wars. In reaching this conclusion, the Court believed that it was reflecting public opinion. According to Arthur Schlesinger, Jr., the Court "registered the sense of Congress and the nation, volubly expressed in the two months since the seizure, that Truman and Vinson had it wrong, and that this was simply not an emergency calling for drastic recourse to inherent presidential power." [117]

In fact, the Court had ample reason to doubt that the immediate "crisis," the need for the continued production of steel, was as acute as the administration described it. Members of the White House staff have stated that they knew before the Supreme Court decision that adequate supplies of steel were on hand, and that the Court knew this too.[118] A White House memorandum discussing a message to Congress on the steel situation observed: "The proposed message stresses once again the need for uninterrupted steel production. Yet the fact is that the public has never believed this contention, and in the face of recent releases of steel for race tracks and bowling alleys, they are even less likely to believe this now." [119] The press had been reporting large inventories of steel. It had become clear that the physical survival of the nation would not be affected by a short interruption in steel production.

In this context, the Court majority could find no reason to sustain an exercise of inherent power when Congress had provided a variety of means—albeit means more cumbersome and less efficacious than seizure—which the President could have used to obtain continued steel production. This was the clearly stated opinion of Justices Clark and Burton. Frankfurter and Jackson declared that the President was precluded from taking independent action when Congress had already enacted legislation designed to meet the emergency confronting him. Why, then, did these Justices (with the exception of Clark, who concurred only in the judgment) join in an opinion for the Court couched in much more sweeping terms?

The favorable reaction to Judge Pine's opinion made it difficult for the Supreme Court to rule on any grounds less broad and less easily understood than the lack of power in the Chief Executive to seize the steel mills. Holmes Baldridge's claim that the President's power was unlimited, and the clamor it produced, and Truman's many statements to the press about the President's duty to take action on the basis of his inherent powers in an emergency undoubtedly contributed to the Court's belief that it had to address this fundamental issue. Whittling down the absolutes contained in the Pine opinion might have been interpreted as a softening of the blow for the President and as reflecting a lack of independence on the Court's part. In this connection, Charles Black, Jr., has noted that the Court when functioning as legitimizer of the national government's power has been much more important in the long run than when acting as invalidator of an assertion of power by Congress or the President. The former function, however, will only be accepted so long as the public has confidence that the latter role will be performed also. That is the value of the Court's occasional use of that power. [120] The Supreme Court evidently thought that the *Steel Seizure* case presented such an occasion and that an authoritative opinion worthy of an equal, coordinate branch of the government was required. The Court also felt it important that its opinion be grounded on clearly delineated principles that would readily be understood by the public. It is for these reasons that Justices like Frankfurter and Jackson joined in the opinion for the Court even though they did not wholly agree with its basic premise.[121]

Furthermore, Frankfurter, Jackson, and Burton were in sympathy with the mood manifested in the Black opinion. They may have thought that a reaffirmation of the first principles of government was not untimely in view of the recent trend of events. The opinions are replete with statements warning of the dangers of concentration of power in the Executive. Justice Brandeis' well-known admonition that the separation-of-powers scheme adopted by the Constitutional Convention was not intended "to promote efficiency but to preclude the exercise of arbitrary power" was quoted by both Douglas and Frankfurter in their opinions.[122] The history of the Court shows that such fundamental principles had seldom stood in the way of upholding presidential action. And the same men on the Court had, in varying degrees, all been supporters of the expansion of presidential power under Roosevelt and Truman. But, absent the existence of a genuine emergency, the Court was unwilling to cloak an exercise of inherent power with the authority of law. The Justices, reflecting the public consensus, believed the time had come to reassert the ordinary limits on presidential power.

eleven

THE CONSTITUTIONAL SIGNIFICANCE OF THE YOUNGSTOWN DECISION

The *Steel Seizure* case is one of the "great" constitutional law cases. It is one of the few which discuss at length the powers of the President. Thus, it would not be unreasonable to expect that *Youngstown* has had a significant influence on the development of constitutional law. Its impact, however, cannot be measured by the number of times it has been cited as a legal precedent.[1] The paramount importance of the *Steel Seizure* decision lies in the fact that it was made. The Supreme Court, by invalidating an act of the President, helped redress the balance of power among the three branches of government and breathed new life into the proposition that the President, like every other citizen, is "under the law." By its ruling in the *Steel* case, the Court created a precedent that future courts and future Presidents could not ignore.[2]

The *Youngstown* decision served as a prelude to a more activist period for the Supreme Court. Traditionally reluctant to deal with constitutional issues if cases could be disposed of on other grounds, the Court rejected its customary stance in the *Steel Seizure* case and in subsequent cases as well. New personnel undoubtedly spurred this change in the Court's attitude. *Brown* v.

Board of Education of Topeka [3] and the other *School Segregation* cases decided in 1954 were only the most spectacular example of the Court's new willingness to face basic constitutional questions. But the Court did not wholly abandon its tradition—one need only read Justice Black's concurrence in *Peters* v. *Hobby* to confirm this [4]—it merely became less rigid in adhering to it.

In the 1960s, however, the impact of *Youngstown* became more evident. Perhaps the most important constitutional initiative in this decade was in legislative apportionment, an area traditionally regarded as off-limits to the courts. Reapportionment had been considered a political question, although the criteria for identifying it as such were unclear.[5] Justice Frankfurter's opinion in *Colegrove* v. *Green,* [6] the leading case on the subject, reflected the widely held view that questions of legislative representation were very complex, involving many considerations, mainly political, which courts were not fit to deal with. In *Colegrove,* a group of Illinois voters had brought suit because, they contended, the divisions of Illinois into congressional districts with unequal population constituted a denial of equal protection of the laws. Voters living in the more populous districts, usually containing the large urban centers, claimed that their votes were worth less than the votes of their fellow citizens who lived in sparsely populated districts. The Supreme Court denied relief. According to Justice Frankfurter, the apportionment issue was "of a peculiarly political nature and therefore not meet for judicial determination." [7] Questions concerning the proper guidelines for apportionment "bring courts into immediate and active relations with party contests. From the determination of such issues this Court has traditionally held aloof." [8] And, Frankfurter warned, "Courts ought not to enter this political thicket. The remedy for unfairness in districting is to secure state legislatures that will apportion properly, or to invoke the ample powers of Congress." [9]

Hence, when the Court considered a challenge to the apportionment of the Tennessee legislature in 1962, in *Baker* v. *Carr,* [10] it had to demonstrate why apportionment was now susceptible of judicial resolution, and *Youngstown* helped it to do so.

In *Baker* the Court reexamined the political question doctrine and fashioned the following criteria:

> It is apparent that several formulations which vary slightly according to the settings in which the questions arise may describe a political question, although each has one or more elements which identifies it as essentially a function of the separation of powers. Prominent on the surface of any case held to involve a political question is found a textually demonstrable constitutional commitment of the issue to a coordinate political department; or a lack of judicially discoverable and manageable standards for resolving it; or the impossibility of deciding without an initial policy determination of a kind clearly for nonjudicial discretion; or the impossibility of a court's undertaking independent resolution without expressing lack of the respect due coordinate branches of government; or an unusual need for unquestioning adherence to a political decision already made; or the potentiality of embarrassment from multifarious pronouncements by various departments on one question.[11]

The Court concluded that the reapportionment of the Tennessee legislature presented a justiciable cause of action, in that the issue was one of "the consistency of state action with the Federal Constitution" and none of the common characteristics of a political question applied. Justice Douglas, in a concurring opinion, noted that many of the cases which had been dismissed by the Court because they involved a political question had been "wrongly decided." A number of other cases in which the Court had ruled on the merits contained questions equally or more "political" than those the Court had refused to consider, Douglas asserted, and he mentioned *Youngstown* as one of them.[12] Thus, the mere fact that the Court had made a decision on the merits in the *Steel* case influenced the Justices' reappraisal of the political question doctrine in *Baker* v. *Carr*.

Five years later, the Court's adjudication of the issue presented in *Powell* v. *McCormack* again demonstrated the importance of the Court's action in the *Steel Seizure* case. Adam Clayton Powell, Jr., the flamboyant congressman from Harlem, claimed that the House of Representatives had unlawfully excluded him from the 90th Congress. The House had refused to seat Powell in March 1967 because its Select Committee had found that Powell had wrongfully asserted a privilege and immunity from the processes of the New York courts, that he had used House funds illegally, and that he had made false reports on expenditures to the Committee on House Administration. Powell maintained that he could not be excluded, since he met the requirements stated in the Constitution for holding congressional office, and he and some of his constituents brought suit against the Speaker of the House and various congressmen and congressional employees. The House of Representatives insisted that it held discretionary power to decide on the pertinent qualifications of its members and urged that the Court dismiss the suit on the ground that it was a political question whose resolution might lead to an embarrassing disagreement between coordinate branches of the government. The district court agreed with the House and held that the court did not have jurisdiction because the issue raised a political question. Consideration of the merits of the dispute and the granting of relief would violate the separation-of-powers doctrine, Judge Hart stated.[13]

The Supreme Court thought differently. A "determination of petitioner Powell's right to sit would require no more than an interpretation of the Constitution," Chief Justice Warren declared for the Court.[14] This was the traditional function of the judicial branch. And the Court had on occasion interpreted the Constitution in a manner at odds with another branch. "The alleged conflict that such an adjudication may cause cannot justify the courts' avoiding their constitutional responsibility," the Chief Justice stated. The *Youngstown* decision was the prime example of such an occasion, Warren wrote.[15] Hence, it was not the holding in *Youngstown*, that the President could not exercise legislative power, so much as the act of ruling against a coordinate branch of the government that had a marked effect on the constitutional doctrine of separation of powers.

Indeed, while the Court has been increasingly unsympathetic to the claim that each branch is the judge of its own constitutional prerogatives, it has been reluctant to apply another aspect of the doctrine of separation of powers, that the Executive cannot exercise legislative power. Whereas the Court seemed to be adhering to that doctrine in *Kent* v. *Dulles*,[16] it drew back in *Zemel* v. *Rusk*,[17] both of them cases which involved a citizen's right to a passport. In *Kent* the issue was whether the Secretary of State could deny a passport to a citizen who, it was alleged, held Communist beliefs and who declined to file affidavits concerning his present or past membership in the Communist party. Justice Douglas held for the Court in 1958 that he could not. Congress had authorized only two grounds for denial of a passport, lack of citizenship or criminal or unlawful conduct. The Secretary of State did not have "unbridled" discretion to set up restrictive regulations but was limited to these two.[18] But in 1965 in *Zemel*, where the Secretary of State's refusal to issue passports for travel to Cuba was examined, the Court declared that the Secretary had the right to impose area restrictions and distinguished the case from *Kent* by pointing out that area restrictions resulted from foreign policy considerations, a field in which the Executive's discretion was unquestioned, and that the proscription on travel to Cuba was not discriminatory since it touched all citizens.[19] Justice Black dissented. All legislative power was lodged in Congress, he noted.

I cannot accept the Government's argument that the President has "inherent" power to make regulations governing the issuance and use of passports. . . . We emphatically and I think properly rejected a similar argument advanced to support a seizure of the Nation's steel companies by the President. . . . And regulation of passports, just like regulation of steel companies, is a law-making—not an executive, law-enforcing—function.[20]

Justice Goldberg also wrote a dissenting opinion in which he stated, "passport restrictions may be imposed only when Congress makes provision therefor in explicit terms," and he cited *Youngstown* as precedent. "I would hold expressly that the Executive had no inherent authority to impose area restrictions in time of peace," Goldberg concluded.[21]

A notable failure of the Supreme Court to follow what some thought were the logical implications of its action in the *Steel Seizure* case occurred in the many attempts to have the constitutionality of the Vietnam War passed on by the Court. The first indication of the Court's attitude came when it considered a request from the state of Massachusetts, whose legislature had passed a law which excused its citizens from fighting in an undeclared war, for leave to file a complaint challenging the constitutionality of United States participation in the Indochina War. Without stating any reasons, the Court denied the motion.[22] Justices Harlan, Stewart, and Douglas dissented. Harlan and Stewart would have set the motion for argument on the issues of standing and justiciability. Justice Douglas would have granted leave to file a complaint and, at the very least, wanted the motion to be argued. In a dissenting opinion, he explained why.

Douglas believed that Massachusetts had standing and that the controversy was justiciable. It was on the question of justiciability that Douglas invoked *Youngstown* as well as *Baker* v. *Carr*. Douglas reviewed the six standards promulgated in *Baker* v. *Carr* for finding that a case involved a political question and concluded that none applied to the question of whether the Vietnam War was constitutional. One of the criteria was "the impossibility of a court's undertaking independent resolution without expressing lack of respect due coordinate branches of government." The solicitor general urged, Douglas wrote, that to examine the authority of the chief executive to act in Indochina would be to show disrespect. That had not stopped the Court in the *Steel Seizure* case, Douglas noted. There the Court found that its duty was to interpret the Constitution and to decide if the President had the authority to seize the mills. Douglas quoted Frankfurter on the necessity for the Court to exercise its power despite its distaste for inquiry into the powers of the President and Congress. "It is far more important," Douglas concluded in *Massachusetts* v. *Laird*, "to be respectful to the Constitution than to a coordinate branch of government." [23]

The power of the Commander in Chief was controlled by constitutional limitations, Douglas continued. That was the crucial point in the *Youngstown* ruling, he noted. As Justice Clark had stated, the President must follow the specific procedures laid down by Congress for dealing with the type of emergency in which the President had acted. "If the President must follow procedures prescribed by Congress," Douglas contended, "it follows *a fortiori* that he must follow procedures prescribed by the Constitution." [24] The members of the Supreme Court had not been afraid to reach the merits of the steel seizure, Douglas declared, and their decision should be "instructive." [25]

Douglas drew one more lesson from the *Steel Seizure* case, and he was to repeat it many times in the next few years, as the Court consistently refused to consider the constitutionality of the war in Indochina. In *Youngstown*, private individuals (the steel companies) had asserted that their property had been taken by the Executive without proper authority. In the Massachusetts case, private citizens were claiming that their lives and liberties were being taken without proper authority. The Constitution does not protect property more than life and liberty, Douglas maintained, and it had always been the Court's duty to determine the legality of the taking of any of them. [26] "If Truman could not seize the steel mills," Douglas once rhetorically asked a group of students, "how can they seize you?" [27]

Several cases disputing the constitutionality of the Indochina War were adjudicated by lower courts, however. Some judges dismissed the cases because a political question was involved, others thought the issue justiciable and found that although Congress had not declared war it had ratified American participation in the conflict by its legislative acts to support the war. *Youngstown* was prominently discussed in many of the cases. In *Berk v. Laird*, District Judge Judd held that Congress had authorized hostilities in Vietnam in a manner sufficient to meet the constitutional requirements. "This is not a case where the President relied on his own power without any supporting action from Congress, as in the Steel Seizure. . . . In that case it was conceded that there was no Congressional authorization for the seizure. . . . The President had reported his action to Congress, and Congress took no action." [28] District Judge Wyzanski construed *Youngstown* in a similar way when he considered *Massachusetts v. Laird*. [29] In a class action suit testing the constitutionality of the war, a three-judge district court divided 2 to 1 over whether the question was political. The majority found it was and distinguished between cases dealing with internal matters as opposed to cases concerned with foreign affairs. According to the court,

> The *Steel Seizure Cases*, because of their supposed relevance to this country's participation in the Korean conflict, perhaps went farthest in adjudicating an issue having potential repercussions beyond the normal judicial sphere. . . . Even though the nationalization and the Court's injunction of the President's action might have had some, although indirect, effect on the foreign relations of this country, such import, if any, would have been clearly minimal compared to the drastic change which nationalization by the President would otherwise have brought about in the free enterprise system.[30]

The dissenting judge, to the contrary, thought *Youngstown* more relevant to the Vietnam War issue, since, in *Youngstown*, steel had been a means of maintaining the troops in Korea and the Court's decision definitely affected the conduct of foreign affairs.[31]

When these cases came to the Supreme Court on petitions for certiorari, the Court refused to consider them. The Court's continued resistance to adjudicating the constitutionality of the war in Indochina evoked anguished cries from Justice Douglas:

> Once again, this Court is confronted with a challenge to the constitutionality of the Presidential war which has raged in Southeast Asia for nearly a decade. Once again, it denies certiorari. Once again, I dissent.
>
> This Court, of course, should give deference to the coordinate branches of the Government. But we did not defer in the *Prize Cases* . . . when the issue was presidential power as Commander in Chief to order a blockade. We did not defer in the *Steel Seizure Case*, when the issue was presidential power, in time of armed international conflict, to order the seizure of domestic steel mills. Nor should we defer here, when the issue is presidential power to seize, not steel, but people.[32]

Youngstown Sheet & Tube Co. v. *Sawyer* notwithstanding, the Supreme Court exercised restraint.[33]

But when, in the presidency of Richard M. Nixon, the Supreme Court was faced with many questions of presidential power in domestic affairs, its reluctance to act disappeared. The *Steel*

Seizure case became a useful precedent, both because the opinion of the Court denied inherent power in the President and because the ruling struck down a President's order. The first of these important cases was the attempt of the government to enjoin the publication of the so-called Pentagon Papers, a history of American involvement in the Vietnam War, which had been surreptitiously copied by Daniel Ellsberg and given to the *New York Times* to print. The government sought an injunction against publication on the ground that the President had inherent power to protect national security. In a per curiam opinion, the Court held that no injunction could be issued against publication because the government had not met the burden of showing justification for prior restraint.[34] For three Justices—Black, Douglas, and Brennan—the command of the First Amendment was conclusive: "Congress shall make no law . . . abridging the freedom of . . . the press." Prior restraints of publication could never be legal, and the Court could not do what Congress was prohibited from doing. For another three Justices—Stewart, White, and Marshall—however, the holding in the *Steel Seizure* case was decisive. Where Congress had legislated in a particular field, as it had in the Espionage Act of 1917—which dealt in part with the communication of government secrets—and had refused to authorize the remedy the government now sought in the Supreme Court, "It would . . . be utterly inconsistent with the concept of separation of powers for this Court to use its power . . . to prevent behavior that Congress has specifically declined to prohibit."[35] When Congress had enacted legislation concerning a certain subject, the President could not go beyond the specific stipulations of the law. He was precluded from an exercise of inherent power.

A similar question arose when the Nixon administration claimed it had the inherent power to order wiretaps in internal security matters without judicial authorization. Title III of the Omnibus Crime Control and Safe Streets Act of 1968 had empowered the Attorney General to tap telephones in certain types of crimes after he had obtained the approval of a court. The government admitted, however, in a case involving the dynamite bombing of a

Central Intelligence Agency office in Michigan which came before Judge Damon Keith of the United States District Court for the Eastern District of Michigan, that it had gotten information about the defendant through a wiretap ordered solely on the authority of the Attorney General. The district court decided that the electronic surveillance violated the Fourth Amendment and ordered the government to disclose all conversations overheard on the taps.[36] The government challenged this order in the Court of Appeals for the Sixth Circuit. Judge Edwards, for the court, upheld the district court and took issue with the Attorney General's contention that the President possesses inherent power to protect the national security. In dealing with the threat of domestic subversion, every official of the executive branch of the government was subject to the limitations of the Fourth Amendment. The Fourth Amendment was adopted, Judge Edwards pointed out, in response to the colonists' experience under an assertion of inherent power in King George III to authorize uncontrolled searches and seizures. The framers of the United States Constitution had deliberately provided checks on this sovereign power; thus, Judge Edwards denied that the President had any inherent power in domestic subversion cases to dispense with the appropriate procedures to gain authorization for a wiretap. Edwards noted that the Supreme Court had "squarely rejected" the inherent power doctrine in the *Steel Seizure* case. Although *Youngstown* did not concern electronic surveillance, the Judge indicated that it was "the authoritative case dealing with the inherent powers of the Presidency—a doctrine which is strongly relied upon by the government in this case." [37]

The Supreme Court emphatically affirmed the court of appeals. Justice Powell, for the Court, observed that if the President had the power to order wiretaps without prior judicial approval, the source of that power had to be in the Constitution. He rejected the government's claim that the wiretap was legal because a section of the Crime Control Act of 1968 stated:

Nor shall anything contained in this chapter be deemed to limit the constitutional power of the President to take such measures as he deems nec-

essary to protect the United States against the overthrow of the Government by force or other unlawful means, or against any other clear and
present danger to the structure or existence of the Government.[38]

The Court interpreted this section as merely a disclaimer of any
congressional intent to define the President's power in this sphere,
rather than as a grant of power to the Chief Executive. Therefore,
in cases of subversion, the President retained whatever constitutional power he had before the Crime Control Act was enacted,
but no more. Hence, the Court had to define the constitutional
limits of the Executive's wiretapping authority in national security
cases.[39]

Implicit in the Fourth Amendment directive that a warrant
shall be obtained is review by a disinterested magistrate of the
request of the investigating officers so as to determine the necessity
for the search or seizure.[40] For the executive branch to be its own
judge of when electronic surveillance should be used would sub-
vert our basic constitutional system. "Individual freedoms will best
be preserved through a separation of powers and division of functions among the different branches and levels of Government,"
Powell asserted.[41] The requirements of prior judicial review, the
Court believed, would in no way obstruct the President in carrying
out his constitutional duties to protect domestic security.[42]

The *Youngstown* decision served as a useful precedent in
challenges to another practice of the Nixon administration: presidential impoundment of congressionally appropriated funds. Cases
testing the impoundment authority raised the question of the
proper division of power between the executive and Congress.
Nixon grounded his impoundment authority on a sweeping claim
to constitutional power stemming from the vesting of the executive
power in the President. Nixon asserted this claim on his own, for
his Justice Department had advised: "With respect to the suggestion that the President has a *constitutional* power to decline to
spend appropriated funds, we must conclude that the existence of
such a broad power is supported by neither reason nor precedent." [43] Unhindered by the Justice Department's opinion, President Nixon, maintaining that he was acting in the interest of

sound fiscal management, proceeded to impound funds from a variety of congressionally approved and funded programs solely on the basis of his own judgment that the expenditure of those monies for those purposes would be harmful to the nation.[44] A host of suits were filed to test the legality of Nixon's impoundment policy. Those decided before Nixon resigned, *Sioux Valley Empire Electric Association, Inc.* v. *Butz,* for example,[45] stressed the point that the appropriation power was the "essence of legislative power" and the President could not "nullify the mandates of Congress." [46] Conceding to the President the authority to impound funds which had been appropriated by Congress in support of programs it had approved, the court declared in the *Sioux Valley* case, would "emasculate" our constitutional system of government.

An even more flagrant abuse of power occurred when the Nixon administration tried to dismantle the Office of Economic Opportunity, which Congress had voted to continue and for whose programs Congress had appropriated funds. Nixon merely neglected to include funds for the OEO programs in his budget message for 1974, and a specially appointed acting director of OEO proceeded to use what funds remained from the previous year to phase out the program. When the employees of OEO brought an action in the United States District Court for the District of Columbia challenging the President's authority to destroy OEO, the government argued that the President's budget message superseded the law enacted by Congress.

Guided by the *Youngstown* precedent, Judge William Jones ruled against the President. Jones held that programs under OEO cannot be terminated before their authorization expires, unless Congress orders such termination either by failing to supply funds or by expressly forbidding the further use of funds for the specific programs. Recalling the *Steel Seizure* case, which he labeled the "leading case on the constitutional division of power between the President and the Congress," Jones declared that the President cannot make law; therefore, the "budget message cannot have the effect of law." The President must carry out laws passed by Congress.[47]

Thus, *Youngstown Sheet & Tube Co.* v. *Sawyer* proved a valuable precedent for courts faced with deciding a variety of cases challenging assertions of executive power. The separation of powers implicit in the Constitution, which *Youngstown* reaffirmed, was vital to many of these decisions. Ironically, in an attempt to escape the mandates of the courts in his Watergate difficulties, President Nixon sought refuge in the separation-of-powers doctrine. He used it, however, in a very different sense. To Nixon, separation of powers meant not only that the Presidency was an independent branch of the government with its own powers and duties but that the President was not subject to judicial review. In declaring Nixon's theory invalid, the courts found the *Steel Seizure* decision crucial to their determination. The first of the Watergate Tapes cases, *In re Grand Jury Subpoena Duces Tecum Issued to Richard M. Nixon,* [48] and, on appeal, *Nixon v. Sirica,* [49] demonstrated *Youngstown*'s importance.

The case of *In re Grand Jury Subpoena* arose as a result of a grand jury investigation of the unauthorized entry into the Democratic National Committee headquarters at the Watergate office building on June 17, 1972, by men subsequently found to be connected with the Committee for the Re-election of the President and with the White House. After the conviction, in January 1973, of seven individuals whom the grand jury had indicted (some pleaded guilty), James McCord, one of the defendants, wrote to Chief Judge John Sirica, of the United States District Court for the District of Columbia, who had presided over the trial. McCord alleged that his superiors at the Re-election Committee and high officers of the government had taken part in the original planning of the break-in and wiretapping of the Democratic National Committee headquarters. Moreover, he stated that perjury had been committed at the trial and that these same government and Re-election Committee officials had conspired to cover up their participation in the crime and to hinder the investigation of the Watergate burglary. The grand jury then began a thorough examination of the allegations. On July 16, 1973, it was revealed, at a Senate hearing on the Watergate break-in and other presidential campaign activities, that conversations in the Oval Office of the White House had been routinely tape recorded. The grand jury then requested the White House to produce specific tapes which the grand jury, from other evidence, believed would be helpful to its investigation of the Watergate crimes. When the White House refused to produce the tapes, the grand jury directed the Watergate special prosecutor to subpoena the materials. A subpoena *duces tecum* (to produce documents) was issued on July 23, 1973, which required Richard Nixon or any subordinate officer with custody or control to furnish nine tape recordings to the grand jury. The White House did not comply with the subpoena because, it said, "the President is not subject to compulsory process from the courts." [50] The grand jury instructed the special prosecutor to seek a court order requiring compliance with the subpoena. Chief Judge Sirica, who had administered all the court proceedings arising from the criminal charges brought after the Watergate burglary, entered an order, returnable August 7, 1973, to show cause why the President should not be compelled to produce the tape recordings.

The heart of the President's response to the show cause order was the Nixon theory of separation of powers. The President declared that if the Court forced him to disclose the contents of private presidential conversations, "the total structure of government—dependent as it is upon a separation of powers—will be impaired." [51] A ruling that the President is "personally subject to the orders of a court would effectively destroy the status of the Executive Branch as an equal and coordinate element of government." No precedent could be found in the entire history of the development of constitutional law which would "justify or permit such a result." [52] Many cases existed where subordinate executive branch officials were held to be subject to court orders. But none was applicable to the tapes case, the brief contended, because the President had taken personal custody of the tapes and the court's process would be directed to him alone.

It was within the President's power, the brief argued, to refuse to furnish the tapes to the grand jury if he thought disclosure of those conversations would not be in the public interest. The President, and only the President, could balance the interest in prosecuting a wrongdoer (without the evidence required from the President, this might not be possible) against the interest of confidentiality of all presidential conversations.[53] The absolute discretion of the President to claim executive privilege could not be reviewed by the courts, for the judiciary lacked power to force compliance by the President. The Chief Executive is "answerable to the Nation but not to the courts." [54] The President did not intend to imply, his lawyers asserted, that he was above the law, but merely that he could be held accountable only by the method prescribed in the Constitution: impeachment.[55] Moreover, since the courts were without the physical power to enforce a decision striking down a claim of executive privilege such a ruling would be meaningless.[56]

Judge Sirica, in an opinion released on August 29, 1973, found the President's arguments unpersuasive. The court recognized that there were executive privileges which barred the production of evidence when a need to protect the confidentiality of presidential discussions was shown. Sirica rejected, however, the President's contention that it is the executive alone who determines whether the privilege is properly employed. The judiciary must decide "the availability of evidence including the validity and scope of privileges." Courts had not been reluctant in the past to rule on claims of executive privilege. "Executive fiat is not the mode of resolution," Sirica declared.[57] Therefore, he ordered the President to produce the tape recordings for judicial examination *in camera*, after which the unprivileged portions of the tapes would be furnished to the grand jury.

Since this ruling amounted to an order which enforced the subpoena, Judge Sirica discussed the reasons which led him to conclude that the court had the authority to compel the President's compliance with a grand jury subpoena of unprivileged evidence. "To argue that it is the constitutional separation of powers that bars compulsory court process from the White House . . . overlooks history," the chief judge declared.[58] Courts had always required the performance of nondiscretionary acts by the executive branch when necessary. Although these court orders were directed to subordinate officials, Sirica asserted, they were still applicable precedents. The *Steel Seizure* case effectively invalidated the proposition that the President was not subject to the court's orders. To insist, after the *Youngstown* ruling, that the President could not be reached by court process, Sirica wrote, "would seem to exalt the form of the *Youngstown Sheet & Tube Co.* case over its substance. Though the court's order there went to the Secretary of Commerce, it was the direct order of President Truman that was reversed." [59]

Moreover, Judge Sirica found the President's conception of the separation of powers to be inaccurate. The President's letter to the court refusing to comply with the subpoena had stated: "It would be wholly inadmissible for the President to seek to compel some particular action by the courts. It is equally inadmissible for the courts to seek to compel some particular action from the President." [60] Relying heavily on the analysis in the special prosecutor's brief, Sirica explained that the President's view of separation of powers did not comport with either the structure of the government set up in the Constitution or its actual practice. The framers of the Constitution lodged the powers of government in three branches, but "a 'watertight' division of different functions was never their design." Citing various examples of interaction between two branches, such as the executive's power to veto legislation, Sirica stressed the relationship between the judiciary and the executive. "The executive appoints judges and justices," Sirica noted, "and may bind judicial decisions by lawful executive orders. The judiciary may pass on the constitutionality of executive acts." [61] Here the chief judge quoted Justice Jackson for support: "While the Constitution diffuses power the better to secure liberty, it also contemplates that practice will integrate the dispersed powers into a workable government. It enjoins upon its branches separateness but interdependence, autonomy but reciprocity." [62] As for the President's contention that the court lacked the physical force necessary to ensure compliance with its order and therefore

should not rule, Sirica thought the whole argument "immaterial."
"The court," Sirica declared, "has a duty to issue appropriate orders." Throughout history, the courts had relied on the good faith
of the executive branch, "even in such dire circumstances as those
presented by *Youngstown Sheet & Tube Co.* v. *Sawyer*," Sirica observed, and he saw no reason why the courts' expectation should
change.[63] If the tapes were important to the grand jury's investigation, "would it not be a blot on the page which records the judicial
proceedings of this country," Sirica concluded, "if, in a case of
such serious import as this, the court did not at least call for an
inspection of the evidence in chambers?" [64]

The United States Court of Appeals for the District of Columbia Circuit, in *Nixon* v. *Sirica,* wholly agreed with the principles set down by Judge Sirica.[65] There was no doubt in the court's
mind that the President must obey an order of the court. The
court of appeals cited *Youngstown* as "only the most celebrated instance of the issuance of compulsory process against Executive officials." [66] This court, too, made short shrift of the argument that
because the President had taken personal custody of the tapes,
these precedents were inapposite. Although it had been the custom
to address court orders to subordinate executive branch officials,

To rule that his case [*Nixon* v. *Sirica*] turns on such a distinction would
be to exalt the form of *Youngstown Sheet & Tube* over its substance. Justice Black, writing for the *Youngstown* majority, made it clear that the
Court understood its affirmance effectively to restrain the President.
There is not the slightest hint in any of the *Youngstown* opinions that the
case would have been viewed differently if President Truman rather than
Secretary Sawyer had been the named party. If *Youngstown* still stands, it
must stand for the case where the President has himself taken possession
and control of the property unconstitutionally seized, and the injunction
would be framed accordingly. The practice of judicial review would be
rendered capricious—and very likely impotent—if jurisdiction vanished
whenever the President personally denoted an Executive action or omis-
sion as his own.[67]

The court of appeals interpretation of the *Steel Seizure* decision terminated the long-held belief that the President could not
be a party to a suit. Ever since the Supreme Court decision in

Mississippi v. *Johnson* this had been a common assumption. But in that case the Court had specifically declared:

> We shall limit our inquiry to the question presented by the objection, *without expressing any opinion* on the broader issues discussed in argument, whether, in any case, the President of the United States may be required, by the process of this court, to perform a purely ministerial act under a positive law, or may be held amenable, in any case, otherwise than by impeachment for crime. [Emphasis added.] [68]

The Court held that it did not have jurisdiction to enjoin the President from enforcing the Reconstruction acts. It is clear from a subsequent case that the Court declined jurisdiction because *Mississippi* v. *Johnson* involved a political question and not because the President was a party. In *Georgia* v. *Stanton* [69] the Court would not take jurisdiction of a similar injunction bill even though a subordinate official was the named defendant. The court of appeals in *Nixon* v. *Sirica* noted the misinterpretation of *Mississippi* v. *Johnson*, and its reading of the case soon gained approval in other lower court decisions. [70]

It remained for the Supreme Court to put its imprimatur on this new departure in constitutional law, and it did so in *United States* v. *Nixon*. [71] This case dealt with many of the same issues present in *Nixon* v. *Sirica*. But *Nixon* v. *Sirica* had never been appealed to the Supreme Court. Instead, in an effort to escape compliance with the district court's order, President Nixon on October 20, 1973, fired Watergate Special Prosecutor Archibald Cox, who had refused to agree to any compromise of Sirica's order. The discharge of Cox aroused a storm of protest, however, and under pressure from Congress and the public Nixon announced that he would comply with the subpoena. [72] To avoid congressional legislation establishing an independent Watergate special prosecutor, the President on November 12, 1973, appointed a new special prosecutor, Leon Jaworski, who was given the same powers as Cox had had and who was specifically authorized to resort to judicial process to contest an assertion of executive privilege as to any evidence needed by the prosecution. On April 16, 1974, Jaworski moved for the issuance of a subpoena *duces tecum* to produce tapes and documents relating to sixty-four presidential conversations which the special prosecutor believed necessary in the trial of seven associates of the President indicted by the grand jury in the Watergate matter for conspiracy to defraud the United States and to obstruct justice.[73] The litigation resulting from Jaworski's motion culminated in *United States* v. *Nixon.*

The district court, by Judge Sirica, ordered the subpoena to issue on April 18, 1974. Counsel for the President filed a formal claim of privilege and a motion to quash. Sirica, after argument, denied the motion, relying on the holding in *Nixon* v. *Sirica.* Moreover, he rejected a new contention put forth by the President that the court lacked jurisdiction because the litigation over the tapes was an intra-executive dispute. The court found that the special prosecutor, representing in this case the sovereign, the United States, had sufficient independence and adverse interests from the President as to provide the court with a concrete legal controversy over which the court had jurisdiction under Article III of the Constitution.[74] The President sought review in the court of appeals, but the special prosecutor decided to go directly to the Supreme Court because he needed the subpoenaed evidence quickly. He filed a petition for certiorari before judgment and the Supreme Court granted that petition. In Chief Justice Burger's subsequent opinion, he cited *Youngstown* as a precedent for the granting of certiorari before judgment in the court of appeals when a prompt resolution of important public questions was necessary.[75]

In a unanimous decision,[76] the Supreme Court affirmed the district court's ruling and in effect ratified the holdings in *In re Grand Jury Subpoena* and *Nixon* v. *Sirica.*[77] *Youngstown* influenced the result in *United States* v. *Nixon,* as it had in the previous tapes case; it provided a recent and dramatic example of the Court's authority to rule against the President. The most substantial part of Chief Justice Burger's opinion dealt with the President's claim of privilege.[78] The President's major argument, Burger wrote, had been that the separation-of-powers doctrine "precludes judicial review of a President's claim of privilege." But ever since *Marbury* v. *Madison,*[79] Burger declared, it had been the duty of the courts to state what the law is. Though the Supreme Court had never before ruled specifically on the scope of the judicial power to enforce a subpoena for confidential presidential conversations for use in a criminal prosecution, it had certainly held other exercises of executive authority unconstitutional, and he cited *Youngstown* as precedent. In *Baker* v. *Carr*, the Court had asserted that it was the final interpreter of the Constitution; therefore, it fell to the Supreme Court to decide if a particular matter had been committed by the Constitution to another department of government or if the action of that department was beyond the authority given to it.[80] Although each branch of government had to respect the others, Burger wrote, federal courts must occasionally interpret the Constitution in a different way from another branch. The Court could not share the judicial power of the United States, assigned to the federal courts by Article III, section 1, of the Constitution, with the executive branch. It was exclusively the function of the courts to state what the law is; therefore to permit the President to be his own judge of executive privilege would be contrary to the doctrine of separation of powers, Burger concluded.[81]

As neither the separation-of-powers doctrine nor the need for confidentiality was enough to sustain the Chief Executive's position that he was immune from judicial process, to accede to the President's claim of an absolute privilege, Burger wrote, "would plainly conflict with the function of the courts under Art. III." [82] Burger pointed out that in the American system of government, powers were divided among three branches; but the framers intended these branches to interact, and none was given unqualified independence. Here Burger quoted Justice Jackson's concurring opinion in *Youngstown* for its statement on the interdependence of the branches,[83] just as Judge Sirica had in *In re Grand Jury Subpoena*. The *Youngstown* precedent led Burger to conclude that

To read the Art. II powers of the President as providing an absolute privilege as against a subpoena essential to enforcement of criminal statutes on no more than a generalized claim of the public interest in confidentiality of nonmilitary and nondiplomatic discussions would upset the constitutional balance of "a workable government" and gravely impair the role of the courts under Art. III.[84]

The President's claimed privilege, the Court ruled, "must yield to the demonstrated, specific need for evidence in a pending criminal trial." [85] Hence, the district court had been correct when it ordered the President to comply with the subpoena by submitting the materials for *in camera* inspection. [86]

The decision in *United States* v. *Nixon* endorsed the principle that the President is subject to judicial review, not only by legal challenge to the actions of his subordinates but, if necessary, by suits against the President himself. The *Steel Seizure* case had a critical role in making this decision possible. *Youngstown Sheet & Tube Co.* v. *Sawyer* thus has had a twofold effect on the doctrine of separation of powers. It defined more precisely the division of functions between the Executive and Congress, by holding that the President could not exercise the seizure power without legislative authority. At the same time, by invalidating an executive order of the President, *Youngstown* dealt a telling blow to the interpretation of the separation-of-powers doctrine which held that each branch of the government was the arbiter of its own powers and responsibilities. When the *Youngstown* decision was made, it seemed dramatic, but more an aberration than the culmination or beginning of a legal trend. It has proved, however, to be an important foundation for the reaffirmation of the proposition that the President is not above the law.

epilogue

The immediate effect of the Supreme Court's decision in *Youngstown Sheet & Tube Co.* v. *Sawyer* was to trigger a strike which shut down the steel industry for fifty-three days. In their opinions the Justices had indicated that the President had available to him three statutory means of averting the strike, and if those were unsuitable, he could request new legislation from Congress. What must the Court have thought, as the walkout dragged on, when neither the President nor Congress took any action to terminate it? [1]

A few hours after the Court's judgment was announced, Truman conferred with some of his advisers: the new Attorney General, James McGranery, Solicitor General Perlman, John Steelman, David Stowe, Secretary of Defense Lovett, Charles Murphy, Joe Short, the press secretary, and Secretary of Commerce Sawyer. They discussed use of the Taft-Hartley Act, but Perlman counseled against it. The union would resent it. Moreover, Perlman was not sure that it would be effective. The union might ignore the injunction and go to court to fight it. As no consensus developed, Truman called another meeting for the next night. [2]

At the second conference—to which had been invited Sam Rosenman, Clark Clifford, Secretary of Labor Tobin, David Bell, a White House staff member, and Henry Fowler, head of the National Production Authority—three courses of action were considered. Secretary Sawyer suggested that the President send a message to Congress asking for new legislation. Sam Rosenman, who introduced his remarks by saying that he would probably be thrown out of the room, urged Truman to employ the Taft-Hartley Act immediately. Rosenman indicated that he sympathized with the union's position, but that the government had to resort to the only remaining legal remedy. The President interrupted to state that he thought that the Steelworkers' agreement to postpone the strike for ninety-nine days precluded invoking Taft-Hartley. Rosenman observed that a legal test of that issue should be welcomed by the government, for if the Court decided that the union was right and an injunction could no longer be employed, the President's position would be strengthened. Clifford and Secretary of Labor Tobin supported Rosenman's view.[3] Steelman believed the union and the industry should be brought together to negotiate before the President took any official measures.[4] After mulling over these options, Truman decided to try the collective bargaining route. Five days of serious negotiations at the White House brought no agreement, however. As a result, he decided on a different course of action.

On June 10, President Truman addressed a joint session of Congress. Briefly describing the steps he had taken since the beginning of the dispute to prevent an interruption in the production of steel, Truman pointed out that for more than 150 days, he had managed to keep the mills operating despite the lack of a contract between the companies and the union. But the Supreme Court ruling had precipitated a strike and now it was Congress' responsibility to deal with the situation. Truman mentioned two possible approaches: Congress could enact a seizure law which would provide fair compensation for both the owners of the industry and its employees or the Taft-Hartley Act could be invoked. The President recommended the first course of action and made clear that he would employ the second only if Congress passed legislation specifically directing him to obtain an injunction. Truman explained that using Taft-Hartley would be "unwise, unfair, and quite possibly ineffective."[5] "Every action I have taken in the dispute in the steel industry," the President declared, ". . . has been based on the paramount necessity for maintaining the production of essential steel products in the present defense emergency." The matter was no longer in his hands, however.

Whatever may have been the intention of the Court's majority in setting limits on the President's powers, there can be no question of their view that the Congress can enact legislation to avoid a crippling work stoppage in the steel industry. . . .

The issue is squarely up to you gentlemen of the Congress. I hope the Congress will meet it by enacting fair and effective legislation.[6]

On the same day as the President's address, the Senate decided upon its response: it voted down three amendments to the Defense Production Act which would have authorized government seizure of the steel mills.[7] Instead, it passed the Byrd amendment, which requested, but did not direct, the President to use the emergency procedures of the Taft-Hartley Act.[8] The House concurred with the Senate request on June 25.[9] These actions constituted the sum of Congress' efforts to end the steel strike.[10] Not surprisingly, Truman did not follow Congress' suggestion.

As the long strike continued, each of the two elected branches urged the other to take action but declined to do anything itself. Time and again President Truman was asked by reporters why he did not invoke the Taft-Harley Act. His answer was always the same: The objectives of the Taft-Hartley Act had already been met; Congress had had plenty of time to think about a solution; another eighty days would just prolong the agony.[11] There was considerable merit to Truman's position. If Congress felt so strongly that invocation of Taft-Hartley was the only appropriate means to terminate the walkout, it could have passed legislation directing the President to use it.

The behavior of Truman and Congress during the 1952 steel strike may not have been the abdication of responsiblity it appeared to be. For the dire consequences of even a short interruption in the production of steel, which had earlier been predicted by the administration, never materialized. As soon as the strike began, the newspapers reported that no shortage of steel existed.

The *New York Times* noted that experts had informed it that "the average stock of steel in the hands of manufacturers exceeded ninety days supply. This did not take into account additional reserves in commercial warehouses . . . or steel products already in the fabricating stage." [12] A government study after the dispute was settled observed: "The enormous inventory of steel on hand at the beginning of the strike became a source of amazement to the entire country—a supply which permitted steel-using industries to operate substantially at full levels for nearly two months with new production virtually shut off." [13]

Meanwhile, the companies and the union met several times in attempts to negotiate an agreement.[14] In the early weeks of the strike, the government and the industry had informally settled on a price increase: $5.20 a ton—the same $4.50 the government had offered in April plus 70 cents to take into account a rise in freight costs. Apparently the government had convinced the industry that this was the highest price adjustment it could obtain. The wage issue was then resolved: rates were increased by 16 cents an hour and fringe benefits by approximately another 5½ cents an hour.[15] By the end of June, the union had hammered out a union shop formula with Bethlehem Steel, but it refused to settle unless the other companies did likewise and their assent was not forthcoming.[16]

Finally, after the industry had been idled for more than fifty days, the administration decided that the country could afford the steel strike no longer. Secretary of Defense Lovett declared at a press conference that the strike had probably caused more damage to the defense production program than any enemy bombing attack could have inflicted.[17] The President determined to bring Philip Murray and Benjamin Fairless to the White House for negotiations. According to David Stowe, if the parties did not reach a settlement quickly, the President was prepared to employ the cumbersome seizure provisions of the Selective Service Act.[18]

On July 24, Murray and Fairless arrived at the White House and by the end of the day concluded an agreement. The terms of the settlement were virtually the same as had been discussed weeks before: an estimated 21½ cents an hour in wage and fringe benefit increases was awarded to the workers; the government allowed the steel companies to charge $5.20 more per ton of steel; [19] and the union shop formula, following the Bethlehem plan, provided that all new employees would be given application cards for union membership at the time of their employment and would become union members after thirty days unless they notified the company in writing of their desire not to become union members. [20] At 4:55 P.M., the President announced with relief that the steel strike was over.

Although pleased that the dispute had ended, price stabilization officials resented the granting of the steel price increase, which they had sought to prevent, and castigated the steel companies for unpatriotic behavior. Roger Putnam wrote to Ellis Arnall that the industry had held "a loaded gun poised at the Government's head." The companies were in a position to derail the entire defense program, Putnam asserted, and they determined to do it unless they got the price adjustment they wanted. Moreover, after studying the terms of the settlement Putnam said that he could find no reason why an agreement had not been reached earlier. "Some day," Putnam observed, "I hope we shall know the real motives behind the mysterious conduct of the steel industry in this tragic incident. Some day, perhaps we will know why this strike was dragged out for 53 days only to be settled on substantially the same wage *and price* terms which the companies could have had nearly four months ago." Putnam noted that he had been going through the steel industry's full-page newspaper advertisements and their other publications "searching for a clue as to the 'principles' involved in their long and costly holdout." And he concluded: "The settlement terms completely expose the hollow character of their pretensions." [21]

Perhaps some clues have been provided by industry leaders. Benjamin Fairless declared, "It became evident as early as April that the roadblock to settlement was the Union's demand for compulsory Union membership." [22] A contract would have been signed before the seizure, Fairless continued, had it not been for the union shop issue.[23] Substantially the same economic terms as finally agreed upon were available before the seizure and also in the early weeks of the strike. It was regrettable that the Steelworkers chose to press the union shop demand and to prolong the strike, Fairless said.[24]

But why did the steel industry, ignoring the costs to itself and to the country, resist a compromise on this issue for such a long time? Edward L. Ryerson, chairman of the board of the Inland Steel Company, revealed that the industry had decided early in the dispute to wage a campaign for public opinion. When the steel mills were seized, Ryerson reported, the companies realized "that we had a responsibility that went beyond our industry, and a remarkable public relations opportunity." [25] The steel companies argued that an unfair alliance between the government and the United Steelworkers, using illegal means (i.e., the seizure), was trying to force the union shop on the industry. The companies' efforts resulted in an impressive public relations victory for the steel industry, Ryerson asserted.

In retrospect the seizure of the steel mills quite possibly was itself responsible for the length of the steel strike. When the Wage Stabilization Board made its recommendations, which provided for a union shop to be negotiated between the parties, the industry determined not to accede to the union shop demand because the Steelworkers were trying to obtain it through government pressure on the companies. While industry leaders were sincere in their resistance to a union shop, they also used this issue to camouflage their more important goal of obtaining a satisfactory price increase from the government. The seizure of the steel plants gave the companies notice, however, that the government had made its final offer on price adjustments. Moreover, the seizure appeared to the industry to be a blatant move to help the union achieve its goals through positive government action. Not only was the seizure a deprivation of property without due process of law, but it also skewed the balance necessary for fair collective bargaining. Hence, when the strike finally occurred, the industry, fresh from its Supreme Court triumph, was in no mood to surrender easily.

The companies had defeated the government in court; the strike was their opportunity for revenge.

The seizure ultimately failed, therefore, to accomplish the goal set for it by the administration: providing a new context for the settlement of the labor dispute. Instead, government intervention had exacerbated the conflict between the industry and the union and thus delayed the final agreement. Of course, had the Court upheld the seizure, as the administration had expected, the result might have been different.

At the end of July 1952, the Truman administration could look back on the sequence of events and ponder whether it should have taken measures other than the ones it had employed. The administration surely must have realized that the lack of coordination among the stabilization agencies and the White House had made the resolution of the steel labor dispute more difficult. The OPS and the WSB may well have individually arrived at economically correct recommendations on steel prices and steel wages, respectively. Taken together, however, in the government's settlement package they appeared to be heavily weighted in the union's favor and were obviously unacceptable to the industry. Before any recommendations were made public, the administration could have studied them and then announced a suggested settlement which all the stabilization officials had approved. This might have eliminated what the industry saw as a good opportunity to shop around for the best price and might have made the administration's case more credible since its own officials would not then have criticized the recommendations as out of line with stabilization regulations. On the other hand, even such a unified front might not have impressed steel industry leaders who, with little concern for stabilization objectives, were intent upon getting the highest possible price increase and willing to defy the government to get it.

In fact, the administration's actions and the steel industry's responses had led to an impasse in the early days of April. The administration had tried to encourage a settlement through collective bargaining. It had offered the industry a reasonable price increase—$4.50—one that was larger than the administration would have liked, but was small enough so that damage to the stabilization program would be minimal. The industry had turned down this offer, and the administration had refused to make a higher one. At this point the President had decided that the nation could not afford a steel strike, and he had chosen seizure of the mills as the means of averting it.

In retrospect, the Truman administration may well have questioned the wisdom of that decision and been tempted to conclude that the strike should have been allowed to occur in April. The recently ended strike had shown that the country could survive an interruption in steel production. The walkout had caused economic dislocation, but not disaster. Nor had the defense program suffered unduly as a result of the fifty-three-day cessation in production. On April 8, however, the administration had not known that this might be the case. Steel strikes had always been thought of as national emergencies, and the Truman administration, already having experienced two major steel strikes, never thought to question this assumption. Given the fact that the United States was involved in the Korean conflict and the administration was responsible for supervising an adequate defense mobilization program, it was not surprising that the administration did not stop to ascertain the exact state of the nation's steel supply but jumped to the conclusion that a strike at that time could not be tolerated. Guided by the advice of several cabinet members who, in their desire to avoid any hitch in the defense program, may have exaggerated the consequences of a steel strike, Truman had decided that he could not gamble with the nation's security.

Hence, President Truman seized the steel mills because he believed that any interruption in the production of steel would be disastrous for the country. Granted, he chose seizure rather than a Taft-Hartley injunction on the basis of political and equitable considerations which reflected his prolabor feelings. In view of the union's cooperation in postponing the strike for more than three months and the industry's unwillingness to make any concession on a price increase, however, these considerations appear justified.

A legitimate argument could also be made that the Taft-Hartley Act was not relevant to the situation existing on April 8, because the dispute between the steel industry and the Steelworkers could no longer be settled by collective bargaining. The industry and the government had first to agree on a price increase. Imposition of a Taft-Hartley injunction would have provided no incentive for the industry to compromise its demands on prices. Of course, had Truman not been convinced that a strike would create a serious emergency, he might not have taken any action—just as he took no action when the walkout was in progress. But the President thought that circumstances required averting the strike. And the seizure, after all, had kept steel production going for an additional two months. Moreover, the seizure had successfully provided a new context for bargaining, since a settlement had been reached before the seizure was one month old. The agreement fell through only because the Supreme Court had decided to hear the case.[26]

The results of the seizure episode, from the administration's point of view, were not all bad. The price increase granted to the industry was larger than the government had wanted but smaller than many of the companies' price demands had been and smaller than the figure reportedly agreed upon by Defense Mobilizer Wilson and the industry. For all practical purposes, the Steelworkers had won the union shop, a concession they might not have gained had the strike taken place before seizure. And, most important, the seizure held organized labor in the Democratic party. The unions were convinced that the Democrats would use government power to assist them;[27] Republicans had shown themselves, in contrast, to be "the servants and errand boys of the steel industry . . . destroying the faith and trust of millions of union members."[28]

With these developments in mind it is tempting, but probably incorrect, to interpret the seizure of the steel mills as a measure indicative of the Truman administration's bias against big business. Although Truman deliberately chose a weapon which would hurt the steel industry—in keeping with his conviction that the companies were acting unreasonably—he picked Charles Sawyer, the Secretary of Commerce, to direct the seizure. The placing of Sawyer, business' representative in the cabinet, as the government's manager of the steel plants was intended to lessen opposition to the seizure in the business community. Moreover, had the President wanted to indulge fully an antibusiness prejudice, he could have ordered an increase in wages for the Steelworkers as soon as the seizure occurred. But Truman knew he could not do this because it would alienate the business sector whose support he needed for the mobilization. A President as politically weak as Truman was in 1952 could hardly afford to undertake an antibusiness crusade.

The seizure episode is more profitably viewed, therefore, as an illustration of the problems faced by an administration in prosecuting a limited war, especially one which has not been declared by Congress. Such a war imposes constraints on presidential action not encountered in a congressionally authorized war. By April 1952, when Truman seized the steel mills, the war was in its second frustrating year. The initial enthusiasm which had greeted the administration's announcement that the United States would come to the aid of the Republic of Korea had long since disappeared. Congressional and popular support for the administration's Korean policy gradually eroded as battles and negotiations continued with no apparent results. Regardless of the lack of public confidence in his policies, however, Truman still considered his responsibility to be no less than Roosevelt's had been during World War II. He, as President, had to ensure a strong domestic economy which could provide for immediate as well as potential military needs. Thus Truman, knowing that he could not depend on cooperation from Congress, prepared to take whatever steps were necessary to avert a strike. After thorough study of a variety of possible courses, the President decided on seizure of the steel industry. He had been advised that his authority would be challenged, but he was willing to take that risk in order to protect the national security.

Moreover, Truman entertained few doubts that his authority would be upheld. Convinced that a strong President was the key to successful government, Truman could not believe that an action taken to preserve the nation's security would be ruled beyond his powers. Throughout his career in the federal government, Truman had witnessed the steady expansion of presidential power, a trend he wholly approved. Determined to maintain that power and to hand it on to his successors intact, Truman was insensitive to the difference between the situation he faced and the circumstances surrounding the occasions when Franklin Roosevelt wielded extraordinary presidential authority. A pragmatist, Truman did not consider the fine legal line between a congressionally declared war and a presidentially instituted one to be of importance when the safety of the nation was at stake. On the eve of the seizure, Truman told Arthur Goldberg that he preferred basing the action on the constitutional power of the President rather than on a statute, because that was the honest thing to do.[29]

Truman failed to perceive that vigorous presidential rule no longer received the public support necessary for it to be effective. Nor did he realize that the infelicitous state of his administration had contributed greatly to the undermining of confidence in the President. Frequent charges of corruption, communism, and incompetence in the executive branch gave evidence of the disreputable position to which the Truman administration had fallen. By the spring of 1952, Truman's political base had evaporated: A mere 26 percent of the nation approved of his leadership. When, in these circumstances, the President seized the steel mills and the Supreme Court ruled against him, it appeared to be a last attempt by a faltering administration to assert power.

And yet, from a historical perspective, the seizure episode, by triggering a constitutional confrontation between the President and the Court, arguably made a positive contribution to the future practice of American government. The braking of the trend toward concentration of power in the executive branch may well have been timely. For twenty years the employment of presidential power had steadily increased under pressure from a succession of crises. As a consequence of this development, Truman believed that he had the authority to seize the steel mills. The industry's challenge to this authority provided an opportunity for the Court to reexamine the doctrine of separation of powers in the light of the nation's experience in the 1930s and 1940s. And the decision in the *Steel Seizure* case gave notice to the President that the limitations on executive power implicit in that doctrine were still to be respected.

Advocates of a strong Presidency have called the ruling in *Youngstown* the "most serious setback" ever experienced by the President,[30] but that judgment is open to question. The decision in the *Steel Seizure* case was certainly a rebuke to President Truman. How the decision affected and will affect the actions of future Presidents is more difficult to determine. Surely no President would seize the steel mills or another major industry again without congressional authorization. Beyond that, however, a President might take many steps, arguably forbidden under the holding or spirit of *Youngstown* but important enough to him so that he would be willing to assume the risks of litigation.[31] Perhaps the President of the United States needs that flexibility. The value of the *Steel Seizure* decision lies in the fact that it ensures that such actions of the President will, when challenged, be reviewed by the courts. A country which suffered through the abuses of power of the Nixon administration has reason to appreciate the importance of the *Steel Seizure* precedent, which reaffirmed the basic principle of the rule of law and its corollary that the President, like every other citizen, is subject to the law.

NOTES AND BIBLIOGRAPHY

ABBREVIATIONS USED IN THE NOTES

CEA	Council of Economic Advisers
DPA	Defense Production Act
ESA	Economic Stabilization Agency
LC	Library of Congress
NA	National Archives
NLRB	National Labor Relations Board
ODM	Office of Defense Mobilization
OF	Official File
OPA	Office of Price Administration
OPS	Office of Price Stabilization
PPF	President's Personal File
RG	Record Group
TL	Harry S. Truman Library
UMW	United Mine Workers
WSB	Wage Stabilization Board

notes

PREFACE

1. John Marshall Harlan, "Thoughts at a Dedication: Keeping the Judicial Function in Balance," *American Bar Association Journal* 49(October 1963):943–44.

2. 343 U.S. 579 (1952).

ONE ★ LIMITED WAR AND THE ECONOMY

1. Harry S. Truman, *Memoirs*, 2:393.

2. Paige, *Korean Decision*, p. 187.

3. Connally, *My Name Is Tom Connally*, pp. 347–48.

Margaret Truman asserts that President Truman's refusal to obtain congressional approval was a matter of principle. See *Harry S. Truman*, pp. 472–73.

For the reaction of Republicans in Congress to the outbreak of hostilities in Korea, see Caridi, *The Korean War and American Politics*, pp. 33–50.

4. The President's News Conference of June 29, 1950, U.S. President, *Public Papers of the Presidents of the United States, Harry S. Truman, 1950*, (Washington, D.C.: 1965), p. 503.

5. *Ibid.*, p. 504.

6. *Ibid.*

7. Statement of Rep. Harold Cooley, Democrat of North Carolina, in U.S. House, 81st Cong., 2d sess., August 10, 1950, *Congressional Record*, 96:12201.

8. June 28, 1950, *ibid.*, p. 9320.

9. June 30, 1950, *ibid.*, p. 9540.

10. Quoted in Paige, *Korean Decision*, p. 266n.

11. For further discussion of the legality of American entry into the Korean War, see Lofgren, "Congress and the Korean Conflict," p. 77; U.S. House, Committee on Foreign Affairs, *Background Information on the Use of United*

States Armed Forces in Foreign Countries, 1951; Binkley, *The Man in the White House,* p. 91; Cabell Phillips, *The Truman Presidency,* p. 399; Pye, "The Legal Status of the Korean Hostilities," pp. 45–60.

A more complete discussion of the decision not to seek specific congressional authorization can be found in Paige, *Korean Decision,* pp. 187, 305–6. See also Dean Acheson, *Present at the Creation: My Years in the State Department* (New York: Norton, 1969), pp. 414–15; Caridi, *The Korean War and American Politics,* pp. 44–48; Schlesinger, *The Imperial Presidency,* pp. 131–33; Rees, *Korea: The Limited War,* pp. 21–35.

12. See, for example, the statements of Rep. Graham Barden, Democrat of North Carolina, August 2, 1950, *Congressional Record,* 96:11626, and Republican Senator Kenneth Wherry of Nebraska, August 11, 1950, *ibid.,* p. 12277. Additional similar statements are scattered throughout the *Congressional Record,* vol. 96, during the debate on the Defense Production Act (which is discussed later). The same point was also made many times during the hearings on that legislation. See, for example, the exchange between Bernard Baruch and Senator Homer Capehart, Republican of Indiana, in U.S. Senate, Committee on Banking and Currency, *Defense Production Act of 1950, Hearings,* p. 135.

13. For the text of the proclamation see U.S. President, *Public Papers 1950,* pp. 746–47.

14. Harry S. Truman, *Memoirs,* 2:476–85. Rossiter commented in *Constitutional Dictatorship,* p. 299: "Few Americans seem to realize that almost all of the President's lengthy catalogue of emergency powers go into operation upon the declaration of an emergency ascertained and proclaimed by himself alone." This became a critical issue in the minds of some Supreme Court Justices when they considered the *Youngstown* case.

15. Neustadt, "Congress and the Fair Deal: A Legislative Balance Sheet," p. 373.

16. Hamby, "The Vital Center, the Fair Deal, and the Quest for a Liberal Political Economy," p. 675. See also Wage Stabilization Program 1950–1953, 1:3, Terminal Reports and Histories, 1953, Wage and Salary Stabilization Boards of the Economic Stabilization Agency, RG 293, NA.

17. Flash, *Economic Advice and Presidential Leadership,* p. 41. See also U.S. President, *The Economic Report of the President Transmitted to the Congress January 12, 1951* (Washington, D.C.: 1951), pp. 1–7.

Some economists disagreed with the Keyserling policy. See Harris, *The Economics of Mobilization and Inflation,* p. 98; and Chandler, "The Basic Economic Problems of War and Defense Programs," in Chandler and Wallace, *Economic Mobilization and Stabilization,* pp. 3–16.

For a brief discussion of the behavior of the American economy during the Korean War years, see Vatter, *The U.S. Economy in the 1950's,* pp. 72–90.

18. Vatter, *The U.S. Economy in the 1950's,* p. 78; Flash, *Economic Advice and Presidential Leadership,* pp. 41, 47.

19. Hamby, "The Vital Center, the Fair Deal, and the Quest for a Liberal Political Economy," p. 675.

20. Special message to the Congress Reporting on the Situation in Korea, July 19, 1950, U.S. President, *Public Papers 1950*, pp. 527–37.

21. The President's News Conference of July 27, 1950, U.S. President, *Public Papers 1950*, p. 563. During congressional hearings on the proposed legislation, spokesmen for the administration supported Truman's contention that price and wage controls were unnecessary. See, for example, the statement of W. Stuart Symington, chairman of the National Security Resources Board, in U.S. House, Committee on Banking and Currency, *Defense Production Act of 1950, Hearings*, p. 26.

22. Flash, *Economic Advice and Presidential Leadership*, p. 42.

23. U.S. President, *Economic Report 1951*, pp. 193–94.

24. U.S. Senate, Committee on Banking and Currency, *Defense Production Act of 1950, Hearings*, p. 98.

25. See the debate in the House of Representatives, August 1, 1950, *Congressional Record*, 96:11506–41, and the remarks in the Senate, August 2 and August 4, 1950, *ibid.*, pp. 11558, 11581, 11824–27. Senator Burnet Maybank, Democrat of South Carolina, chairman of the Committee on Banking and Currency, stated: "After hearing witnesses, and receiving hundreds of thousands of telegrams and letters which came to the members of the committee and the committee itself, it was felt by the committee, in its wisdom that it would be proper to grant stand-by controls." *Ibid.*, p. 12154. See also Flash, *Economic Advice and Presidential Leadership*, pp. 43–44.

Senator Tom Connally, however, issued a statement agreeing with the President's policy: "For the present, general controls are probably not necessary." Press Statement on President's Message to Congress, July 19, 1950, Box 570, Connally Papers, LC.

26. Wage Stabilization Program, 1950–1953, 2:1.

27. Letter to Committee Chairmen on the Defense Production Bill, August 1, 1950, U.S. President, *Public Papers 1950*, p. 567.

28. The House passed its version of the bill on August 10, 1950, by a vote of 383 to 12. *Congressional Record*, 96:12224–25. The Senate bill was passed on August 21, 1950, by a vote of 85 to 3. *Ibid.*, p. 12910. The conference bill was enacted by both houses on September 1 without a recorded vote. *Ibid.*, pp. 14074, 14138.

29. U.S., *Statutes at Large 1950*, "Defense Production Act," September 8, 1950, 64:798.

30. *Ibid.*, p. 799.

31. *Ibid.*, p. 800.

32. U.S. Senate, Committee on Banking and Currency, *Defense Production Act of 1950, Hearings*, pp. 19–20, 35–36, 183–84.

33. *Ibid.*, p. 35.

34. *Congressional Record*, 96:12412.

35. *Ibid.*, p. 11513.

36. U.S., *Statutes at Large 1950*, 64:803.

37. For the Senate Committee on Banking and Currency's reasons for adopting this stance, see U.S. Senate, Committee on Banking and Currency, *Defense Production Act of 1950*, p. 21.

38. U.S., *Statutes at Large 1950*, 64:804.

39. *Ibid.*, p. 807.

40. S. Rept. 2250, pp. 7–8.

41. Sec. 50↑, U.S., *Statutes at Large 1950*, 64:812.

42. S. Rept. 2250, pp. 5–6.

43. U.S., *Statutes at Large 1950*, 64:812.

44. S. Rept. 2250, p. 41. See also "Analysis and Comparison of S. 3936 and H.R. 9176, Being the Defense Production Bills Presently before the Senate and Passed by the House Respectively," Background Material S. 3936 (DPA 1950), Senate Committee on Banking and Currency, Legislative Records, NA; *Congressional Record*, 96:11745, 12906.

45. *Congressional Record*, 96:12905–6.

46. S. Rept. 2250, pp. 41–42.

47. Statement of Rep. Andrew Jacobs, Democrat of Indiana, *Congressional Record*, 96:11631.

48. For a law professor's attempt to ferret out the meaning of Section 503, see Sylvester Garrett, "The Emergency Dispute Privisions of the Taft-Hartley Act as a Framework for Wartime Dispute Settlement," in Industrial Relations Research Association, *Proceedings of the Third Annual Meeting*, pp. 25–29.

49. U.S. President, Executive Order 10161, *Federal Register*, 15:6105.

50. McConnell, *The Steel Seizure of 1952*, p. 4. The White House viewpoint was supported by most industrial relations experts. See, e.g., W. Ellison Chalmers, Milton Derber, and William H. McPherson, "Problems and Policies of Dispute Settlement and Wage Stabilization in World War II," in Chandler and Wallace, *Economic Mobilization and Stabilization*, p. 388.

51. Wage Stabilization Program 1950–1953, 1:17.

52. *Federal Register*, 15:6106. For comments on why this particular structure was chosen, see John H. Kaufmann, "The Problem of Coordinating Price and Wage Programs in 1950–1953: Part I," p. 511.

53. There was some confusion on this point. The administration thought the act clearly directed the President to try to gain the voluntary concurrence of business and labor in the economic objectives of price and wage stabilization. A House committee report on the mobilization program said as much: "It was the over-all policy of the act to encourage voluntary cooperation between business and Government," and specifically cited Section 402 as proof. (U.S. House of Representatives, Committee on the Judiciary, *The Mobilization Program*, H. Rept. 1217, 82d Cong., 1st sess., 1951, p. 6.) But, in 1951, when the administration was accused of not putting mandatory controls into effect immediately after the Defense Production Act was passed, some congressmen denied that the

act plainly intended that voluntary action be attempted before instituting compulsory controls. See the testimony of Michael DiSalle before the House Committee on Banking and Currency. (U.S. House, Committee on Banking and Currency, *Defense Production Act Amendments of 1951, Hearings*, pp. 2219–20.)

54. McConnell, *Steel Seizure of 1952*, p. 5.

55. U.S. President, *The Economic Report of the President Transmitted to the Congress January 16, 1952* (Washington, D.C.: 1952), p. 189.

56. *Ibid.*, p. 188.

57. Wage Stabilization Program 1950–1953, 1:3.

58. McConnell, *Steel Seizure of 1952*, p. 6.

59. Radio and Television Report to the American People on the National Emergency, December 15, 1950, U.S. President, *Public Papers 1950*, p. 744.

60. Wage Stabilization Program 1950–1953, 1:3.

61. See Radio and Television Report to the American People on the National Emergency, pp. 744–45; U.S. President, Executive Order 10193, *Federal Register*, 15:9031; McConnell, *Steel Seizure of 1952*, p. 6.

62. John R. Steelman and H. Dewayne Kreager, "The Executive Office as Administrative Coordinator," p. 704.

63. *Ibid.*

64. See U.S. House, Committee on Banking and Currency, *Defense Production Act Amendments of 1951, Hearings*, pp. 41–42.

65. McConnell, *Steel Seizure of 1952*, p. 6. Also, author's interview with David Stowe, January 9, 1973.

66. See testimony of Eric Johnston in U.S. Senate, Committee on Banking and Currency, *Defense Production Act Amendments of 1951, Hearings*, p. 519.

67. U.S. President, *Economic Report 1952*, p. 188.

68. Flash, *Economic Advice and Presidential Leadership*, pp. 64–70.

69. Wage Stabilization Program 1950–1953, 1:4; Johnston testimony, U.S. Senate, Committee on Banking and Currency, *Defense Production Act Amendments of 1951, Hearings*, p. 519. See also Hamby, *Beyond the New Deal*, pp. 446–48; and Ching, *Review and Reflection*, pp. 92–93.

70. Johnston testimony, U.S. Senate, Committee on Banking and Currency, *Defense Production Act Amendments of 1951, Hearings*, pp. 525–26; Rayback, *A History of American Labor*, p. 412.

71. The Committee, composed of fourteen members from the AFL, the CIO, the Railway Labor Executives Association, and the International Association of Machinists, had been formed in December 1950, to provide a forum for discussion of problems generated by the government's mobilization and stabilization policies. The labor movement hoped to evolve common solutions to these problems and thus, by presenting a united front, strengthen its case before the government. *Steel Labor* 16 (January 1951): 3.

72. Quoted *ibid.*

73. McConnell, *Steel Seizure of 1952*, pp. 7–8.

74. Quoted in *Steel Labor* 16 (April 1951): 2.

75. For an economist's analysis of the stabilization crisis of 1951, see Harris, *Economics of Mobilization and Inflation,* pp. 211–13.

76. Ching, *Review and Reflection,* pp. 91–92.

TWO ★ TRUMAN, LABOR, AND THE POLITICS OF LIMITED WAR

1. Margaret Truman, *Harry S. Truman,* pp. 82, 176; McClure, *The Truman Administration and the Problems of Postwar Labor, 1945–1948,* pp. 36, 81; Cabell Phillips, *Truman Presidency,* pp. 31–32; Hamby, *Beyond the New Deal,* p. 46.

2. John T. Dunlop, "The Decontrol of Wages and Prices," in Warne, *Labor in Postwar America,* pp. 5, 10–11.

3. McClure, *The Truman Administration,* p. 159; Hartmann, *Truman and the 80th Congress,* p. 5; Seidman, *American Labor from Defense to Reconversion,* p. 236; Koenig, *The Truman Administration,* p. 4; Bernstein and Matusow, *The Truman Administration,* p. 46.

4. U.S., *Statutes at Large 1947,* "Labor Management Relations Act, 1947," June 23, 1947, 61:136.

5. Millis and Brown, *From the Wagner Act to Taft-Hartley,* p. 398.

6. See the discussion of these provisions *ibid.,* pp. 489–96.

7. *Ibid.,* p. 363.

8. See ch. 4 in R. Alton Lee, *Truman and Taft-Hartley,* pp. 80–105.

9. Kroll, "Labor's Political Role," p. 118. For a history of the labor movement's increasing participation in partisan campaign work, see Greenstone, *Labor in American Politics,* pp. 39–80.

10. Labor Day Speech, Cadillac Square, Detroit, September 6, 1948, in Bernstein and Matusow, *The Truman Administration,* p. 131.

11. Kirkendall, "Election of 1948," 4:3139.

12. R. Alton Lee, *Truman and Taft-Hartley,* pp. 153, 155. Various analysts credit different voting blocs as being the most important factor in the Truman victory. Lee concludes that a fair statement would be that Truman could not have won without tremendous labor support. *Ibid.,* p. 153. Kirkendall's essay (cited above) is the best source for a balanced view of the election.

13. Author's interview with David Stowe, January 9, 1973.

14. Anderson, *The President's Men,* p. 92. Allen and Shannon, *The Truman Merry-Go-Round,* p. 63, accused Steelman of being unsympathetic to labor, but his associates agree that this was not the case. Author's interview with Charles S. Murphy, September 22, 1972; author's interview with David Stowe, January 9, 1973; author's interview with Arthur Goldberg, September 11, 1972; author's interview with Elliot Bredhoff, September 6, 1972.

15. Cabell Phillips, *Truman Presidency,* p. 114.

16. *Ibid.,* p. 135.

17. Allen and Shannon, *Truman Merry-Go-Round*, p. 75; Flash, *Economic Advice and Presidential Leadership*, p. 19.

18. Richard Neustadt, "Notes on the White House Staff under President Truman," June 1953, p. 25, TL.

19. Margaret Truman, *Harry S. Truman*, p. 449. Allen and Shannon, however, thought Murphy made little of the opportunities his position offered. "Inarticulate," "cautious and slow-moving," "conservative"—these are the adjectives by which they described the President's special counsel. *Truman Merry-Go-Round*, p. 74.

20. Flash, *Economic Advice and Presidential Leadership*, p. 19. See also Neustadt, "Notes on the White House Staff," pp. 10–11.

21. Quoted in McConnell, *Steel Seizure of 1952*, p. 8.

22. *Steel Labor* 16 (March 1951):2. Flexibility in the price stabilization program was the result of the variety of adjustments made after the price freeze of January 26, 1951. Whereas the wage stabilization regulation set forth 10 percent as the maximum wage increase, the price stabilization formulas allowed for open-ended increases dependent upon a number of factors, and these formulas were different for each industry and sometimes different for particular products within an industry. Ceiling prices of agricultural products were flexible because the Defense Production Act of 1950 itself provided a choice of methods for computing these ceilings. U.S., *Statutes at Large 1950*, "Defense Production Act," September 8, 1950, 64:805.

23. Otis Brubaker, "The Role of Direct Controls," in Director, *Defense, Controls, and Inflation*, p. 182.

24. What constituted effective price control, was a subjective judgment. In an economy as large as that of the United States, a total price freeze is impossible to enforce; thus there are always price fluctuations. The administration was pleased with its stabilization program after the freeze. Prices did level off generally in the spring of 1951. But labor thought that prices were frozen at too high a level and resented any increases at all.

25. Director, *Defense, Controls, and Inflation*, pp. 183–84.

26. Wage Stabilization Program 1950–1953, 2:3.

27. Title II of the Taft-Hartley Act, concerned with "national emergency" disputes, created an independent Federal Mediation and Conciliation Service (previously the Conciliation Service had been under the jurisdiction of the Department of Labor), which was authorized to mediate disputes that threatened a "substantial interruption of commerce." If the mediation service was unsuccessful, the President, if the proposed strike or lockout affected "an entire industry or a substantial part thereof [and] if permitted to occur or to continue, . . . would imperil the national health or safety," could appoint a board of inquiry that would investigate the dispute and make a report to the President. The report was to be a statement of facts; it could contain no recommendations for the settlement of the dispute. After receiving the board of inquiry's report, the President was authorized, at his discretion, to "direct the Attorney General to petition any district

court of the United States having jurisdiction of the parties to enjoin such strike or lock-out or the continuing thereof." The court could issue the injunction if the strike or lockout did in fact affect "an entire industry or a substantial part thereof" and if, in the court's view, it imperiled "the national health or safety." The injunction could stand for eighty days, after which time the Attorney General had to move the court to discharge it. If the dispute was not settled during the first sixty days, the National Labor Relations Board was required to take a secret ballot of all "the employees of each employer involved in the dispute on the question of whether they wish to accept the final offer of settlement made by their employer as stated by him." If the employees did not accept the last offer and the dispute was still not settled at the end of the eighty days, the President had to submit a report to Congress and could recommend suitable action. Unless Congress took action, the strike could continue until a settlement was reached by the parties. U.S., *Statutes at Large*, 61:152–56.

28. Statement of Ira Mosher, chairman, Industrial Mobilization Committee, National Association of Manufacturers, U.S. Senate, Committee on Banking and Currency, *Defense Production Act Amendments of 1951, Hearings*, pp. 1220, 1237.

29. Wage Stabilization Program 1950–1953, 2:4.

30. Statement of Ira Mosher, U.S. Senate, Committee on Banking and Currency, *Defense Production Act Amendments of 1951, Hearings*, p. 1236.

31. *Ibid.*, p. 1237.

32. Executive Order 10233, *Federal Register*, 16:3503.

33. This arrangement was a source of consternation to Defense Mobilizer Wilson and caused some difficulties when the steel dispute was being considered. Wage Stabilization Program 1950–1953, 2:25–26.

34. David Stowe, a member of Truman's staff, thinks that business representatives agreed to participate in the board's work because they realized that disputes "go hand in hand with stabilization" and their experience with the War Labor Board during World War II was not an unfavorable precedent. Author's interview with David Stowe, January 9, 1973.

35. Rothe and Lohr, *Current Biography* . . . 1952, pp. 183–85.

36. Statement of George W. Taylor, U.S. Senate, Committee on Banking and Currency, *Defense Production Act Amendments of 1951, Hearings*, p. 853. Ching, *Review and Reflection*, pp. 96–97.

37. Executive Order 10233, *Federal Register*, 16:3503.

38. Truman, in 1949, set up the Steel Industry Board instead of using the Taft-Hartley procedure to avert a strike in the steel industry. The board was authorized to investigate the issues in the dispute, make a public report, and recommend settlement terms. While the board performed its duties, the President requested the union and the industry to continue working for sixty days under the old contract. Steel labor and management cooperated with the President, but the industry declared it would not be bound by the board's recommendations. Livernash, *Collective Bargaining in the Basic Steel Industry*, p. 263.

39. Harold Enarson, "The Politics of an Emergency Dispute: Steel, 1952," in Industrial Relations Research Association, *Emergency Disputes and National Policy* (New York: 1955), pp. 52–53n. In other words, the President's power to ask advice from whom he chose enabled him to set up the WSB.

40. U.S. Senate, Committee on Labor and Public Welfare, *The Disputes Functions of the Wage Stabilization Board, 1951*, p. 4.

41. *Ibid.* Taft was the only committee member to register a dissent to the report. For other views on the President's authority to set up the new WSB, see U.S. Senate, Committee on Banking and Currency, *Defense Production Act Amendments of 1951, Hearings*, p. 1267; U.S. House, Committee on Banking and Currency, *Defense Production Act Amendments of 1951, Hearings*, p. 47; speech of Senator Wayne Morse, U.S. Senate, 82d Cong., 1st sess., April 18, 1951, *Congressional Record*, 97:4063.

42. *Congressional Record*, 97:7692.

43. *Ibid.*, p. 8395. The lengthy debate on this amendment begins on p. 8390 and ends on p. 8415.

44. *Ibid.*, p. 8415. Senator Taft intended to offer a similar amendment in the Senate but, at the behest of Senator Maybank, Taft withdrew it. Note, Burnet Maybank to Walter Reuther, July 9, 1951, U.S. Senate, Committee on Banking and Currency, Defense Production Act of 1951, RG 46, NA.

45. Harry S. Truman, *Memoirs*, 2:529; and Harbison and Spencer, "The Politics of Collective Bargaining," p. 713n.

46. Blackman, *Presidential Seizure in Labor Disputes*, p. 210.

47. U.S., *Statutes at Large 1951*, "Amendments to Defense Production Act of 1950," July 31, 1951, 65:135.

48. Testimony of Leon Keyserling, U.S. Senate, Committee on Banking and Currency, *Defense Production Act Amendments of 1951, Hearings*, pp. 288–89.

49. See, for example, U.S. Senate, Committee on Banking and Currency, *A Bill to Amend and Extend the Defense Production Act of 1950 and the Housing and Rent Act of 1947, as Amended*, Senator Irving Ives Individual Views, pp. 38–39. See also Bernard Baruch's Army War College Speech, April 18, 1952, p. 6, Public Papers, vol. 5, Baruch Papers, Princeton University; and "Baruch Links Ills to Controls Delay," *New York Times*, May 15, 1952, p. 20.

50. Hamby, "The Vital Center, the Fair Deal, and the Quest for a Liberal Political Economy," pp. 676–77. See too the President's Special Message to the Congress Recommending Extension and Broadening of the Defense Production Act, April 26, 1951, in U.S. President, *Public Papers 1951*, pp. 247–48.

51. *Public Opinion Quarterly* 15 (Summer 1951):395.

52. *Congressional Record*, 97:9148.

53. *Ibid.*, pp. 9025–29, 9141–59.

54. Statement by the President upon Signing the Defense Production Act Amendments, July 31, 1951, in U.S. President, *Public Papers 1951*, p. 435.

55. *Ibid.*, p. 436. All the stabilization officials were disturbed by the Capehart amendment and worked to have the amendment changed.

56. Special Message to the Congress after Further Review of the Defense Production Act Amendments, August 23, 1951, in U.S. President, *Public Papers 1951*, pp. 480–81.

57. S. Rept. 470, Sen. Irving M. Ives Individual Views, p. 39.

58. Earl Latham has noted, "One of the prime deterrents to the development in the United States of an adequate federal civil service is the manifest hostility, relentless and unceasing, which Congressmen pour out upon officials of the executive establishment. One former official of the federal government said that it was 'like being nibbled to death by ducks,' and former Secretary of the Interior Krug, when asked whether he would return to Washington to mobilize industry at the start of the Korean War in July 1950 (as he had in the Second World War) replied to the effect that he would seek a painless death by joining the military forces this time." *The Group Basis of Politics: A Study in Basing-Point Legislation*, p. 42.

59. Quoted in Rothe and Lohr, *Current Biography* . . . 1952, p. 482.

60. *Ibid.*, pp. 482–83.

61. Quoted in obituary, Washington *Post*, Nov. 26, 1972, p. D14. See also *New York Times*, December 2, 1951, p. IV–12.

62. U.S. Senate, Committee on Banking and Currency, *Nomination of Ellis G. Arnall, Hearings*, 82d Cong., 2d sess., 1952, p. 1.

63. Anna Rothe and Helen Demarest, *Current Biography: Who's News and Why 1945* (New York: H. W. Wilson, 1946), pp. 12–13.

64. See Ellis Arnall's own book, *The Shore Dimly Seen* (Philadelphia: Lippincott, 1946), for a heartfelt recital of his beliefs and his hopes for the South's future.

65. Arnall saw this issue as the most important in his campaign. *Ibid.*, pp. 39–50.

66. Rothe and Demarest, *Current Biography* . . . 1945, pp. 13–14; Arnall, *The Shore Dimly Seen*, pp. 186–207, 263–65.

67. *Christian Science Monitor*, June 29, 1950, p. 1, quoted in Paige, *Korean Decision*, p. 194. For other favorable responses to Truman's decision, see *ibid.*, pp. 212–13.

68. *Congressional Record*, 96:9537.

69. Fenton, *In Your Opinion*, p. 91. By August 1950, the percentage was down, but still a substantial majority approved. Roper, *You and Your Leaders*, p. 145.

70. Fenton, *In Your Opinion*, p. 90.

71. See the polls in *Public Opinion Quarterly* 14 (Winter 1950–1951):814, 818.

72. Neustadt, "Congress and the Fair Deal: A Legislative Balance Sheet," p. 373. See also Lubell, *The Future of American Politics*, pp. 230–31; poll on price

regulation, *Public Opinion Quarterly* 15 (Summer 1951):395. Of those polled, 75 percent were dissatisfied with the administration's price control measures.

73. See the polls in *Public Opinion Quarterly* 15:386–87.

74. Athan Theoharis, "The Rhetoric of Politics: Foreign Policy, Internal Security, and Domestic Politics in the Truman Era, 1945–1950," in Bernstein, *Politics and Policies of the Truman Adminsitration*, p. 217.

75. Cornwell, *Presidential Leadership of Public Opinion*, pp. 162–63.

76. Lofgren, "Congress and the Korean Conflict," pp. 21–24.

77. Fenton, *In Your Opinion . . .* , p. 92.

78. *Ibid.*, pp. 73, 92.

79. Quoted in Chicago *Tribune*, March 24, 1952, p. 13.

80. Bernstein and Matusow, *The Truman Administration*, p. 357; Hamby, "The Vital Center, the Fair Deal, and the Quest for a Liberal Political Economy," p. 672.

81. Roper, *You and Your Leaders*, pp. 146–47; Hamby, *Beyond the New Deal*, pp. 460–64.

82. *New York Times*, January 11, 1952, p. 1; Hamby, *Beyond the New Deal*, pp. 464–66.

83. *New York Times*, January 30, 1952, p. 1.

84. *Ibid.*, April 4, 1952, p. 1.

85. U.S. Senate, 82d Cong., 2d sess., May 20, 1952, *Congressional Record*, 98:5554.

86. "Notes on the White House Staff," p. 24.

87. Marquis Childs, "Upheaval in the Cabinet," Washington *Post*, April 8, 1952, p. 12.

88. Riddick, "The Eighty-Second Congress: Second Session," pp. 632–33.

89. Cabell Phillips, "Congress Falls Behind in Election-Year Lag," *New York Times*, June 1, 1952, p. E7.

90. Truman's remarks at a dinner given in New York City on May 8, 1954, in Koenig, *The Truman Administration*, p. 17.

91. *Ibid.*, pp. 17–18.

THREE ★ THE STEEL INDUSTRY AND THE UNITED STEELWORKERS OF AMERICA

1. Urofsky, *Big Steel and the Wilson Administration*, p. xxix.

2. *Ibid.*, pp. xxvii–viii.

3. Livernash, *Collective Bargaining in the Basic Steel Industry*, p. 21; Walter S. Tower, "Fifty Years of Steel," in American Iron and Steel Institute, *Year Book of American Iron and Steel Institute 1950* (New York: American Iron and Steel Institute, 1950), pp. 25–26; David Brody, *Labor in Crisis: The Steel Strike of 1919* (Philadelphia: Lippincott, 1965), p. 14.

4. Brody, *Labor in Crisis*, pp. 21–22.

5. Livernash, *Collective Bargaining in the Basic Steel Industry*, pp. 20–21; American Iron and Steel Institute, *Annual Statistical Report 1950* (New York: American Iron and Steel Institute, 1951), p. 30.

6. American Iron and Steel Institute, *Annual Statistical Report 1950*, p. 8. Two hundred thousand additional employees were engaged in other facets of the industry: mining, transportation, sales, etc.

7. Livernash, *Collective Bargaining in the Basic Steel Industry*, p. 20.

8. Henry David, "Upheaval at Homestead," in Daniel Aaron, ed., *America in Crisis* (New York: Knopf, 1952), p. 137. Brody's estimate differs from David's: Brody states the union had organized "about two thirds of the skilled men eligible to hold a union card." *Labor in Crisis*, p. 17.

9. David, "Upheaval at Homestead," pp. 136–37.

10. Brody, *Labor in Crisis*, p. 13.

11. *Ibid.*, pp. 17–18.

12. David, "Upheaval at Homestead," p. 170.

13. Brody, *Labor in Crisis*, p. 26.

14. Brooks, *As Steel Goes*, pp. 28–29.

15. *Ibid.*, pp. 31–32.

16. For an incisive discussion of developments during this period see Brody, *Labor in Crisis*, pp. 24–44.

17. Quoted in Brody, *Labor in Crisis*, p. 54.

18. *Ibid.*, pp. 54–55, 60.

19. *Ibid.*, pp. 64–76.

20. Wakstein, "The Origins of the Open-Shop Movement, 1919–1920," pp. 466–67.

21. Brody, *Labor in Crisis*, pp. 87–95.

22. *Ibid.*, pp. 95–102.

23. *Ibid.*, pp. 174–75. Urofsky, *Big Steel and the Wilson Administration*, pp. 288–90.

24. *United States* v. *U.S. Steel Corporation*, 251 U.S. 417 (1920).

25. Urofsky, *Big Steel and the Wilson Administration*, pp. 338–39. See also Schroeder, *The Growth of the Major Steel Companies, 1900–1950*, pp. 293–305.

26. Quoted in Brody, "The Rise and Decline of Welfare Capitalism," p. 147.

27. *Ibid.*, p. 151.

28. Quoted *ibid.*, pp. 154–55.

29. In this period, steel wages kept pace with wages in most durable goods industries and increased faster than in many nondurable goods and other industries. Livernash, *Collective Bargaining in the Basic Steel Industry*, pp. 133–35.

30. U.S. Steel, however, did not organize a company union until the enactment of the National Industrial Recovery Act of 1933.

For the development of the employee representation plans and the institution of the eight-hour day, see Hogan, *Economic History of the Iron and Steel Industry in the United States,* 3:866–74.

31. Brody, "The Rise and Decline of Welfare Capitalism," p. 158. Obviously, other factors also militated against the growth of the organized labor movement in the 1920s: The development of mass production methods and the concept of scientific management, xenophobia and antiradicalism, the disinterest of the federal government, "yellow-dog contracts," injunctions, and poor union leadership were some of these. See, generally, Mark Perlman, "Labor in Eclipse," pp. 103–45.

32. Brody, "The Rise and Decline of Welfare Capitalism," pp. 165, 171–73.

33. Livernash, *Collective Bargaining in the Basic Steel Industry,* pp. 74–75; Galenson, "The Unionization of the American Steel Industry," pp. 8–16.

34. Quoted in Anna Rothe, *Current Biography: Who's News and Why 1949* (New York: H. H. Wilson, 1950), p. 445.

35. *Ibid.*

36. Interview with Elliot Bredhoff, September 6, 1972. (In 1952, Bredhoff was associate general counsel of the United Steelworkers of America.)

37. Livernash, *Collective Bargaining in the Basic Steel Industry,* p. 76; Brooks, *As Steel Goes,* p. 179.

38. Galenson, "The Unionization of the American Steel Industry," p. 21; Brooks, *As Steel Goes,* p. 108. For further discussion of the conditions which favored the recognition of the SWOC, see Livernash, *Collective Bargaining in the Basic Steel Industry,* p. 62; Galenson, "The Unionization of the American Steel Industry," pp. 22–26; Brody, *Labor in Crisis,* pp. 180–85. Also, for an evaluation of the support given the organizing drive by the La Follette committee, see Auerbach, "The La Follette Committee," pp. 443–44.

39. Brooks, *As Steel Goes,* p. 109.

40. Galenson, "The Unionization of the American Steel Industry," pp. 27–29; Auerbach, "The La Follette Committee," pp. 447–48.

41. Galenson, "The Unionization of the American Steel Industry," pp. 30–36; Seidman, *American Labor from Defense to Reconversion,* pp. 47–48; Hogan, *Economic History of the Iron and Steel Industry,* 3:1178–84.

42. Livernash, *Collective Bargaining in the Basic Steel Industry,* pp. 22–24, 110. Mills, in *The New Men of Power,* p. 44, relates that polls showed Murray to be one of the best-known and best-liked labor leaders in the country.

43. Livernash, *Collective Bargaining in the Basic Steel Industry,* pp. 238–40; Addison T. Cutler, "Price Control in Steel," in U.S. Office of Temporary Controls, Office of Price Administration, *Studies in Industrial Price Control* (Washington, D.C.: 1947), pp. 41–42.

44. Cutler, "Price Control in Steel," pp. 37–48. The excess profits tax also reduced the steel companies' gains during the war years.

It should be noted that after the War Labor Board, in 1942, promulgated the "Little Steel" formula, which gave the steelworkers a 15 percent wage increase over January, 1941, levels (the raise was in line with settlements in other industries and took into account the rise in the cost of living), wages were held at that level for the remainder of the war. Livernash, *Collective Bargaining in the Basic Steel Industry,* pp. 242–46.

45. Livernash, *Collective Bargaining in the Basic Steel Industry,* p. 246. The industry reaped some benefits from the war, however. The companies were permitted to write off plants installed for war purposes on a five-year basis, and after the war some firms acquired government-built steel facilities at very reasonable prices. Hogan, *Economic History of the Iron and Steel Industry,* 3:1193; Schroeder, *Growth of the Major Steel Companies,* p. 104.

46. McCabe, "Union Policies as to the Area of Collective Bargaining," p. 124. Also see Livernash, *Collective Bargaining in the Basic Steel Industry,* pp. 4–5.

There were some exceptions—most notably in 1949, when the Steelworkers settled first with Bethlehem Steel, which set the pattern for that year. Livernash, *Collective Bargaining in the Basic Steel Industry,* p. 269.

47. Livernash, *Collective Bargaining in the Basic Steel Industry,* p. 110.

48. The outbreak of the Korean War ended the debate. Broude, *Steel Decisions and the National Economy,* pp. 20, 217–18, 229–30, 222–23, 242, 247–48; Hogan, *Economic History of the Iron and Steel Industry,* 4:1447.

49. Livernash, *Collective Bargaining in the Basic Steel Industry,* p. 110. See also Harbison and Spencer, "The Politics of Collective Bargaining," pp. 705–6.

50. Author's interview with David Stowe, January 9, 1973. Stowe stated that although there was greater rapport between the steel union and the White House than between other unions and the White House, the Steelworkers received no special treatment as a result.

51. Stephen K. Galpin, "Steel Strike Threat: Prospects of Averting Shutdown Seem Dim as Panel Readies Report," *Wall Street Journal,* March 15, 1952, p. 4. At this time, the CIO claimed it had nearly 7 million members. Rayback, *History of American Labor,* p. 406.

52. Charles Moritz, *Current Biography Yearbook 1961* (New York: H. W. Wilson, 1961, 1962), pp. 178–79.

53. Memorandum and Draft of April 3, Gardner Ackley to Arnall, April 7, 1952, Steel, Decentralized Files of Various Directors of OPS, RG 295, NA.

54. Note, Moreell to Roger L. Putnam, July 14, 1952, Wage Stabilization Steel Industry, General Subject Files 1951–1953, Records of the Office of the Administrator, ESA, RG 296, NA.

55. Quoted in *Steel Labor* 16 (March 1951):3. McDonald was referring to the inflexible 10-percent formula for wage increases adopted by the original WSB.

56. *Ibid.*

FOUR ★ THE GOVERNMENT SEIZES THE STEEL MILLS

1. Benjamin Fairless, Speech at Cincinnati, Ohio, November 15, 1951. Quoted in *Steel Labor* 17 (April 1952):1.

2. *Steel* 130 (January 1952):492.

3. Wage Stabilization Program 1950–1953, 2:78–79. Terminal Reports and Histories, 1953, Wage and Salary Stabilization Boards of the Economic Stabilization Agency, RG 293, NA.

4. Quoted in U.S. Senate, Committee on Banking and Currency, *Defense Production Act Amendments of 1952, Hearings*, p. 2039.

5. Memorandum, "Prices and Wages in the Steel Industry," Council of Economic Advisers to the President, November 28, 1951, OF 342, September 1951–February 1952, TL.

6. Quoted in McConnell, *Steel Seizure of 1952*, p. 16.

7. Ching, *Review and Reflection*, p. 99.

8. Statement by the President on the Labor Dispute in the Steel Industry, December 22, 1951, U.S. President, *Public Papers 1951*, p. 651.

9. *Ibid.*, p. 652.

10. *Ibid.* Truman details his reasons for certifying the steel dispute to the WSB in *Memoirs*, 2:528–30.

11. Transcript of speech in Box 726, Taft Papers, LC.

12. Harold L. Enarson, "Politics of an Emergency Dispute," in Industrial Relations Research Association, *Emergency Disputes and National Policy* (New York, 1955), p. 54n.

13. Author's interview with Arthur Goldberg, September 11, 1972.

14. Telegram, December 24, 1951, Wage Stabilization Steel Industry, General Subject Files 1951–1953, Records of the Office of the Administrator, Economic Stabilization Agency, RG 296, NA.

15. Wage Stabilization Program 1950–1953,2:78.

16. *Ibid.*, pp. 53–54.

17. *Ibid.*, pp. 85–86.

18. Panel Report in Case D-18-C of the WSB in the matter of United Steelworkers of America-CIO and Various Steel and Iron Ore Companies (Xerox copy sent to me by the union), pp. 2–6. This report is printed in U.S. Senate, Committee on Labor and Public Welfare, *Disputes Functions of the Wage Stabilization Board, Hearings*, pp. 64–122.

A discussion of the union shop proposal does not appear in the steel panel report because the panel was instructed by the WSB that it would review the original evidence on this matter itself. *Ibid.*, p. 1.

19. *Ibid.*, p. 2.

20. Indeed, by a liberal interpretation of the wage board's regulations, OPS officials found that the Steelworkers were entitled to as much as 22 cents an hour

as well as additional fringe benefits. Memorandum. John K. Meskimen to Michael V. DiSalle, Subject: Steel Wage Negotiations, January 5, 1952, Labor, Decentralized Files of Various Directors of OPS, OPS, RG 295, NA.

21 Memorandum, Gardner Ackley to DiSalle, January 8, 1952, Subject: Steel Wage Increase, Steel, Decentralized Files of Various Directors of OPS, OPS, RG 295, NA.

22. McConnell, *Steel Seizure of 1952*, p. 16.

23. Gardner Ackley, Memorandum: Steel Position Paper, February 27, 1952, Steel, Decentralized Files of Various Directors of OPS, OPS, RG 295, NA.

24. Statement of Arnall, April 3, 1952, Steel Press Releases, Decentralized Files of Various Directors of OPS, OPS, RG 295, NA. See also McConnell, *Steel Seizure of 1952*, p. 20.

25. Memorandum, "Steel—again," Enarson to Stowe and Murphy, January 30, 1952, Box 4, Enarson Papers, TL. See also *New York Times*, January 30, 1952, p. 33; McConnell, *Steel Seizure of 1952*, p. 20.

26. Memorandum, Enarson to Stowe and Murphy, January 30, 1952.

27. The President's News Conference at Key West, March 20, 1952, U.S. President, *Public Papers 1952–1953*, p. 209.

28. "Looming in Steel: A Strike over Prices," *Business Week*, March 8, 1952, p. 22. See also A. H. Raskin, "Steel Case Highlights Wage Board's Dilemma," *New York Times*, March 23, 1952, p. E10; "Price Boosting Formula Given Steel Firms," Chicago *Tribune*, March 13, 1952, part 6, p. 7; "Steel Heads Jittery with CIO Silent," Pittsburgh *Post-Gazette*, March 18, 1952, pp. 1, 4; "Strike Plans Forcing Steel Plant Slow-Up," New York *Herald Tribune*, March 17, 1952, p. 1; Stephen K. Galpin, "Steel Strike Threat: Prospects of Averting Shutdown Seem Dim as Panel Readies Report," *Wall Street Journal*, March 15, 1952, pp. 1, 4.

29. Memorandum, Enarson to Files, March 5, 1952, Box 5, Enarson Papers, TL.

30. Steel Companies in the Wage Case, "These Are the Facts in the Steel Controversy," pamphlet in papers of the American Iron and Steel Institute, New York, p. 10. The steel companies joined together under the name of "Steel Companies in the Wage Case" in order to coordinate most effectively their propaganda efforts.

31. U.S. Senate, Committee on Labor and Public Welfare, *National Emergency Labor Disputes Act*, pp. 17–18.

32. The transcript of the testimony is over 3,100 pages long; more than 145 special exhibits were included as evidence. Panel Report in Case D-18-C, pp. 1, 70.

33. Apparently the WSB had hoped to return the union shop issue to the parties but still maintain jurisdiction if the parties failed to agree. Neither the union nor the industry found this an acceptable solution. The board subsequently discovered that a favorable union shop recommendation was the key to obtaining

the union's assent to the rest of the WSB package. Livernash, *Collective Bargaining in the Basic Steel Industry*, pp. 106–7. See also Murray Kempton, "Steel Talkathon," New York *Post*, March 20, 1952, pp. 3, 33.

The WSB's recommendations are printed in U.S. Senate, Committee on Labor and Public Welfare, *Disputes Functions of the Wage Stabilization Board, Hearings*, pp. 38–63.

34. McConnell, *Steel Seizure of 1952*, p. 24.

35. Magnitudes Involved in Possible Steel Settlements, Memorandum, Gardner Ackley to Ellis Arnall, April 7, 1952, Steel, Decentralized Files of Various Directors of OPS, OPS, RG 295, NA.

36. *New York Times*, March 22, 1952, p. 1; *ibid.*, March 27, 1952, p. 17; New York *Herald Tribune*, March 22, 1952, p. 1; Chicago *Tribune*, March 30, 1952, part 1, p. 14. At this time, the steel companies were focusing primarily on the issue of price relief for the suggested wage increase. Later, when the battle for public opinion raged, other recommendations—e.g., the union shop provision— would be the focal point of the controversy.

37. Statement of Industry Members, March 20, 1952, in U.S. House, 82d Cong., 2d sess., March 24, 1952, *Congressional Record*, 98:2784.

38. U.S. Senate, Committee on Labor and Public Welfare, *Disputes Functions of the Wage Stabilization Board, Hearings*, p. 60.

39. Ching, *Review and Reflection*, pp. 99–100.

40. See, for example, editorials in the following newspapers: Washington *Post*, April 3, 1952, p. 10; New York *Herald Tribune*, March 22, 1952, p. 8; Pittsburgh *Post-Gazette*, March 24, 1952, p. 16; Cleveland *Plain Dealer*, March 22, 1952, p. 10; Wheeling *Intelligencer*, March 25, 1952, p. 4.

41. Apparently this was true. Harry Shulman, chairman of the steel special panel, observed: "I had some talks with Feinsinger. . . . My impression was that he thought of the problem as one of inducing the Union not to strike—i.e., of making a recommendation finally acceptable to the Union. I believe, though I do not know, that he did not take very seriously the possibility of a rejection by the Companies." Note, to Felix Frankfurter, August 20, 1952, Box 102, Frankfurter Papers, LC. See also Ching, *Review and Reflection*, pp. 99–100.

42. Editorial, April 21, 1952, p. 20.

43. *Congressional Record*, 98:2784. See also remarks by Rep. H. Carl Anderson of Minnesota, *ibid.*, p. 2849, and statement by Sen. John W. Bricker of Ohio, *ibid.*, p. 3678.

44. *Ibid.*, pp. 3219–20.

45. U.S. House, Committee on Rules, *Directing the Committee on Education and Labor to Conduct an Investigation of the Wage Stabilization Board*, p. 1.

46. H. Res. 532, *Congressional Record*, 98:1283. See also Wage Stabilization Program 1950–1953, 2:10.

47. Memorandum, Enarson to Files, March 5, 1952.

48. *Congressional Record,* 98:3139, 4398.

49. U.S. Senate, Committee on Labor and Public Welfare, *Disputes Functions of the Wage Stabilization Board, Hearings,* pp. 1–37.

50. U.S. Senate, Committee on Labor and Public Welfare, *Staff Report on WSB Recommendations in Steel Dispute,* pp. 1–2.

51. "Labor," *Fortune,* May, 1952, pp. 77–78; editorial, *New Republic,* April 7, 1952, p. 7; "Pay's Climb Nears Top," *U.S. News & World Report,* March 28, 1952, p. 58; Harry C. Bates, W. C. Birthright, Elmer E. Walker, and Lee W. Minton, "The Truth about the Steel Case," *American Federationist,* April, 1952, p. 6.

For additional defenses of the wage board's recommendations, see A. H. Raskin, "Steel Case Highlights Wage Board's Dilemma," *New York Times,* March 23, 1952, p. E10; Marquis Childs, "Power to Buy Must Be Held," *Washington Post,* April 4, 1952, p. 20; Robert R. Nathan, letter to the editor, *Washington Post,* April 5, 1952, p. 8; Dewey Anderson, letter to the editor, *Washington Post,* April 17, 1952, p. 16; editorial, *New York Post,* March 28, 1952, p. 45; Murray Kempton, "Payment Deferred," *New York Post,* March 24, 1952, p. 23.

A recent study of the history of collective bargaining in the steel industry questions the whole concept of steel as a pattern-setter in the American economy and concludes that steel has not been a leader in its wage settlements. Specifically regarding 1952, the study noted: "Viewed in the light of the wage trends which had developed since the prior steel settlement in late 1950, and considering the nature of underlying economic trends as they existed at the time, there appears to be little ground for the view that the 1952 steel contract [which was virtually the same as the WSB's recommendations] was out of line, or that it contributed to a new and higher pattern of wage-fringe benefits." Livernash, *Collective Bargaining in the Basic Steel Industry,* p. 146.

52. Enarson, "Politics of an Emergency Dispute," p. 57. Richard Neustadt has commented: "The Wage Board's public members . . . surely were guilty of too broad a view of their role and too narrow a conception of the Government's, in rendering their famous Steel decision without sounding out the White House." "The Presidency at Mid-Century," *Law and Contemporary Problems* 21 (Autumn 1956):621.

53. Analysis of Report and Recommendations of the Wage Stabilization Board in the Steel Case, March 27, 1952, Box 28, Murphy Files, TL. Memorandum, "Current Issues in the Steel Industry," CEA to the President, March 28, 1952, Box 4, Stowe Papers, TL.

The CEA was less convinced than the staff that the wage recommendations were strictly within stabilization guidelines: "While the wage recommendations may have been higher than the most desirable recommendations would have been, our preliminary analysis indicates that these wage recommendations are by no means as extreme as has been portrayed in some quarters. Viewing these wage recommendations in their long-range impact, they may be somewhat out of line, but they are not very seriously out of line, with the broad principles of wage policy that have been set forth in various Economic Reports of the President." But the council was more inclined to consider the wage settlement acceptable because, in purely economic terms, it necessitated no price increase other than what was allowed by the Capehart amendment. "Current Issues in the Steel Industry."

54. Analysis of Report and Recommendations of the Wage Stabilization Board, pp. 2, 4.

55. *Ibid.*, pp. 2–3.

56. *Ibid.*, p. 3.

57. *Ibid.*, p. 4. Roger Putnam supported this conclusion: "In the steel dispute our policies on both wages and prices have been based entirely on policies adopted by this Agency as long as a year ago. The steel companies knew what the rules were before they ever started the wage negotiations with the union. I saw to that. Their complaint is not one of lack of coordination between the two, but rather that they just don't like those policies as they apply to the steel industry. They have asked for special treatment of a kind not given to any other industry." Note, Putnam to Glenn E. Burress, May 23, 1952, Wage Stabilization, Steel Industry, General Subject Files 1951–53, Records of the Office of the Administrator, ESA, RG 296, NA.

58. McConnell, *Steel Seizure of 1952*, pp. 24–26.

59. Memorandum, Enarson to Files, February 19, 1952, Box 4, Enarson Papers, TL; Memorandum, Enarson to Files, March 5, 1952.

60. McConnell says Sunday, March 24, but it was the 23d. *Steel Seizure of 1952*, p. 25. "Calendar," *Encyclopedia Americana*, 1972, 5:188–89.

61. Memorandum, Enarson to Files, March 25, 1952, Box 5, Enarson Papers, TL.

For a discussion of why the Truman-Wilson meetings at Key West were unsuccessful, see Neustadt, *Presidential Power*, pp. 21–22.

62. U.S. House, Committee on Education and Labor, *Investigation of the Wage Stabilization Board, Hearings*, p. 7.

63. *New York Times*, March 25, 1952, p. 1.

64. Memorandum, Enarson to Files, March 25, 1952, Box 5, Enarson Papers, TL. *Steel Labor* 17 (April 1952):3. The labor adviser to the OPS apprised Arnall of the reaction of the labor members of the WSB to Wilson's statement: "The Labor groups are agreed that, so far as dispute cases are concerned, they are responsible only to the President. In their view, Mr. Wilson is entitled to his personal opinion but it carries no weight with them. They consider the recommendations of the Board to be no more and no less than the basis for negotiations between the parties. . . .

"There are thus far no indications that the Labor representatives will assume responsibility for wrecking the stabilization program and embarrassing the Administration by walking out." Memorandum, Subject: Steel Wage Negotiations, John K. Meskimen to Ellis Arnall, March 26, 1952, Labor, Decentralized Files of Various Directors of OPS, OPS, RG 295, NA.

65. *Steel Labor* 17 (April 1952):3.

66. Wilson later denied that he had been consulted before the board's recommendations were announced. Bullen, in his testimony before the House Committee on Banking and Currency (*Defense Production Act Amendments of 1952, Hearings*, p. 203), replied to Wilson's denial by describing the meeting in question. An article in *Business Week* (March 29, 1952, p. 174) also gave an account of this meeting and reported that Wilson expressed no objections to the projected recommendations, "a fact that explains Feinsinger's genuine incredulity four days later, when Wilson blasted the recommendations." For additional proof that meetings between the WSB chairman and Wilson had been held, see Wage Stabilization Program 1950–1953, 2:25–26.

It should be noted, however, that the WSB was under no obligation to clear its recommendations with the highest stabilization officials, i.e., Wilson and Putnam. As explained by Putnam to Congressman Walter Mumma, "the Wage Stabilization Board—although a part of my Agency—was acting on a special assignment from the President in recommending a proposed series of terms for settling this dispute. Its recommendations on disputes cases do not clear through me but rather go direct to the parties and to the President. On the economic issues, it was under instructions from the President to make its recommendations conform to stabilization policy; on the non-economic issues it was completely on its own." Note, April 3, 1952, Wage Stabilization Steel Industry, General Subject Files 1951–53, Records of the Office of the Administrator, ESA, RG 296, NA.

67. Memorandum, Enarson to Files, March 25, 1952.

68. McConnell, *Steel Seizure of 1952*, p. 27.

69. *Ibid.*

70. Memorandum, "Events Relating to Charlie Wilson's resignation," Enarson to Files, April 1, 1952, Box 5, Enarson Papers, TL.

71. *Ibid.* For President Truman's account of the conference see his *Memoirs*, 2:531.

72. James L. Sundquist, Oral History Interview No. 13, p. 18, TL.

73. *New York Times*, March 31, 1952, p. 30.

74. Letter accepting Wilson's resignation, March 30, 1952, U.S. President, *Public Papers 1952–1953*, p. 226. For Wilson's version of his resignation, see "Charles E. Wilson's Own Story of Break with Truman," *U.S. News & World Report*, May 2, 1952, pp. 11–14.

75. Editorial, *Nation*, April 5, 1952, p. 311; editorial, *New Republic*, April 7, 1952, pp. 6–7.

76. See editorials appearing in the following newspapers on April 1, 1952: Washington *Post*, p. 14; *New York Times*, p. 28; New York *Herald Tribune*, p. 22; New York *Daily News*, p. 39; *Wall Street Journal*, p. 8; Chicago *Tribune*, p. 20; Chicago *Sun-Times*, p. 23; Cleveland *Plain Dealer*, p. 16; Pittsburgh *Post-Gazette*, p. 10; Los Angeles *Times*, part II, p. 4; and the Washington Evening *Star*, March 31, 1952, p. A-10. These editorials accuse Truman of playing politics with the nation's stabilization program, i.e., of making special exceptions for labor but not for industry. As the Washington *Post* pointed out, "it is always popular to oppose big business." April 1, 1952, p. 14.

77. McConnell noted (*Steel Seizure of 1952*, p. 23): "Steelman's role accounted for the fact that Secretary of Labor Lewis Schwellenbach [*sic*, Maurice Tobin] did not play a prominent part in the steel crisis." Examination of Department of Labor sources confirms that the department was not significantly involved in the steel dispute.

78. David Stowe, in a 1973 interview, commented that Feinsinger was a poor choice to mediate the dispute at that point. As chairman of the WSB, Feinsinger had expended all his energy and goodwill during the steel hearings. By the time of the collective bargaining sessions he was "burnt out." Author's interview with David Stowe, January 9, 1973.

79. McConnell, *Steel Seizure of 1952*, p. 30.

80. *New York Times*, April 3, 1952, p. 38. One small company, Kaiser Steel Corporation, found it possible to sign a contract without any price guarantee. Kaiser and the United Steelworkers of America reached agreement on the basis of the WSB recommendations. Kaiser already had a union shop, however. Henry J. Kaiser commented, "We concluded production should be continued during this critical defense period while price deliberations are continued with the Government." Quoted in *Daily Metal Reporter*, April 4, 1952, p. 1.

Other small firms that reached agreement with the United Steelworkers of America on the basis of the WSB recommendations were Detroit Steel Co., Ohio River Steel Corp., Belmont Iron Works, and Alan Wood Steel Co. But these companies employed only some 15,000 of the 650,000 employed in the basic steel industry. *CIO News*, April 7, 1952, p. 3.

81. McConnell, *Steel Seizure of 1952*, p. 30. See also *Steel*, April 7, 1952, p. 91, in which a caption under a photograph of Arnall and Fairless read, "Mr. Fairless was upholding the companies' claim that a $12 a ton price increase is needed to compensate for wages recommended by Wage Stabilization Board."

The $12-a-ton figure was commonly reported as the price adjustment the steel industry desired. But in later testimony before congressional committees the companies denied that they had ever asked for $12 a ton. U.S. Senate, Committee on Labor and Public Welfare, *National and Emergency Labor Disputes, Hearings*, p. 222.

82. McConnell states that the government's offer was devised by Arnall and was a revelation to the White House staff. *Steel Seizure of 1952*, p. 31. Other sources say the $4.50 figure originated with Steelman and the President; Arnall was "ordered" to offer it to the steel companies. Edwin A. Lahey, "The Steelworkers State Their Case," *New Republic*, May 26, 1952, p. 13; G. H. Baker, " 'Secret' Steel Price Offer Admitted," *Iron Age*, May 8, 1952, p. 95.

83. Author's interview with Charles Murphy, September 22, 1972; author's interview with David Stowe, January 9, 1973.

84. David Stowe remembered the Secretary of Defense saying that even a three-day strike could not be tolerated. Interview, January 9, 1973.

85. The President's military budget for 1952–1953 called for a slowdown of the rearmament program because of a scarcity of strategic materials. *Newsweek*, Jan. 28, 1952, p. 22. With specific reference to steel, John D. Small, chairman

of the Munitions Board, forecast, "On steel, refiguring of schedules should have a moderate good effect in terms of reducing take of high temperature alloys. However, shortages are so severe that no easing in civilian use is probable for many months." Minutes of Defense Mobilization Board Meeting No. 22, January 23, 1952, General Subject File, Secretary Tobin, 1952—Office of Defense Mobilization, NA.

86. Truman, *Memoirs*, 2:531–32. See also Questions and Answers on the Steel Case, from Dick Neustadt, May 1952, Box 53, Elsey Papers, TL.

87. Truman, *Memoirs*, 2:532. Upon his departure from the government, Charles Wilson had commented that a thirty-day steel strike would retard the defense program sixty days. *New York Times*, April 1, 1952, p. 20.

88. Truman, *Memoirs*, 2:532.

89. *Ibid.*

90. Professor Edward R. Livernash has pointed out that the effects of steel strikes on the American economy have been exaggerated. *Collective Bargaining in the Basic Steel Industry*, pp. 31–49.

91. McConnell, *Steel Seizure of 1952*, p. 31. The government, under Title II as amended in 1951, was authorized to institute ordinary condemnation proceedings for any property it needed. This required the government to file in court a declaration of taking for each property it wanted and to deposit the amount of payment it considered just compensation before the government could take over the property and operate the facilities. (U.S., *Statutes at Large 1951*, "Amendments to Defense Production Act of 1950," July 31, 1951, 65:132–33.) In the case of the steel industry, the amount of money involved was staggering and clearly acted as a deterrent to the use of this act. Also, it was unclear whether this power was applicable to property wanted to avoid a production stoppage as a result of a labor dispute. *Ibid.*, p. 32. See too Memorandum to Assistant Attorney General Baldridge re: President's Power to Seize the Steel Industry, p. 5, Lloyd Files, Box 6, Truman Papers, TL.

92. Interview with David Stowe, January 9, 1973; Memorandum Re: Steel Developments, Enarson to Files, April 7, 1952, Box 5, Enarson Papers, TL.

93. Proponents of the Taft-Hartley procedure pointed out that the President could appoint a board of inquiry several days before the strike deadline. But the White House staff realized that appointment of a board of inquiry while collective bargaining was continuing would give steel management even less incentive to settle: an 80-day strike delay could only work to the industry's advantage. Questions and Answers on the Steel Case, Box 53, Elsey Papers, TL.

94. *Ibid.* See also Clarence H. Osthagen, "The 1952 Government Seizure of the Basic Steel Industry," July 1952, Department of Commerce Records, Washington, D.C.

95. U.S., *Statutes at Large 1948*, "Selective Service Act of 1948," June 24, 1948, 62:626.

96. Memorandum, "Meeting with Justice Department Staff on 'Section 18 Seizure Authority'," Kayle to Dave Stowe, April 1, 1952, p. 1, Box 4, Stowe

Papers, TL. Arthur Goldberg submitted to David Stowe an assessment of the President's power to seize the steel mills in which he, too, stated that "It seems plain that the President could seize the steel mills under this authorization [Section 18]." Statutory Authorization for Seizure, p. 3, Box 4, Stowe Papers, TL. (I am not certain that Goldberg was the author of this memorandum. His name, with "I believe?" penciled after it, appeared on the title page.)

On the other hand, John W. Davis, the senior partner in the large Wall Street law firm Davis, Polk, Wardwell, Sunderland, and Kiendl, stated categorically, in an opinion for Republic Steel Corporation, that Section 18 seizure authority was not applicable to the steel dispute. Opinion of John W. Davis, March 31, 1952, p. 1, Papers of John H. Pickering, Washington, D.C.

97. Legislative History of Section 18 of the Selective Service Act of 1948, Box 1, Baldridge Papers, TL.

98. Government officials had previously examined the question of seizing, under Section 18, plants closed by strikes. During the hearings on the Defense Production Act of 1950, Senator John Sparkman discussed with W. Stuart Symington, chairman of the National Security Resources Board, the proposed power to requisition property:

"Sen. Sparkman: Under Title II, if the plant that was needed in the war effort should be strike-bound, would the President be enabled to take it over? If there should be a breakdown because of labor relations?

"Mr. Symington: I don't think this act gets into the question of labor relations. I think it gets into the question of requisitioning.

"Sen. Sparkman: I know. But the statement is made that when the President determines that this plant is needed in the war effort, he can requisition it. Does it mean that he could requisition it for any of those reasons?

"Mr. Symington: I think he could requisition, under Selective Service, a strike-bound plant."

Senator Sparkman accepted that statement. Senate, Committee on Banking and Currency, *Defense Production Act of 1950, Hearings*, p. 24.

99. Memorandum, Kayle to Stowe, April 1, 1952, p. 1.

100. David Lloyd, an administrative assistant to the President, sent a memorandum to the Justice Department stating: "Although each of the statutes considered above is of at least arguable applicability, it is not desirable to rely on them even as first lines of defense because of their clear commitment to the principle of just compensation. This principle may involve the Government in the payment of damages of considerable magnitude. Basing the seizure solely upon the inherent powers—while it does not assure freedom from this principle—at least leaves some leeway for maneuver." Memorandum to Assistant Attorney General Baldridge re: President's Power to Seize the Steel Industry, p. 7.

101. On June 9, 1941, the North American Aviation plant was seized by Executive Order 8773 (*Federal Register*, 6:2777); on August 23, 1941, the Federal Shipbuilding & Drydock Co. was seized by Executive Order 8868 (*Federal Register*, 6:4349); and on October 31, 1941, Air Associates, Inc. was seized by Executive Order 8938 (*Federal Register*, 6:5599).

102. Memorandum, Kayle to Stowe, April 1, 1952, pp. 1–2.

103. Memorandum, Kayle to Stowe, April 3, 1952, pp. 1–2, Box 5, Enarson Papers, TL. It is interesting to speculate whether Baldridge, at the time of this meeting, had already received the memorandum from David Lloyd on the President's power to seize the steel industry (the memorandum was undated). Lloyd clearly favored basing a seizure order on the inherent constitutional powers of the President (p. 39).

104. Interview with Arthur Goldberg, September 11, 1972. Interview with Elliot Bredhoff, associate general counsel of the United Steelworkers of America, September 6, 1972.

105. "The discussion above indicates that ample seizure power is vested in the President for use in emergencies. It should be clear that the present emergency warrants invocation of this power.

"It is true that several of the opinions and cases referred to above dealt with emergencies in time of war. However, the President himself has proclaimed that we are in a grave national emergency at present, and that we are faced not only with hostilities in Korea, but with the spectre of possible war. The President has inherent powers to take all steps necessary to wage war. But these powers also run to the steps necessary to prepare to wage war. The acuteness of the emergency differs today only in degree, rather than in kind, from a war situation. The safety, health and welfare of our troops, the nation, and of our allies require all out mobilization of our productive facilities. These facilities would be crippled by a steel strike forced by the steel industry." Goldberg, Statutory Authorization for Seizure, p. 10.

106. *Ibid.*

107. John W. Davis, in his opinion for Republic Steel Corporation, came to the opposite conclusion: the President had neither statutory nor inherent power to seize the steel industry. Opinion, March 31, 1952, pp. 1–6.

Newspaper editorials appearing at this time were inclined to the Davis view; for example, an editorial in the Washington *Post*, April 7, 1952, p. 8, stated:

"The current talk about governmental seizure of the steel industry is in striking contrast to President Truman's assertion two years ago, when the coal miners refused to obey a court injunction, that he had no authority to seize the mines. It appears that the Presidential power in this sphere blooms and withers in accord with the political sympathies of the White House in the dispute. In our opinion, however, the President was right when he told the press that he lacked power to seize the mines in an emergency. And we know of no law that has since given him power to take over steel plants because of a strike or potential strike."

See too editorial, *New York Times*, April 4, 1952, p. 24. In an article in the *New York Times*, A. H. Raskin described the views of government lawyers: "The President has not received any new legal authority to seize the mills since December, but the consensus among Federal lawyers is that it would take so long for anyone to get a definitive ruling on his powers that the steel crisis would have been passed successfully before any court said, Stop." "Steel Seizure Proposal Raises Many Serious Questions," April 6, 1952, p. E10.

108. Neustadt, *Presidential Power*, pp. 59–60.

A review of the *Congressional Record* for the week before the strike deadline shows that Congress generally agreed that the President must employ the Taft-Hartley Act. Representative Jacob Javits of New York, however, introduced a bill on April 8, 1952, giving the President seizure authority, and urged Congress to pass it, because existing legislation did not meet the demands of the steel crisis. *Congressional Record*, 98:3778.

109. Memo to All Employees, National Office, from Max Hall, Director of Public Information, May 1, 1952, Steel Case File, Decentralized Files of Various Directors of OPS, OPS, RG 295, NA.

110. The steel industry claimed the direct employment costs of the WSB proposals would be $6 a ton; the OPS said the costs would be, at most, $4.50 a ton. "Randall Puts Price Tag on Industry Steel Offer," Daily Labor Report No. 83, April 25, 1952, pp. A-6, A-7, in Exhibits, Publications and Newspaper Clippings Relating to the 1952 Steel Dispute, General Subject File Maintained by Sidney L. Feiler, ESA, RG 296, NA; Magnitudes Involved in Possible Steel Settlements, Memorandum, Gardner Ackley to Ellis Arnall, April 7, 1952, Steel, Decentralized Files of Various Directors of OPS, OPS, RG 295, NA.

111. Lahey, "The Steelworkers State Their Case," p. 13; Memorandum, "Events Preceding the Presidential Speech on April 8 and the Resort to Seizure," Enarson to Files, Box 5, Enarson Papers, TL.

112. Neustadt, *Presidential Power*, p. 29.

113. Radio and Television Address to the American People on the Need for Government Operation of the Steel Mills, April 8, 1952, U.S. President, *Public Papers 1952–1953*, pp. 246–50.

114. Harold Enarson had suggested that the President invoke a Taft-Hartley board of inquiry at the same time that he seized the steel mills. The advantages of doing this were many: It would relieve the President of criticism for using vague seizure authority when the Taft-Hartley Act was the law of the land. A board of inquiry would draw public attention to the price problem instead of concentrating it on the question of the legality of the seizure and the avoidance of the Taft-Hartley law. An impartial board of inquiry would point out the reasons for not seeking a Taft-Hartley injunction. While the board of inquiry did its fact finding, seizure would ensure continued production. Handwritten memorandum of Harold Enarson with "For Goldberg" scrawled at the top, Box 5, Enarson Papers, TL.

As most of the members of the White House staff were receptive to this idea, it was included in several drafts of the President's speech. But it was agreed that Arthur Goldberg should be consulted about the probable response of the union to the use of the first step of the Taft-Hartley procedure. Goldberg, who could not be reached until about two hours before the President's scheduled speech, vetoed the board of inquiry idea on the grounds that the steelworkers would not understand it and wildcat strikes might result. Enarson asserts that Murphy and Stowe made the decision to exclude the first step of the Taft-Hartley procedure and did not check it with the President. Memorandum, "Events Preceding the President's

Speech on April 8 and the Resort to Seizure," pp. 2–5. Goldberg, however, thinks the President made the final decision. Interview with Arthur Goldberg, September 11, 1972. See also Neustadt, *Presidential Power*, p. 25. The reasons for not using Taft-Hartley were substituted in the speech for the sections on the board of inquiry. "Steel Strike Speech—April 8, 1952," Files of David D. Lloyd, Box 27, TL.

115. Radio and Television Address, April 8, 1952, p. 247.

116. *Ibid.*, pp. 248–49.

117. *Ibid.*, p. 249.

118. *Ibid.*, p. 250.

FIVE ★ THE COUNTRY REACTS

1. *Federal Register*, 17:3139.

2. *Ibid.*, p. 3141.

3. A successful lawyer and businessman himself, Sawyer justified his defense of industry in the preface to his memoirs. See *Concerns of a Conservative Democrat*, p. viii.

4. Quoted *ibid.*, pp. 259–60.

5. Interview with Charles Sawyer, August 16, 1971. See also Confidential Dictation, May 3, 1952, p. 8, Sawyer Papers, Cincinnati, Ohio; Memorandum, "Events Preceding the Presidential Speech on April 8 and the Resort to Seizure," Enarson to Files, Box 5, Enarson Papers, TL.

6. Sawyer, *Concerns of a Conservative Democrat*, pp. 257–58; Memorandum, "Re: Steel Developments," Enarson to Files, April 7, 1952, Box 5, Enarson Papers, TL.

7. Interview with Arthur Goldberg, September 11, 1972. The Department of Defense apparently was considered for the job, but Secretary Lovett objected. Memorandum, "Re: Steel Developments," Enarson to Files, April 7, 1952, Box 5, Enarson Papers, TL.

Roger Putnam noted, "Secretary Sawyer . . . is a perfect man for the job because his whole duties will be to find a settlement and get the mills back to their owners in short order. Many newspapers and certain numbers of the opposition will undoubtedly talk as though we are interested in Government ownership or control. There is absolutely no truth in this but we are interested in keeping intact our stabilization rules which the steel industry was determined to break." Note, Putnam to Harold Crowther, April 9, 1952, Wage Stabilization Steel Industry, General Subject Files 1951–53, Records of the Office of the Administrator, ESA, RG 296, NA.

8. Copies of the telegrams may be found in "The 1952 Government Seizure of the Basic Steel Industry: Significant Documents," Records of the Department of Commerce, Washington, D.C.

9. New York *Herald Tribune*, April 9, 1952, pp. 1, 20.

10. The President's News Conference of April 10, 1952, U.S. President, *Public Papers, 1952–1953*, p. 259. See too Sawyer, *Concerns of a Conservative*

Democrat, p. 380. Joseph Rayback mistakenly asserts that the government put the WSB recommendations into effect after seizure. *History of American Labor*, p. 413. There also are other inaccuracies in his treatment of the steel seizure.

11. Interview with Arthur Goldberg, September 11, 1972; interview with Elliot Bredhoff, September 6, 1972; Enarson, "Politics of an Emergency Dispute," in Industrial Relations Research Association, *Emergency Disputes and National Policy*, pp. 62–63; A. H. Raskin, "Murray Hurls 'Bias' Charge at Sawyer as Steel 'Boss'," *New York Times*, May 17, 1952, p. 1.

12. Press Release, April 15, 1952, Doc. #20, "The 1952 Government Seizure of the Basic Steel Industry: Significant Documents," Department of Commerce Records, Washington, D.C.

13. Press Release, April 18, 1952, Doc. #26, *ibid*. At this April 17 news conference, President Truman had made it clear that the wage question would be decided by him and not by Secretary Sawyer (U.S. President, *Public Papers 1952–1953*, p. 274). Richard E. Neustadt asserts (*Presidential Power*, pp. 22–24) that within a week after the seizure, Truman ordered Sawyer to put the following adjustments into effect: The union would receive a wage increase equal to the amount to which it was automatically entitled under the stabilization guidelines (a cost-of-living adjustment) and the industry would be granted the price increase to which it was entitled under the Capehart amendment. This settlement would be so "unattractive" to both the union and the industry that they would make an extra effort to come to better terms through collective bargaining. According to Neustadt, "The President's intent was clear, . . . But Sawyer did not act." (*Presidential Power*, p. 23) Instead, the Secretary of Commerce engaged in dilatory tactics because he, as the representative of business in the cabinet, did not want to put "his signature on wage orders and price requests committing the steel industry." (*Presidential Power*, p. 23)

I can find no evidence that this order was ever given. Sawyer vehemently denies ever receiving such an order (*Concerns of a Conservative Democrat*, pp. 274–77). Charles Murphy and David Stowe both stated during interviews that if wages were not raised it was because the President had not ordered the changes. The White House staff was closely coordinated and, in their opinion, they would have known about the order if it had been given. Furthermore, both men said that Sawyer would have followed an order given to him by the President. Arthur Goldberg also indicated in an interview that he thought President Truman did not want to increase wages by government action.

The tactic described by Neustadt, moreover, would not seem to be calculated to produce a quick settlement. Indeed, steel industry leaders might have concluded that they had gotten a good deal. The government's best previous offer had been a $4.50-a-ton price increase in return for the full WSB recommendation of approximately 26 cents an hour plus the union shop. According to Neustadt, the administration was planning to institute the price adjustment allowed by the Capehart amendment (about $3 a ton) and an automatic cost-of-living wage increase (8 or 9 cents an hour allowed by the WSB's General Wage Regulation #8). The steel companies might have reasoned that this was a very attractive settlement, and there would have been little impetus for the industry to bargain with the Steelworkers any further. It was the union that would have lost most by the proposed tactic mentioned by Neustadt. Surely the Steelworkers would not have been willing to settle on those terms. Thus the industry would have been able to claim that it was satisfied with the government's settlement and the union would have become the uncooperative party—decidedly not a result desired by the administration. It therefore seems unlikely that President Truman ordered Sawyer to put these adjustments into effect.

Furthermore, there were other considerations that might have counseled a delay in changing the terms of employment. Judge Holtzoff had indicated in his opinion (see chapter 6) that if the possibility of wage changes by the government became an actuality, he would entertain a new motion by the steel companies for a temporary restraining order (U.S. House, *Steel Seizure Case*, p. 265). If the administration's primary concern was the production of steel, it probably would not have done something that increased considerably its chances of having a temporary restraining order issued against the seizure. Rather, it pursued the more prudent course of using changes in the terms and conditions of employment as a threat to encourage the steel industry to settle. The seizure speech and the act of seizure itself gave the administration enough credit with the union so that the government could afford to wait a few weeks before taking action on the wage front. In addition, the union knew that the administration was discussing making any wage increases it ordered retroactive to April 8. Memorandum, James McI. Henderson to Roger L. Putnam, April 21, 1952, 1952 Steel Dispute: Plant Seizure—Key Papers, Statements, Speeches, Reports, General Subject File maintained by Sidney L. Feiler, ESA, RG 296, NA. See also Statement by Sawyer (not released to the press), April 21, 1952; Notes on phone conversation, Sawyer to Steelman, April 18, 1952; Notes on phone conversation, Sawyer to Steelman, April 22, 1952, all in Sawyer Papers, Cincinnati, Ohio.

In a letter to the President written on May 20, 1952, Sawyer alluded to reports that union officials had been told that he had refused to put new wage rates into effect immediately after the government had seized the mills. Sawyer informed the President: "I have no apologies to make for my action in this matter; I am certain that I acted impartially and wisely and, of course, with your full knowledge of what I was doing." It seems unlikely that the Secretary of Commerce would have written this if the President had actually ordered him to raise the steelworkers' wages within a week after the seizure had commenced. The remainder of Sawyer's letter deals with his handling of the wage issue until the Supreme Court took the matter out of his hands. His account appears to be accurate. In answering Sawyer's letter, Truman did not take issue with any of the statements the Secretary of Commerce had made. This correspondence is contained in a Memorandum for Sawyer from the President, May 26, 1952, Box 136, President's Secretary's Files, Truman Papers, TL.

14. *New York Times*, April 21, 1952, p. 1.

15. Note, Sawyer to Roger Putnam, April 23, 1952, Doc. #28, "The 1952 Government Seizure of the Basic Steel Industry: Significant Documents"; Press Release, April 29, 1952, Doc. #33, *ibid.*; Note, Roger Putnam to Ellis Arnall, April 23, 1952, Decentralized Files of Various Directors of OPS, OPS, RG 295,

NA; Note, Ellis Arnall to Roger Putnam, April 25, 1952, Decentralized Files of Various Directors of OPS; *New York Times*, April 24, 1952, p. 1.

Secretary Sawyer did not impose any price increases on the steel companies. The industry simply was told that these increases were now available to it if it wanted them.

Putnam did not deliver the ESA's recommendations to Sawyer until May 3. U.S. Dept. of Commerce, Records, Osthagen Study, pp. 48–49.

16. *Daily Labor Report*, April 9, 1952, p. A-2.

17. Interview with Arthur Goldberg, September 11, 1972; Interview with Elliot Bredhoff, September 6, 1972.

18. Quoted in Pittsburgh *Press*, April 9, 1952, p. 2.

19. Quoted in Murray Kempton, "Harry's Night," New York *Post*, April 9, 1952, p. 37.

20. *Ibid*. See also Pittsburgh *Press*, April 9, 1952, pp. 1–2; "Steelworker Joy," *Wall Street Journal*, April 10, 1952, p. 1.

21. Randall, *Over My Shoulder: A Reminiscence*, p. 214.

22. *Ibid*., pp. 215–16.

23. New York *Post*, April 9, 1952, p. 14.

24. Pittsburgh *Press*, April 9, 1952, p. 1. Admiral Moreell, a naval engineer before he retired from active duty in September, 1946, to work in private industry, spent the last year of his government service managing petroleum plants and soft coal mines seized by President Truman. *Ibid*. See also Anna Rothe, *Current Biography: Who's News and Why 1946* (New York: H. W. Wilson, 1947), pp. 409, 411.

25. April 26, 1952, p. 393.

26. April 10, 1952, p. 41.

27. See, for example, editorials in the following newspapers: on April 9, 1952, Washington *Evening Star*, p. 12 and New York *Herald Tribune*, p. 26; on April 10, *New York Times*, p. 28; Washington *Post*, p. 16; Chicago *Sun-Times*, p. 27; Chicago *Tribune*, p. 16; Washington *Times-Herald*, p. 18; Wheeling *Intelligencer*, p. 4; Pittsburgh *Post-Gazette*, p. 12; Cleveland *Plain Dealer*, p. 12; *Wall Street Journal*, p. 6; on April 11, *Wall Street Journal*, p. 4 and Washington *News*, p. 42.

28. William Henry Chamberlain, "Steel Seizure," *Wall Street Journal*, April 16, 1952, p. 8.

29. May 10, 1952, p. 10. See too Pittsburgh *Post-Gazette*, April 10, 1952, p. 12.

30. See, for example, Washington *Post*, April 10, 1952, p. 16.

31. See *Saturday Evening Post*, May 10, 1952, p. 10; *Christian Science Monitor*, April 10, 1952, p. 18; Arthur Krock, "Issues of Law and Politics in Seizure," *New York Times*, April 11, 1952, p. 22.

32. *Iron Age*, April 10, 1952, p. 7.

33. April 21, 1952, p. 23.

34. Pittsburgh *Press*, April 9, 1952, p. 18; *Newsweek*, April 21, 1952, p. 37.

35. Pittsburgh *Press*, April 11, 1952, p. 16.

36. April 10, 1952, p. 25. The *New Republic* was the only significant periodical to support the President.

37. Editorial, Wheeling *Intelligencer*, April 10, 1952, p. 4. See also editorials in the following: St. Louis *Post-Dispatch*, April 9, 1952, p. 2C; Los Angeles *Times*, April 10, 1952, II, p. 4; *Steel*, April 14, 1952, p. 59.

38. Radio and Television Address, April 8, 1952, U.S. President, *Public Papers 1952–1953*, p. 248. The $6.69 figure was reported in the *Wall Street Journal* as the one computed by the Federal Trade Commission (April 10, 1952, p. 3). Other newspapers and journals used slightly different figures; the Washington *Post*, for example, mentions $6.83 a ton as net profit (April 10, 1952, p. 16).

Although $19.50 was the figure used by OPS for its computations, Truman erred in not stating frankly that this figure represented profits before taxes. People were suspicious (judging from congressional mail) of the largeness of the figure and thought Truman had exaggerated it. This led to the belief that perhaps other points in his address had been exaggerated also.

39. Washington *Post*, April 19, 1952, p. 6; "Steel Earnings Hit the Skids," *Iron Age*, April 10, 1952, p. 80; *Wall Street Journal*, April 10, 1952, p. 3. The *Wall Street Journal* pointed out that sales and operating revenue increased from 1950 to 1951 but that net profit was lower because of higher taxes in 1951.

40. *New Republic*, May 5, 1952, p. 6. In a letter to the editor of the *New York Times*, James Tobin, an economist from Yale University, advanced similar observations and asserted that if the profits-after-taxes figure were to be used by other industries and economic groups, "it would seriously imperil the prospects for economic stability. The argument is the more dangerous because it is superficially so plausible." April 21, 1952, p. 20.

Throughout the month of April, newspapers and periodicals kept up the barrage of editorials criticizing the President. See, for instance, Washington *Post*, April 19, 1952, p. 6; April 26, 1952, p. 6; *Nation*, April 26, 1952, p. 393; *Steel*, April 21, 1952, p. 47. David Lawrence, in *U.S. News & World Report*, April 18, 1952, p. 96, even called for Truman to be impeached, but no serious impeachment campaign materialized.

41. Memorandum, Harold Enarson to Dr. Steelman, April 15, 1952, Box 28, Murphy Files, TL. Some of the letters received by the White House can be found in President's Address re Steel Strike, 4/8/52, PPF 200, TL; Box 1214 and Box 1215, OF 407 B, TL.

42. *Congressional Record*, 98:4386.

43. *Ibid.*, p. 4562.

44. See, for example, William Langer Papers, Chester Fritz Library, University of North Dakota; Herbert H. Lehman Papers, Columbia University; Robert A. Taft Papers, LC (many letters addressed to Taft urged the impeachment of the President); Tom Connally Papers, LC (most of his constituent mail was antiseizure). Many letters on the seizure were also sent to the Senate Committee on

Banking and Currency, which was responsible for the Defense Production Act. These can be found in Correspondence, U.S. Senate Committee on Banking and Currency, Defense Production Act, 1952, 82d Cong., 2d sess., RG 46, NA.

45. Telegrams, April 17 and April 18, 1952, Charles Sawyer Papers, Cincinnati, Ohio. Most of these telegrams are from businessmen. An industry journal, *Iron Age*, however, editorially berated businessmen for their lack of interest in the seizure issue (May 1, 1952, p. 7).

46. U.S. Dept. of Justice, *Youngstown Sheet & Tube Co.* v. *Sawyer*, Files.

47. Statement of William J. Grede, president of NAM, in *Congressional Record*, 98:A2252. The Buffalo Chamber of Commerce passed a resolution critical of the seizure. Copy of the resolution, April 26, 1952, Departmental File #19, Drawer 109, Lehman Papers, Columbia University.

48. *CIO News*, April 21, 1952, pp. 2–3.

49. The Public Opinion Index for Industry, Special Report, enclosed in note, Claude Robinson to Senator Robert A. Taft, April 30, 1952, Box 720, Taft Papers, LC.

50. *Congressional Record*, 98:A3430.

51. Gallup, *The Gallup Poll, 1935–1971*, 2:1065. The fact that some of the interviewing took place after Federal Judge David A. Pine had declared the seizure illegal should be taken into account when weighing the results of this poll. Undoubtedly, many people who might have supported the seizure were influenced against it by the judge's decision.

52. See the poll cited in Lowi, *The Politics of Disorder*, p. 94. Lowi suggests that the poll taken after the seizure also came a few weeks after Truman announced his retirement and this may have affected the results. It should be noted that although the President's popularity increased it was still at a very low point: In November 1951, only 23 percent had approved the way Truman was handling his job; in January 1952, 28 percent thought the President was doing a good job. By June 1952, the approval rate had gone up to 32 percent.

53. For Randall's account of his experiences before and after giving this speech see *Over My Shoulder*, pp. 216–29.

54. "These Are the *Facts*, Mr. President," reprint of Randall's speech, April 9, 1952, p. 2, in papers of the American Iron and Steel Institute, New York. Excerpts from Randall's speech may be found in Westin, *The Anatomy of a Constitutional Law Case*, pp. 18–20.

55. Quoted in "These Are the *Facts*, Mr. President," p. 5.

56. *Ibid.*, pp. 7–8. Randall omitted the fact that the steel industry received more rapid tax amortization allowances than any other industry during the Korean defense buildup. By reducing a company's income tax payments for the first five years of operation of a new plant, the government's fast tax amortization allowances encouraged industries whose products were needed for defense to expand their facilities, thereby increasing production. According to the *Wall Street Journal* (April 9, 1952, p. 5), as of February 25, 1952, the accelerated tax amor-

tization allowance for the steel companies amounted to about $2.1 billion. See, too, the editorial in the *New Republic*, May 5, 1952, p. 6.

57. See collection of these reprints, Papers of the American Iron and Steel Institute, New York.

58. "Basic Facts in the Steel Wage Case," April 14, 1952, Papers of the American Iron and Steel Institute, New York.

59. Westin, *Anatomy of a Constitutional Law Case*, p. 46.

60. Memorandum for Sawyer from Charles S. Murphy, April 21, 1952, Sawyer Papers, Cincinnati, Ohio.

61. U.S. Senate, Committee on Labor and Public Welfare, *National and Emergency Labor Disputes, Hearings 1952*, pp. 588–89. For other comments on press coverage of the seizure situation see *ibid.*, p. 586.

62. Special Message to the Congress Reporting on the Situation in the Steel Industry, April 9, 1952, U.S. President, *Public Papers 1952–1953*, p. 251. See too "Speech on Steel Situation; Executive Order and Message to Congress Relative to Steel Mill Seizures," April 8 and 9, 1952, Box 14, Murphy Files, TL.

63. Special Message to the Congress, April 9, 1952, p. 251. In his memoirs, Truman stated that the message to Congress had said that "*unless* there was congressional action I would naturally have to take the responsibility myself." In fact, Truman's message was not as emphatic an ultimatum as he remembered. He merely said he would follow Congress' lead if it passed a bill, but he did not think legislation was necessary. *Memoirs*, 2:535.

64. *Newsweek*, April 21, 1952, p. 37.

65. *Congressional Record*, 98:3909.

66. See *Congressional Record*, 98:3794ff. Throughout April, Senators took advantage of every opportunity to voice their opinions on the seizure. See also Westin, *Anatomy of a Constitutional Law Case*, p. 45.

67. See, for example, the remarks of Senators Hoey and Mundt. *Congressional Record*, 98:3797, 3956. Whether a negative vote on a proposal constitutes a positive prohibition against the action considered became a key issue when the steel seizure case was reviewed by the Supreme Court.

68. *Ibid.*, p. 4102.

69. *Ibid.*, p. 3789.

70. *Ibid.*, pp. 3820, 3825, 3840.

71. Note, Langer to A. C. Bjerken, April 29, 1952, Langer Papers, Chester Fritz Library. There are several letters in this collection defending the seizure.

72. See, for example, Senator Aiken's comments, *Congressional Record*, 98:4133. With more time for consideration, speeches became less emotional and more concerned with the legal issues involved in the steel seizure. For examples, see *ibid.*, pp. 4142, 4192–95, 4332, 4469, 4562. The congressmen, of course, made their feelings known in other ways too. Insertion into the appendix of the *Congressional Record* of editorials critical of the President, as well as extensions of congressmen's remarks, became epidemic after the steel seizure (see *Congres-*

sional Record, vol. 98 Appendix, beginning April 9, 1952). Speeches were given around the country. Senator Taft addressed many different audiences, condemning the steel seizure and urging the House to commence impeachment proceedings (see "Seizure of the Steel Industry," Address, Pittsburgh, Pa., April 15, 1952, Box 1286, Robert A. Taft Papers, LC; Note, Taft to Dr. James E. Schaal, April 26, 1952, Box 1011, Taft Papers, LC). And letters were written to constituents and others: See "This Week in Washington," George Bender to his constituents, April 19, 1952, Bender Papers, Western Reserve Historical Society; Note, Sen. Pat McCarran to Mrs. E. D. Biathrow, May 21, 1952, Files on Hearings on S. Res. 306, Senate Committee on the Judiciary, NA; Note, Sen. Lester Hunt to O. A. Knight, May 10, 1952, Box A4-114, Philip Murray Papers, Catholic University of America; Note, Sen. James Duff to O. A. Knight, Box A4-114, Murray Papers.

73. See, for example, various bills introduced at *Congressional Record*, 98:4222, 4417, 4421, 4518–19, 4539.

74. *Ibid.*, pp. 4012–13. See also Stenographic Transcript of Hearings to Inquire into the Legal Authority of the President to Seize and Operate Certain Steel Plants and Facilities (S. Res. 306), April 24, 25, May 1, 1952, Records of the Senate Committee on the Judiciary, NA. Senator Hugh Butler submitted a bill which ordered the Secretary of Commerce to return the steel companies to their owners. It was referred to the Judiciary Committee, but Senator McCarran took no action on it because Judge Pine's decision had been handed down and McCarran wanted to wait for further developments. Note, McCarran to Hon. Hugh Butler, May 13, 1952, Files on S. 3106, NA.

75. U.S. House, Committee on Education and Labor, *Investigation of the Wage Stabilization Board, Hearings.*

76. U.S. Senate, Committee on Labor and Public Welfare, *National and Emergency Labor Disputes, Hearings.*

77. *Congressional Record*, 98:4016.

78. U.S. Senate, Committee on Banking and Currency, *Defense Production Act Amendments of 1952, Hearings*; U.S. House, Committee on Banking and Currency, *Defense Production Act Amendments of 1952, Hearings.* The administration had the additional burden of requesting another year's extension of the Defense Production Act when it appeared to many that controls were no longer necessary. As an administration official noted: "One sophisticated attack upon the need for continuation of price controls is that, 'It's the market, not OPS, that is keeping prices in line.' "

He concluded that price controls must be continued because defense expenditures were going to rise, not fall; because a truce in Korea did not appear imminent; and because consumer demand might not remain slack—an inflationary buying spree could be set off at any time. Memorandum, Charles Stewart to Ewan Clague, April 25, 1952—Labor Statistics—General (January–June), Secretary Maurice J. Tobin, General Subject File 1952, Office of the Secretary, Department of Labor Papers, NA.

79. H.R. 7647. *Congressional Record*, 98:4639.

80. There was so much opposition to the Smith bill from organized labor and the administration that the committee never reported it to the House. See, for example, Statement of Joseph Curran (representing the CIO before the House Committee on Armed Services) on H.R. 7647, May 8, 1952, and Note, Ralph Wright (Acting Secretary of Labor) to Hon. Carl Vinson, Chairman, Committee on Armed Services, May 6, 1952, Smith Antistrike Bill, Secretary Tobin, General Subject File, 1952, Office of the Secretary, Department of Labor Papers, NA.

81. See, for example, the testimony of Ellis Arnall, one of the most effective administration spokesmen, Senate, Committee on Labor and Public Welfare, *National and Emergency Labor Disputes, Hearings*, pp. 127ff., and the testimony of Nathan Feinsinger, *ibid.*, pp. 27ff., 511ff.

82. U.S. Senate, Committee on Labor and Public Welfare, *National and Emergency Labor Disputes, Hearings*, p. 388.

83. See, for example, the testimony of John C. Bane, Jr., *ibid.*, pp. 264ff., 424ff.; testimony of John A. Stephens, *ibid.*, pp. 164ff.; testimony of Charles E. Wilson, U.S. House, Committee on Education and Labor, *Investigation of the Wage Stabilization Board, Hearings*, pp. 2ff.; testimony of Ben Moreell, U.S. Senate, Committee on Banking and Currency, *Defense Production Act Amendments of 1952, Hearings*, pp. 2178ff.

84. See Philip Geyelin, "Steel Seizure," *Wall Street Journal*, April 16, 1952, p. 8; John P. Roche, "Executive Power and Domestic Emergency: The Quest for Prerogative," p. 613.

85. For text of the amendment see *Congressional Record*, 98:4026. The amendment passed 44 to 31: 11 southern Democrats joined 33 Republicans to form the majority (*ibid.*, p. 4155). See also Westin, *Anatomy of a Constitutional Law Case*, pp. 47–52; McConnell, *Steel Seizure of 1952*, p. 38.

86. Note, Connally to Hon. Alvin H. Lane, April 28, 1952, Connally Papers, Box 307, LC.

President Truman had hoped to convince more Senators of this argument when he sent a letter to Vice-President Barkley, in his capacity as president of the Senate, on April 21. The letter, which was read to the Senate, stressed the President's willingness to abide by Congress' actions but admonished the Senators: "I do not believe the Congress can meet its responsibilities simply by following a course of negation. The Congress cannot perform its Constitutional functions simply by paralyzing the operations of the Government in an emergency. . . . The Congress should do more than simply tell me what I should not do. It should pass affirmative legislation to provide a constructive course of action looking toward a solution of this matter which will be in the public interest" (*Congressional Record*, 98:4131). Truman concluded with a strong statement against invoking the Taft-Hartley Act. Murphy and Steelman, the President's advisers, had urged him to send the letter (Memorandum, Charles Murphy to Admiral Dennison, April 20, 1952, OF 272, Truman Papers, TL). Whether they informed him of Secretary Sawyer's comment that it would be better to send no message than to send such a provocative one, is unknown (Notes for Files, April 21, 1952, Sawyer Papers, Cincinnati, Ohio).

87. *Congressional Record*, 98:4467.

88. Congress, on April 9, 1952, did pass a sixty-day extension of certain provisions of the War Powers Act that would have terminated with the final deposit of the peace treaty with Japan. (The treaty became effective on April 28, 1952. Until that date the United States was technically at war with Japan.) The administration needed this statutory authority for many reasons, chief among them the necessity of continuing the railroad seizure. The railroads were seized by the government in the summer of 1950 (Executive Order 10141, July 8, 1950, *Federal Register*, 15:4363; Executive Order 10155, August 25, 1950, *Federal Register* 15:5785) to avert a strike that threatened to interrupt the transportation system during the Korean crisis. The President cited as authority for the seizure the Constitution and laws of the United States and specifically mentioned the act of August 29, 1916, which gave the President the power to take possession, in time of war, of any transportation system for whatever purpose necessary during the emergency. (U.S., *Statutes at Large*, 39:619 at 645. Westin incorrectly calls this statute the Railway Labor Act of 1916. See *Anatomy of a Constitutional Law Case*, p. 94. The August 29, 1916, law was an army appropriations act with the seizure provision thrown in, apparently because it might relate to the transportation of troops.) This provision remained in effect only during a declared war; thus, in 1950, when the United States was still technically at war with Japan, the President could use it as authority for the railroad seizure. The sixty-day extension of the President's war powers was for the purpose of continuing various provisions of many acts that were valid only in wartime, until the Congress could make a more thorough investigation of which statutes were still necessary. Where the administration was inconsistent was in maintaining that constitutional authority was sufficient for the steel seizure, whereas it had claimed it needed statutory authority for the railroad seizure.

When Congress passed the extension on April 9, 1952, it approved the joint resolution with a proviso that had been added by the Senate Judiciary Committee on April 8: "Nothing contained herein shall be construed to authorize seizure by the government, under authority of any act herein extended, of any privately owned plants or facilities which are not public utilities." (See *Congressional Record*, 98:3816ff.) The proviso was passed by the Senate Judiciary Committee because there had been talk of an imminent steel seizure. President Truman seized the mills on April 8, one day before the proviso was approved by Congress and weeks before he actually signed the bill. According to the chairman of the Judiciary Committee, Senator McCarran, the presence of this proviso as law probably would not have deterred the steel seizure, since the President had based the seizure on constitutional authority. (*Congressional Record*, 98:3816; New York *Herald Tribune*, April 9, 1952, p. 19.)

89. Arnall radio speech, April 18, 1952, Decentralized Files of Various Directors of OPS, OPS, RG 295, NA.

90. Address at the National Convention Banquet of the Americans for

Democratic Action, May 17, 1952, U.S. President, *Public Papers 1952–1953*, p. 343.

91. *Ibid.*

92. The President's News Conference of April 17, 1952, U.S. President, *Public Papers 1952–1953*, pp. 272–73.

93. The President's News Conference of April 24, 1952, *ibid.*, pp. 290–91.

94. *Ibid.*, p. 291.

95. *Ibid.*, pp. 293–94.

96. *Ibid.*, p. 294.

SIX ★ THE STEEL COMPANIES GO TO COURT

1. Interview with David Stowe, January 9, 1973; interview with Charles Murphy, September 22, 1972. Truman *Memoirs*, 2:535.

2. Westin, *Anatomy of a Constitutional Law Case*, pp. 17–18.

3. Pittsburgh *Press*, April 9, 1952, p. 2.

4. Quoted in Westin, *Anatomy of a Constitutional Law Case*, p. 18.

5. Motions for temporary restraining orders and injunctions are heard by the court sitting as a court of equity. In the United States, the same federal courts discharge both equity and common law functions. According to Alexander M. Bickel, "The mainstream of Anglo-American legal development has been the common law, administered by judges who evolved and reasoned from principle. But there soon flowed alongside the common law another stream, the equity jurisdiction, whose headwaters were in the discretionary royal prerogative. Equity was a more flexible process, more unprincipled, initially quite ad hoc. It often worked the accommodation that made the rigorous principles of the common law fit to live with." *The Least Dangerous Branch*, p. 250.

A court of equity deals with justiciable controversies in which money damages would not be suitable compensation for the wronged party. A showing must be made by the complaining party that the action or threatened action it is contesting would result in irreparable injury. When the judge (in equity proceedings a jury is not required) has made this finding, he can issue one of three orders: (1) a temporary restraining order, in force for a maximum of ten days under federal law, which prevents the contemplated action from being put into effect until a full hearing with all parties attending can be held; (2) a preliminary injunction, which enjoins the action, after a hearing, until a final disposition of the case is made (at this stage, the party asking for relief has only to show that irreparable injury will be done to him and that he has a substantial likelihood of success when the legal merits of the controversy are determined in a court of law; a full appraisal of the legal merits of the case is not necessary although a judge may elect to determine them); (3) a permanent injunction, which will stem from a final decision on the rights of the parties involved. Gregory, *Labor and the Law*, p. 96; Westin, *Anatomy of a Constitutional Law Case*, p. 27n.

It should be noted that when the questioned action will affect the public, a different balancing of interests is indicated: "Courts of equity may, and frequently do, go much further both to give and withhold relief in furtherance of the public interest than they are accustomed to go when only private interests are involved." *Virginian Ry.* v. *System Federation,* 300 U.S. 515, 552 (1937).

The complete record in these cases may be found at the U.S. District Court for the District of Columbia under the following numbers: Civil Action No. 1550–52: Youngstown Sheet and Tube Company, Youngstown Metal Products Company, Plaintiffs, v. Charles Sawyer, individually and as Secretary of Commerce, Defendant; Civil Action No. 1539–52: Republic Steel Corporation, Plaintiff, v. Charles Sawyer, individually and as Secretary of Commerce, Defendant; Civil Action No. 1549–52: Bethlehem Steel Company, et al., Plaintiffs, v. Charles Sawyer, individually and as Secretary of Commerce, Defendant.

A partial transcript of the record beginning with these cases and continuing through the Supreme Court hearing may be found in U.S. House, *Steel Seizure Case.*

6. Judge Holtzoff had been in private law practice in New York City from 1911 until 1924, when he became special assistant to the attorney general of the United States. Holtzoff was the author of several books on federal practice and procedure and was an expert in the field of tort claims against the federal government.

7. Oral argument of John J. Wilson, representing the Youngstown Sheet & Tube Co., U.S. House, *Steel Seizure Case,* pp. 221–22.

8. *Ibid.,* pp. 225–29.

9. *Ibid.,* p. 222.

10. *Ibid.,* p. 234. See also Bruce Bromley's argument on this point, *ibid.,* pp. 238–39.

11. *Ibid.,* p. 236.

12. *Ibid.,* p. 246.

13. *Ibid.,* p. 247.

14. *Ibid.,* p. 249.

15. *Ibid.,* p. 250.

16. *New York Times,* April 27, 1952, p. 51.

17. Baldridge cited the Krug-Lewis agreement of 1946 between the government and the UMW as the only example of a contract concluded during a period of seizure. This seizure of the coal mines, however, was not based on the President's constitutional powers alone but also on the War Labor Disputes Act of 1943. See Blackman, *Presidential Seizure in Labor Disputes,* p. 278.

18. U.S. House, *Steel Seizure Case,* p. 253.

19. *Ibid.,* pp. 253–54. Baldridge was unable to explain satisfactorily what he meant by a legal taking subject to just compensation under the Fifth Amendment. The Fifth Amendment refers to the power of eminent domain: "nor shall private property be taken for public use without just compensation." By conceding that the steel companies could bring an action under the Fifth Amendment, Baldridge seemed to be saying that the seizure was authorized by the eminent domain power. But Baldridge admitted that the government had no

intention of instituting eminent domain proceedings; the steel companies would have to go to the Court of Claims with an action against the government for damages. In addition, Judge Holtzoff reminded Baldridge that eminent domain was a power "exercised pursuant to an Act of Congress." At that point Baldridge declared that the steel seizure was not a taking under the eminent domain powers. But he could not specify what kind of taking the seizure did represent. *Ibid.*, pp. 253, 255.

20. *Ibid.*, p. 256.

21. Baldridge, knowing he was not ready to argue this point, asked Judge Holtzoff if he could submit a brief before discussing it. Holtzoff replied that a request for a TRO was an urgent matter which allowed for no delay in decision. He expected to give his opinion soon after the oral argument was concluded. *Ibid.*, p. 254.

22. *Ibid.*, p. 255.

23. *Home Building & Loan Association* v. *Blaisdell*, 290 U.S. 398, 425–26 (1934). In that opinion Chief Justice Hughes observed:

"Emergency does not create power. Emergency does not increase granted power or remove or diminish the restrictions imposed upon power granted or reserved. The Constitution was adopted in a period of grave emergency. Its grants of power to the Federal Government and its limitations of the power of the States were determined in the light of emergency and they are not altered by emergency. What power was thus granted and what limitations were thus imposed are questions which have always been, and always will be, the subject of close examination under our constitutional system.

"While emergency does not create power, emergency may furnish the occasion for the exercise of power. 'Although an emergency may not call into life a power which has never lived, nevertheless emergency may afford a reason for the exertion of a living power already enjoyed.' *Wilson* v. *New*, 243 U.S. 332, 348. The constitutional question presented in the light of an emergency is whether the power possessed embraces the particular exercise of it in response to particular conditions."

24. U.S. House, *Steel Seizure Case*, pp. 257–58.

25. *Ibid.*, pp. 260–61.

26. *Ibid.*, p. 264.

27. Holtzoff quoted Chief Justice Chase in the case of *Mississippi* v. *Johnson*, 4 Wall. 475 (1867): "The Congress is the legislative department of the Government. The President is the executive department. Neither can be restrained in its action by the judicial department, though the acts of both when performed are in proper cases subject to its cognizance." U.S. House, *Steel Seizure Case*, p. 265. *Mississippi* v. *Johnson* will be discussed in subsequent chapters in greater detail. But it should be noted that, at the time of the steel seizure litigation, this was the only case involving an action for an injunction specifically against the President, and it has been taken as conclusive by many authorities. See, for example, Schubert, *The Presidency in the Courts*, pp. 320–21. There are some who

disagree, however. Judge Charles Fahy (U.S. Court of Appeals, District of Columbia Circuit), a past solicitor general of the United States, indicated in an interview that he thought the President, in certain circumstances, could be enjoined, because, under the Constitution, the powers of the President are limited. Interview, March 10, 1972. In 1973, the Watergate tapes case renewed the controversy.

28. U.S. House, *Steel Seizure Case*, pp. 265–66.

29. A trial on the merits involves a determination of the legal rights of the concerned parties. Because all relevant information and arguments must be brought to the court's attention before a decision is made, a more leisurely pace prevails. In the steel case, the companies wanted the Justice Department to agree to an accelerated schedule for the filing of briefs so that the hearing on the merits could be held as soon as possible.

Harbaugh, in *Lawyer's Lawyer: The Life of John W. Davis*, p. 470, mistakenly asserts that the steel companies appeared in court on April 10 with motions for temporary and permanent injunctions. Harbaugh later states, p. 471, that the trial on the merits began on April 24, when, in fact, there never was a trial on the merits in the steel seizure case. What Harbaugh refers to was the hearing before Judge Pine on the companies' motions for a preliminary injunction.

30. U.S. House, *Steel Seizure Case*, p. 269.

31. *Ibid.*, p. 272.

32. Pine, who was a Democrat, lost his Justice Department position when the Harding administration came in.

33. Westin, *Anatomy of a Constitutional Law Case*, p. 55n.

34. Pittsburgh *Post-Gazette*, April 30, 1952, p. 7.

35. Editorial, Washington *Post*, June 15, 1970, p. 22.

36. *New York Times*, April 27, 1952, p. 51.

37. Pittsburgh *Post-Gazette*, April 30, 1952, p. 7; obituary, Washington *Post*, June 12, 1970, p. C4. Pine had gained some publicity as the judge who had presided at the trial of movie writer Dalton Trumbo, one of the "Hollywood 10" given a jail sentence for declining to inform the House Un-American Activities Committee of his possible Communist affiliations. Pittsburgh *Post-Gazette*, April 30, 1952, p. 7. In legal circles, Judge Pine had also acquired a bit of a reputation for absentmindedness when he almost sentenced a lawyer instead of his convicted client to jail. Pittsburgh *Post-Gazette*, April 30, 1952, p. 7.

38. Baldridge was not unlike Pine in both personality and demeanor and the press speculated that the two would come to no "dramatic, climactic clashes of temperament." *New York Times*, April 27, 1952, p. 51.

39. U.S. House, *Steel Seizure Case*, pp. 272–73.

40. *Ibid.*, p. 276. The Federal Rules of Civil Procedure for the United States District Courts, as Amended to April 30, 1951, (Rule 12a) provide that when a complaint is served on a private individual, he must file his answer to the complaint within twenty days after service upon him. The United States, or an

agency or officer of the United States, however, has sixty days to file an answer if it is a party to a suit. The judge can set a case down for trial only after the answer to the complaint is received. In the steel seizure case, there was a question of whether Charles Sawyer was sued as a private individual or as an officer of the United States. The steel companies addressed suits against him in both capacities. The government maintained that Sawyer was an officer of the United States and thus entitled to sixty days to file his answer. Judge Pine thought it unnecessary to decide that issue at the April 10 hearing. *Ibid.*

41. *Ibid.*

42. *Ibid.*, pp. 277–78.

43. The seven steel companies were Youngstown Sheet & Tube Co., U.S. Steel Co., Bethlehem Steel Corp., Republic Steel Corp., Jones & Laughlin Steel Corp., Armco Steel Corp., and E. J. Lavino & Co. Ten suits were filed, however. Some of the plaintiff companies chose to sue Sawyer individually and as Secretary of Commerce in one suit. Other plaintiffs filed separate suits, one against Sawyer personally and another in his official capacity. The moving papers were virtually the same in all the suits. The case numbers in the U.S. District Court for the District of Columbia were Civil Action Numbers 1550-52, 1539-52, 1549-52, 1581-52, 1624-52, 1625-52, 1635-52, 1647-52, 1700-52, and 1732-52.

44. Civil Action No. 1635-52: The Youngstown Sheet & Tube Co. v. Charles Sawyer, Secretary of Commerce, U.S., Complaint for Injunction and for a Declaratory Judgment—Filed April 14, 1952, U.S. House, *Steel Seizure Case*, p. 3. E. J. Lavino & Co. filed a like complaint but claimed additionally that its three plants which had been seized should not have been taken over by the government because they did not produce steel despite the fact that their workers belonged to the Steelworkers union. *Ibid.*, pp. 167–98.

45. *Ibid.*, pp. 4–5.

46. Notice of Special Appearance—Filed April 24, 1952, *ibid.*, pp. 210–11.

47. Defendant's Opposition to Plaintiffs' Motion for a Preliminary Injunction—Filed April 25, 1952, *ibid.*, p. 19.

48. *Ibid.*, pp. 19–20.

49. For a concise discussion of the relevant factors in considering preliminary relief, see *Virginia Petroleum Jobbers Ass'n. v. FPC*, 259 F.2d 921 (D.C. Cir., 1958).

50. The first part of the Justice Department's brief, a total of eighteen pages, was devoted to the equity problems involved in the case. The second part, a total of thirty-five pages, discussed the constitutional question in depth.

51. The importance in a constitutional law case of what arguments are presented, how they are presented, and by whom should not be underestimated. See Freund's lecture on " 'Judge and Company' in Constitutional Law" in *On Understanding the Supreme Court*, pp. 77–116.

Ironically, the Justice Department was probably able to brief the merits of the case in detail in the short time (less than two weeks) available only because it had a ready-made brief from a previous case, *United States* v. *Montgomery Ward*

& *Co.*, 150 F.2d 369 (7th Cir., 1945), which it could use. The brief in the *Youngstown* case repeated, almost word for word (with some changes in organization) an elaborate discussion of the President's constitutional power to seize private property in wartime. Compare Case No. 8765, Brief for the United States—Plaintiff-Appellant, *United States* v. *Montgomery Ward & Co., Inc.* (in the United States Circuit Court of Appeals for the Seventh Circuit), pp. 37–55, 63–68, 76–80, with the Memorandum of Points and Authorities in Opposition to Plaintiffs' Motion for Preliminary Injunction, *Youngstown Sheet and Tube Co., et al.* v. *Charles Sawyer* (in the U.S. District Court for the District of Columbia), pp. 40–60, Records, Department of Justice, Washington, D.C. (I am indebted to Stanley Temko, a member of the firm of Covington & Burling, which also represented U.S. Steel in the seizure litigation, for suggesting that Baldridge may have used the *Montgomery Ward* brief.)

Baldridge did, however, file a number of affidavits from government officials that stressed the country's need for steel and the possible harmful consequences to the war effort if a strike occurred. He did not make sufficient use of these affidavits. In the preparation of its brief and in its oral argument before the district court, the Justice Department apparently suffered from a lack of leadership and confidence. Various attorneys were responsible for different sections of the brief, and no one played an authoritative role in its formation. One argument, that the steel companies had an adequate remedy at law (i.e., they could be compensated for the seizure despite the fact that the government had not followed the precise statutory procedures prescribed before private property could be taken), based on *Hurley* v. *Kincaid*, 285 U.S. 95 (1932), for example, was prepared but was not going to be used in the brief until Paul Freund wrote to the department urging it to make that argument. (Interview with Herman Marcuse, Attorney, Department of Justice, May 12, 1971. See Memorandum in Opposition to Plaintiff's Motion for Preliminary Injunction, pp. 17–18.) It became apparent during the hearing that Baldridge had still not decided which was his strongest argument and, as a result, presented neither the equitable nor the constitutional points well.

52. U.S. House, *Steel Seizure Case*, p. 285.

53. See Judge Holtzoff's opinion, *ibid.*, p. 266.

54. Following the holding in *Hooe* v. *United States* (218 U.S. 322 (1910)), the taking of private property by an officer of the government without proper authorization from Congress cannot be construed as an act of the government, Kiendl stated. Moreover, in the subsequent case of *United States* v. *North American Transportation and Trade Co.* (253 U.S. 330 (1920)), the Supreme Court ruled that the government cannot be held liable for an unauthorized taking by one of its officials. Clearly, if the steel seizure was "wholly illegal and wholly unconstitutional, there can be no remedy under that line of cases," Kiendl declared. U.S. House, *Steel Seizure Case*, p. 287. The question of an adequate remedy at law was discussed, although no cases were mentioned, in the argument before Judge Holtzoff.

55. *Ibid.* According to the Supreme Court, Kiendl observed, Executive Orders come under the meaning of regulations in that provision.

56. Sawyer had appeared on "Meet the Press" on Sunday, April 20, 1952, and had categorically stated that there would be some wage increases made.

57. U.S. House, *Steel Seizure Case*, pp. 288–99.

58. *Ibid.*, pp. 302–5. The Supreme Court decisions relied on by Kiendl were *Larson* v. *Domestic & Foreign Commerce Corp.*, 337 U.S. 682 (1949); *Land* v. *Dollar*, 330 U.S. 731 (1947); *Ickes* v. *Fox*, 300 U.S. 82 (1937); *Williams* v. *Fanning*, 332 U.S. 490 (1947); *Marbury* v. *Madison*, 1 Cranch 137 (U.S. 1803).

59. U.S. House, *Steel Seizure Case*, p. 305.

60. *Ibid.*, pp. 308–9. Judge Pine was unimpressed with Kiendl's learned exposition: "Do you think this argument is serving a useful purpose?" he inquired. *Ibid.*, p. 308.

61. 330 U.S. 258 (1947).

62. U.S. House, *Steel Seizure Case*, p. 311.

63. *Ibid.*, p. 313.

64. "I can't understand Mr. Kiendl's position when he asks me to find the act illegal, and yet he wants to continue the illegality. That is the reason I was astonished when he told me that that was all you were asking, because it seems inconsistent to me," Judge Pine exclaimed. *Ibid.*, p. 314.

65. *Ibid.*, pp. 315–16.

66. *Ibid.*, p. 318. Kiendl's inflexibility surprised the other attorneys representing the steel industry. It was obvious to them that Kiendl should have changed to a request for full relief after Judge Pine so clearly indicated in what direction the court was headed.

67. *Ibid.*, pp. 327–28.

68. *Ibid.*, p. 331.

69. *Ibid.*, pp. 338–40.

70. 272 U.S. 52 (1926).

71. *Ibid.*, pp. 163–64.

72. U.S. House, *Steel Seizure Case*, pp. 340–41. Although the *Myers* case dealt with the President's removal power, Chief Justice Taft did make certain general statements on executive power in his opinion for the Court which gave credence to the theory of inherent power. Taft, for example, endorsed the Hamiltonian construction of Article II of the Constitution, that all executive power was granted to the President except for the express limitations enumerated throughout the document. 272 U.S. 52, 118 (1926).

In his argument Tuttle did not mention *Humphrey's Executor* v. *United States* (295 U.S. 602 (1935)) which qualified *Myers*. For a discussion of these cases see William E. Leuchtenburg, "The Case of the Contentious Commissioner," pp. 276–312.

73. Article I, Section 8, Clause 18.

74. U.S. House, *Steel Seizure Case*, pp. 343–45.

75. *Ibid.*, p. 346.

76. John J. Wilson, who had intended to discuss the difference between inherent and implied power, agreed to hold his argument for rebuttal. Randolph Childs, counsel for E. J. Lavino & Co., observed that his argument was specific to his company; it would be better, therefore, to hear both sides on the general issues before he presented Lavino's case. *Ibid.*, pp. 358–59.

77. *Ibid.*, p. 363.

78. *Ibid.*, p. 364.

79. See discussion, *ibid.*, pp. 364–66.

80. Affidavits were filed by the following officials: Robert A. Lovett, Secretary of Defense; Gordon Dean, chairman of the Atomic Energy Commission; Manly Fleischmann, administrator of the Defense Production Administration; Henry Fowler, administrator of the National Production Authority; Oscar Chapman, Secretary of the Interior; Jess Larson, administrator of General Services and defense materials procurement administrator; Homer King, acting administrator of the Defense Transport Administration; Charles Sawyer, Secretary of Commerce. For the texts of the affidavits see *ibid.*, pp. 27–59.

81. Many years later, Baldridge complained that Judge Pine would let him talk about nothing but the question of the constitutional power to seize. Pine was not interested in the equitable issues and would not allow reference to the government's affidavits, according to Baldridge. (Note, Holmes Baldridge to Nancy Dixon Bradford, October 9, 1964; Note, Holmes Baldridge to Nancy Dixon Bradford, undated, Box 1, Baldridge Papers, TL.) Although his recollection is faulty about the discussion of the affidavits, it was accurate as to Pine's overriding interest in the constitutional aspects of the steel seizure case.

Baldridge also unjustly accused the steel industry's attorneys of concentrating, to the exclusion of all else, on the powers of the executive. (Note, Baldridge to Nancy Dixon Bradford, undated.)

82. U.S. House, *Steel Seizure Case*, p. 371. It should be noted that Judge Pine always referred to Secretary Sawyer's power to seize while Baldridge discussed the *President's* power. Pine obviously was determined to stick with the legal fiction that the steel industry's suits ran against Sawyer.

83. *Ibid.*

84. *Ibid.*, pp. 372–74. The cases Baldridge cited were *United States* v. *Pewee Coal Co.*, 341 U.S. 114 (1951), and *Employers Group of Motor Freight Carriers, Inc., et al.* v. *National War Labor Board, et al.*, 143 F.2d 145 (1944).

Newspaper articles describing the first day of the hearing noted Baldridge's lack of preparation to discuss these issues. See, for example, New York *Herald Tribune*, April 25, 1952, pp. 1, 13.

85. 17 Fed. Cas. No. 9487 (1861).

86. U.S. House, *Steel Seizure Case*, pp. 374–76. See also Rossiter, *The Supreme Court and the Commander in Chief*, pp. 20–26. According to Rossiter, "The one great precedent is what Lincoln did, not what Taney said. Future Presidents will know where to look for historical support. So long as public opinion sustains the President, as a sufficient amount of it sustained Lincoln in his shad-

owy tilt with Taney and throughout the rest of the war, he has nothing to fear from the displeasure of the courts. . . . The law of the Constitution, as it actually exists, must be considered to read that in a condition of martial necessity the President has the power to suspend the privilege of the writ of habeas corpus. The most a court or judge can do is read the President a lecture based on *Ex parte Merryman.*" *The Supreme Court and the Commander in Chief,* p. 25.

Although Rossiter's interpretation might have sustained Baldridge's position in regard to the hypothetical situation posited by Judge Pine, it offered little support in the circumstances of the steel case. Rossiter's view is dependent upon two factors, martial necessity and favorable public opinion (including that of Congress), both of which were absent in 1952. Baldridge's argument that the Korean War could be equated with the Civil War or World War II did not impress the courts or the public, which did not suffer during the Korean hostilities as they had during previous wars. In addition, Congress eventually ratified Lincoln's suspension of the writ (Habeas Corpus Act of 1863) and gave him the power to continue to suspend it; and, during World War II, Congress ratified Roosevelt's seizures based on constitutional authority only, by passing the War Labor Disputes Act giving him statutory power to seize. In 1952, however, there was little chance of congressional approval of the steel seizure.

Moreover, Baldridge's choice of *Ex parte Merryman,* emphasizing as it did executive defiance of the judiciary, could hardly have been expected to win favor in the eyes of Judge Pine.

87. U.S. House, *Steel Seizure Case,* p. 376.

88. *Ibid.,* p. 377.

89. Baldridge later realized, however, the error he had made. Consequently, he filed a supplemental memorandum with the court in order to clarify the government's position with regard to the question of executive power: "We feel that a further statement is justified and perhaps necessitated by misunderstandings which may have arisen during the course of oral argument.

"At no time have we urged any view that the President possesses powers outside the Constitution, and our brief . . . is clear on that point. On the contrary we have urged that the President must act within the Constitution, specifically Article II. We have shown, by historic executive usage, legislative recognition, and judicial precedent that the President possesses the constitutional power and duty to take action in a grave national emergency such as existed here. Beyond this, of course, we do not go. If the Court understood us to say more, we respectfully ask that this memorandum be accepted as the accurate statement of our views." Supplemental Memorandum of Defendant, *Youngstown Sheet & Tube Co. and Other Steel Companies* v. *Charles Sawyer,* Box 1, Baldridge Papers, TL.

The Justice Department filed the memorandum only a few hours before Judge Pine handed down his opinion. *New York Times,* April 30, 1952, p. 19.

90. 4 Wall. 475 (1867).

91. U.S. House, *Steel Seizure Case,* pp. 380-82.

92. *Ibid.,* p. 382. Secretary Sawyer's memoir, *Concerns of a Conservative Democrat,* contains a misleading account of Baldridge's unwillingness to commit

the government to maintenance of the status quo. Sawyer noted that "During the hearing of this case, Judge Pine requested the Attorney General [sic, Assistant Attorney General] to secure a statement from me, agreeing to make no changes in wage schedules before the case was decided. I tried several times to get such assurance to the judge. The Solicitor General, Phillip [sic, Philip] Perlman, acting as my counsel, refused to pass this word on. When I mentioned this to the President, he said he was surprised and would talk to Perlman. I was in Cincinnati on April 26 when my solicitor [in the Commerce Department], C. Dickerman Williams, phoned to say it was of the greatest importance that I get this message to Judge Pine; on three separate occasions he had asked for such a commitment. I told Williams to tell Holmes Baldridge . . . to assure the judge that I would make no change while his decision was pending. But the Solicitor General refused to let Baldridge deliver this message. I then instructed Williams to set forth my position in a letter to Baldridge. Only then did Perlman get word to the judge" (*Concerns of a Conservative Democrat*, p. 261).

This description of events is marred by errors of fact and interpretation, which appear to be chargeable to Baldridge. As has been seen from the discussion in the text of the oral argument before Judge Pine, the question of maintenance of the status quo did come up three times but never in the manner reported by Sawyer. Once Pine asked Baldridge if the government would agree to preserve the status quo, since that was all the relief U.S. Steel was requesting. The judge clearly did not expect an affirmative answer and Baldridge explained to the judge that if the government made such an agreement U.S. Steel would be in the position of having its cake and eating it too. The seizure would be just like a Taft-Hartley injunction: no wage increases could be made but the workers could not go on strike. The seizure could go on indefinitely; the steel companies would have nothing to lose in that situation (see U.S. House, *Steel Seizure Case*, pp. 359–61). But Baldridge, in a letter written many years after the steel seizure, gave more importance to this request for maintenance of the status quo than it deserved. By claiming that the other steel companies followed the lead of U.S. Steel, Baldridge made it sound as if the whole case could have been concluded instantly if he had had the authority to commit the government to preservation of the status quo (Note, Baldridge to Nancy Dixon Bradford, undated). Sawyer follows Baldridge in this interpretation (*Concerns of a Conservative Democrat*, p. 373 n8).

Twice more during the hearing the subject of maintenance of the status quo was discussed but only in reply to requests from Baldridge for more time to file additional briefs. Judge Pine never asked Baldridge to get a statement from Sawyer that wages and prices would not be increased while the judge was considering the case. Pine simply said that if Baldridge wanted more time before the court ruled, he would have to promise to preserve the status quo (U.S. House, *Steel Seizure Case*, pp. 364–66, 380–82).

Apparently what occurred was that after the hearing was adjourned on Friday, April 25, Baldridge decided that strategically it would be a good idea to inform Judge Pine that the government would not make any changes in terms and conditions of employment while he was considering the case. Thus, Baldridge called C. Dickerman Williams to see if he could get such a commitment from Secretary Sawyer. Williams, after his conversation with Baldridge, informed Sawyer that Baldridge thought it would be a serious mistake, bordering on contempt of court, to grant a wage increase before Pine ruled (Note, C. D. Williams to Holmes Baldridge, April 26, 1952, Sawyer Papers). Secretary Sawyer said Baldridge was free to give such assurance to the judge. Williams gave this information to Baldridge. Later in the day, Baldridge talked with Williams again and said that Solicitor General Perlman, as head of the Justice Department, had refused to give Judge Pine this commitment but that Perlman had given permission to Baldridge to do so if he wished. Baldridge had not yet decided what to do. Williams called him again the next morning and Baldridge told him that he would not convey Secretary Sawyer's statement to the judge.

Sawyer, when told about these conversations, ordered Williams to write a letter to Baldridge making clear the Secretary's willingness to give Judge Pine the assurance he apparently wanted and announcing that, in view of Baldridge's suggestion and advice, he would not make any changes in the terms and conditions of employment until Judge Pine ruled (Note, C. D. Williams to Baldridge, April 26, 1952; Confidential Dictation, April 28, 1952, Sawyer Papers).

Judge Pine never was told that Secretary Sawyer had agreed to preserve the status quo. Sawyer's statement to the contrary (in the passage quoted from *Concerns of a Conservative Democrat*, p. 261) is difficult to comprehend because in Sawyer's own confidential dictation he had noted that, "Perlman refused to give the Judge the assurance which I had authorized him to give and that the matter was therefore left, so far as the Judge was concerned, in a state of complete uncertainty" (Confidential Dictation, April 28, 1952, Sawyer Papers). Newspaper accounts corroborate that Sawyer's assurance was not passed on to Judge Pine (*New York Times*, May 9, 1952, p. 16; May 11, 1952, p. E9). William H. Harbaugh's *Lawyer's Lawyer: The Life of John W. Davis* contains a chapter on the steel seizure case in which Harbaugh follows in toto Sawyer's misleading description of the Pine hearing with regard to the requests for maintenance of the status quo and subsequent events (see *Lawyer's Lawyer*, p. 471).

When this series of events was reported in the *New York Times*, reporters at first could not identify who was responsible for ordering Sawyer's assurance withheld from Judge Pine. Subsequent articles, however, stated that "the White House had countermanded Secretary Sawyer" (May 9, 1952, p. 16; May 11, 1952, p. E9; Westin, *Anatomy of a Constitutional Law Case*, p. 68). Sawyer, however, claimed that he had told President Truman about Perlman's refusal to give the assurance and Truman had agreed with Sawyer's position (Confidential Dictation, April 28, 1952).

There is evidence that some members of the White House staff were anxious to implement some of the WSB recommendations regardless of when Judge Pine ruled. A memorandum from Harold Enarson to Charles Murphy, dated April 29, 1952, discussed a draft of a memorandum addressed to Secretary Sawyer stating that the WSB recommendations must be put into effect and Secretary Sawyer should start the machinery rolling (Box 28, Murphy Files, TL). In addition, during the afternoon that Judge Pine handed down his decision, some of the

White House staff were meeting with officers of ESA to plan for institution of parts of the WSB's proposed settlement (McConnell, *Steel Seizure of 1952*, p. 42).

93. 135 U.S. 1 (1890).

94. 158 U.S. 564 (1895).

95. U.S. House, *Steel Seizure Case*, p. 387.

96. 341 U.S. 114 (1951).

97. U.S. House, *Steel Seizure Case*, pp. 387–89. Edward S. Corwin supports Baldridge's view of this case. See U.S. Senate, *The Constitution of the United States of America*, pp. 494–95.

98. U.S. House, *Steel Seizure Case*, pp. 390–91. Even Senator Taft recognized that too much time elapsed between the appointment of a fact-finding board and the issuance of an injunction under the Taft-Hartley Act. In 1949, Taft supported an amendment that, if enacted, would have given the Attorney General the power to seek an injunction immediately after the President proclaimed that a strike would endanger national health or security. The Attorney General would not have had to wait for a report from the emergency fact-finding board. See note, Thomas E. Shroyer [Senator Taft's administrative assistant] to Senator H. Alexander Smith, May 29, 1952, Box 1013, Robert A. Taft Papers, LC.

99. U.S. House, *Steel Seizure Case*, p. 392. In support of Baldridge's argument, see Jerre Williams, "The Steel Seizure: A Legal Analysis of a Political Controversy," p. 38.

100. Theodore Kiendl, however, reiterated his position that an injunction prohibiting any wage increases was the relief U.S. Steel was seeking until a trial on the merits could be held. U.S. House, *Steel Seizure Case*, p. 411. Despite all that had transpired during the two-day-long hearing, Kiendl gave no evidence of having understood Judge Pine's approach to the case.

101. 295 U.S. 602 (1935); U.S. House, *Steel Seizure Case*, pp. 402–4. The rule of stare decisis stipulates that the holding in a specific case creates a precedent which should be followed in subsequent cases concerning similar questions. The holding in a case, however, is usually established on the narrowest possible legal grounds, and any further opinions expressed by the Justices in discussion of the case are not held to create a precedent and are called dicta.

To the consternation of the other attorneys representing the steel companies, Wilson also discussed, in great detail, Justice McReynolds' dissent in the *Myers* case, which rejected Chief Justice Taft's interpretation of the Constitution in regard to the executive power. Wilson concluded that in the *Humphrey* case, the Supreme Court, in a unanimous decision, "must have accepted the theories of Mr. Justice McReynolds as he expressed them in the *Myers* case in his dissent" (U.S. House, *Steel Seizure Case*, p. 406). Since McReynolds' views were hardly in vogue in 1952, the steel industry lawyers could not imagine why Wilson had invoked him as authority so often. Later they learned that Judge Pine had been a special assistant to McReynolds when the latter was Attorney General. Interview with John Pickering, August 19, 1972; interview with Stanley Temko, January 16, 1974.

102. U.S. House, *Steel Seizure Case*, pp. 416–17. The case cited is *House v. Mayes*, 219 U.S. 270, 281 (1911).

103. 13 Wall. 623 (1871).

104. 120 U.S. 227 (1887).

105. U.S. House, *Steel Seizure Case*, p. 420. The *Wall Street Journal* commented editorially on Baldridge's response: "It is not the first time—nor will it be the last—when a man under pressure told the naked truth." April 29, 1952, p. 8.

106. Before the hearing adjourned, Judge Pine heard argument on behalf of E. J. Lavino & Co. The thrust of Randolph Childs' presentation was that no labor controversy existed at the seized plants of the Lavino Company. Discussion of this point was the occasion for an amusing but troublesome, from the government's point of view, exchange between Childs and Judge Pine:

"Mr. Childs: Suppose, for example, that there had been no controversy at all with any employer, and the President had made a proclamation, or statement saying there was a controversy. He could not manufacture one.

The Court: He could not what?

Mr. Childs: I say he could not manufacture a controversy, because none ever existed.

The Court: The Attorney General claims that the President has unlimited powers.

Mr. Childs: Well, with all due respect, I think that argument is something like my suit. I could not get a taxicab at noon and it is all wet." U.S. House, *Steel Seizure Case*, p. 424.

107. Harold Enarson, "The Steel Strike—A Modern Tragedy," July 9, 1952, p. 6, Box 5, Enarson Papers, TL.

Baldridge later confessed that he had made a terrible argument, but he blamed the press and Judge Pine for distorting the Justice Department's position by the use of ridiculous hypothetical situations, so that instead of insisting only that the steel seizure was justified under the general welfare clause of the Constitution, Baldridge appeared to be arguing that the President's power was unlimited. Note, Baldridge to Nancy Dixon Bradford, October 9, 1964.

108. Enarson, "The Steel Strike—A Modern Tragedy," p. 6, Box 5, Enarson Papers, TL.

109. Editorial, New York *Post*, April 29, 1952, p. 27. See also editorial, *New York Times*, April 29, 1952, p. 26.

110. *Congressional Record*, 98:4469–79.

111. *Ibid.*, pp. 4472–73.

112. Enarson, "The Steel Strike—A Modern Tragedy," July 9, 1952, p. 6, Box 5, Enarson Papers, TL.

113. Letter to C. S. Jones in Response to Questions on the Steel Situation, April 27, 1952, U.S. President, *Public Papers 1952–1953*, p. 301. The letter begins on p. 299.

Truman later complained that few newspapers in the country printed the whole letter or even described its main points for their readers. Note, Harry S. Truman to Vic H. Householder, May 26, 1952, OF 407B, Truman Papers, TL.

114. 103 F. Supp. 569; U.S. House, *Steel Seizure Case*, p. 66.

115. *New York Times*, April 30, 1952, p. 19. See also *Time*, May 12, 1952, p. 19. Compare the exciting but highly imaginative accounts in Westin, *Anatomy of a Constitutional Law Case*: "To a packed courtroom on April 29, Judge Pine read his opinion" (p. 68), and Harbaugh, *Lawyer's Lawyer*: "In a courtroom jammed with newspapermen poised to run for the telephone, Pine declared . . ." (p. 472).

116. U.S. House, *Steel Seizure Case*, p. 68. At this point Pine made no reference to the government's argument that the legal merits of a case should not be reached on a motion for preliminary injunction. Further along in his opinion, he noted: "As to the necessity for weighing the respective injuries and balancing the equities, I am not sure that this conventional requirement for the issuance of a preliminary injunction is applicable to a case where the Court comes to a fixed conclusion, as I do, that defendant's acts are illegal. On such premise, why are the plaintiffs to be deprived of their property and required to suffer further irreparable damage until answers to the complaint are filed and the cases are at issue and are reached for hearing on the merits? Nothing that could be submitted at such trial on the facts would alter the legal conclusion I have reached" (*ibid.*, p. 74). Pine did not cite any cases to support his departure from the "conventional requirement," whereas the Justice Department's brief cited many important precedents for its point of view (*Youngstown Sheet & Tube Co., et al.* v. *Charles Sawyer*, Memorandum of Points and Authorities in Opposition to Plaintiffs' Motion for Preliminary Injunction, pp. 7–8).

117. In fact, there were two statutes giving the President seizure power in order to ensure that the defense needs of the nation were met, although neither mentioned seizure as a remedy in labor disputes: the Selective Service Act of 1948 and the Defense Production Act of 1950 as amended in 1951. These acts described certain procedures to be followed before the seizure could be consummated. Because these procedures were not followed by the administration, the Justice Department did not rely on these statutes. The existence of these acts, however, arguably put the steel seizure in a different light. There had been previous cases where the end action of the government had been sanctioned by Congress but the specific administrative procedures prescribed had not been followed, and the Supreme Court had upheld the government. See especially *Hurley* v. *Kincaid*, 285 U.S. 95 (1923).

118. U.S. House, *Steel Seizure Case*, p. 69.

119. William Howard Taft, *Our Chief Magistrate and His Powers*, pp. 139–40. Quoted in U.S. House, *Steel Seizure Case*, p. 70.

120. U.S. House, *Steel Seizure Case*, p. 70.

121. *Ibid.*, p. 72. Pine did not mention *United States* v. *Midwest Oil Co.*, 236 U.S. 459 (1915), where the Supreme Court had approved an assertion of

executive power not expressly authorized by the Constitution or by statute, because the power had been used repeatedly by Presidents and had never been repudiated by Congress. This case was a source of difficulty for the steel companies, both because of its holding and because John W. Davis, who would present the industry case before the Supreme Court, had been the solicitor general who argued and won the *Midwest Oil* case for the government.

122. U.S. House, *Steel Seizure Case*, p. 73.

123. *Ibid.*, pp. 73–74.

124. *Ibid.*, p. 75.

125. *Ibid.*, p. 74. One of the attorneys for Bethlehem Steel Company, Bruce Bromley, later admitted that he was surprised that Judge Pine was persuaded by the irreparable injury argument. In fact, the seizure was token; the steel companies were in full control of their operations and, if wages had been increased without appropriate price relief, the damage would have been compensable by money. Interview with Bruce Bromley, June 30, 1972.

126. U.S. House, *Steel Seizure Case*, pp. 75–76.

127. *Ibid.*, p. 76.

128. *Ibid.* Counsel for U.S. Steel typed a new motion in the courthouse immediately after Judge Pine released his opinion. From that point on, U.S. Steel's case was joined with the other plaintiff steel companies. Interview with Stanley Temko, December 11, 1973.

SEVEN ★ THE STEEL SEIZURE CASE IN THE COURT OF APPEALS

1. Quoted in Beverly Smith, "What a Spanking He Gave Truman," *Saturday Evening Post*, August 2, 1952, p. 27.

2. *Ibid.*, p. 65.

3. Editorial, April 30, 1952, p. 26.

4. April 30, 1952, p. 16.

5. Editorial, May 1, 1952, p. 4. For additional editorials praising Judge Pine, see on April 30, Los Angeles *Times*, Part II, p. 4; New York *Herald Tribune*, p. 18 (this was one of a few editorials which cautioned that Pine's decision was not final and that experts disagreed on the points of law involved); *Wall Street Journal*, p. 8; Washington *News*, p. 38; and Washington *Evening Star*, p. 14; and on May 1, Pittsburgh *Post-Gazette*, p. 10; Cleveland *Plain Dealer*, p. 10; *Wall Street Journal*, p. 10; Chicago *Sun-Times*, p. 29; Chicago *Tribune*, p. 14; Washington *Times-Herald*, p. 16; and Cincinnati *Enquirer*, p. 4.

Arthur Krock, in his *New York Times* column of May 1, 1952, p. 28, declared that Pine's opinion contained the "most precise and firmest restraints on Executive power that have been stated by a federal court in our history." The Pine decision was vital to the operation of the American government, Krock maintained, and therefore the desire was "general here that the Supreme Court will not contract the broad issue."

6. April 30, 1952, p. 8.

7. "Cry Wolf? Claim of 'Emergency' to Justify Seizure of Steel Plants Was Open to Questioning," *Wall Street Journal,* April 30, 1952, p. 8.

8. *Wall Street Journal,* April 30, 1952, p. 8.

9. New York *Post,* April 30, 1952, p. 48.

10. John P. Roche, another critic of Pine's ruling, in an article written later in 1952, observed that Pine's opinion read "more like a Liberty League tract than a realistic appraisal of the duties and responsibilities of the President of the United States in an era of permanent crisis." "Executive Power and Domestic Emergency," p. 613.

11. See, for example, the Washington *Post* editorial, April 30, 1952, p. 16. The *Post* thought the Steelworkers were unpatriotic to call a strike. The editorial writer apparently was persuaded by the argument in Pine's opinion that if an emergency existed the union would not order a work stoppage. When Judge Pine was informed that the Steelworkers had walked off the job, he was greatly surprised. See McConnell, *Steel Seizure of 1952,* p. 42.

12. Editorial, April 30, 1952, p. 2C. See, too, the New York *Post,* editorial, April 30, 1952, p. 49.

13. May 1, 1952, p. 24.

14. Two newspapers, the New York *Post* and the St. Louis *Post-Dispatch,* rejected the employment of Taft-Hartley on the ground that every objective of the Taft-Hartley Act had already been met and no further gain could be expected from its imposition.

15. See *New York Times,* April 30, 1952, p. 20, and *Congressional Record,* 98:4614–15, 4644–49. The most vocal of the President's defenders, Senator Morse, announced that he would wait for the Supreme Court to rule before he gave up his and President Truman's theory of executive power, since in his view, Judge Pine's opinion lacked a thorough, scholarly discussion of the legal problems. *Congressional Record,* 98:4678.

16. *New York Times,* May 1, 1952, p. 24.

17. See the remarks of Rep. Jacob Javits, *Congressional Record,* 98:4630, and Senator Morse, *ibid.,* p. 4682.

18. Memorandum to Files, April 30, 1952, Box 5, Enarson Papers, TL.

19. *Ibid.*

20. *New York Times,* April 30, 1952, p. 19.

21. Memorandum to Files, April 30, 1952, Box 5, Enarson Papers, TL. See also Murray Kempton, "Bitter and Weary," New York *Post,* April 30, 1952, p. 47.

22. Memorandum to Files, April 30, 1952, Box 5, Enarson Papers, TL.

23. Sawyer, *Concerns of a Conservative Democrat,* p. 262. Sawyer also informed Truman that he was aware that if Taft-Hartley were invoked, it would be an admission by the administration of the usability of that act, but in the circumstances the production of steel took precedence over other considerations.

24. *Ibid.,* p. 263.

25. Memorandum to Files, April 30, 1952, Box 5, Enarson Papers, TL.

26. Attending the meeting, which began when news of the Pine ruling was received, were Enarson, Stowe, Kayle, Murphy, Bell, Lloyd, and Neustadt of the White House, Ellis Lyons and some unidentified attorneys from the Justice Department, and, later that evening, Philip Perlman and Holmes Baldridge. *Ibid.*

27. *Ibid.* See also *New York Times*, April 30, 1952, pp. 1, 19.

28. Memorandum to Files, April 30, 1952. Charles Sawyer, who would again be in charge of the mills if the stay were granted, was upset that his advice on the Taft-Hartley Act had not been taken and that decisions that affected him were being made without consulting him. Sawyer complained to John Steelman and asked to be included in any further discussions so that he would not be presented with a fait accompli. Notes on telephone conversation between Steelman and Sawyer, April 30, 1952, 1:15 P.M., Sawyer Papers.

29. The Federal Rules of Civil Procedure (Rule 62(c)) require that the party granted a preliminary injunction submit a bond to insure the losing party against damages if he should ultimately win the case. Before Pine signed the orders, he had to decide the amount of the bond each steel company would submit. Industry attorneys argued that, to comply with the rules, the bond need only be nominal, for, in the unlikely event that Pine's decision were to be reversed, there would be no damages to Secretary Sawyer. Holmes Baldridge maintained, on the contrary, that the damage resulting from Judge Pine's opinion was incalculable to the President and to the public because of the steel strike. Pine reminded Baldridge that the President and the public were not parties to the case. In the judge's opinion no damage would be caused to Secretary Sawyer, and the judge therefore set the bond for each company at $100. U.S. House, *Steel Seizure Case*, pp. 429–34.

30. *Ibid.*, pp. 434–39. While the government's appeal was pending, it could petition the Supreme Court for a writ of certiorari before judgment in the court of appeals.

31. *Ibid.*, p. 439. It is clear that the government had no expectation that Judge Pine would grant a stay. The sole ground of the Justice Department's application was the need for steel so as to maintain the national security, and the application cited the union's threat to continue the strike if possession of the steel plants were not returned to the government. This was hardly an argument that could be expected to win sympathy from Judge Pine.

32. *Ibid.*, pp. 442–43. The ten suits filed in the district court were identified in the court of appeals as Nos. 11, 404–13.

33. Usually a case is heard in the court of appeals by a three-judge panel. But Chief Judge Stephens had met with the associate judges at 12:30 P.M. on April 30, to discuss the seizure case and they had voted 5 to 4 to hear argument and to rule on the stay application *en banc*. Judges Edgerton, Proctor, Bazelon, Fahy, and Washington approved the *en banc* hearing; Stephens, Clark, Miller, and Prettyman opposed it. Daily Time Sheets, April 30, 1952, Box 2, Stephens Papers, LC; Prettyman Diary, May 1952, p. 5, Box 1, Prettyman Papers, LC.

The May 1952 diary is a special segment of the diaries containing Judge Prettyman's observations on the *Steel Seizure Case*.

34. The presence of Perlman, Acting Attorney General of the United States and its solicitor general, was unusual although not unprecedented. Clearly, the administration had determined to improve its courtroom performance.

35. Holmes Baldridge took the position that, for the purpose of obtaining a stay, the validity of the President's executive order was not a proper subject for discussion. But on application for a stay it is normal practice for the petitioner to try to convince the court that he will ultimately prevail on the merits. See Stenographic Record of the Proceedings before the United States Court of Appeals for the District of Columbia Circuit, April 30 and May 1, 1952, p. 23. I am indebted to Stanley Temko, of Covington and Burling, Washington, D.C., for a copy of the transcript of this oral argument, which is not printed.

36. *Ibid.*, pp. 51–54. Judge Prettyman explained in his diary that the fact that, in the district court, U.S. Steel had asked only for an injunction against any change in employment terms led him to question the irreparable injury argument of the companies: If the companies were really being harmed by the seizure, lawyers of the caliber of those representing U.S. Steel would never have asked for relief that did not enjoin the seizure. Prettyman Diary, May 1952, p. 14, Box 1, Prettyman Papers, LC.

37. Rothe and Lohr, *Current Biography* . . . 1952, pp. 465–67; obituary, *New York Times*, August 1, 1960, p. 23.

38. Stenographic Record of the Proceedings before the United States Court of Appeals, p. 88.

39. *Ibid.*, p. 89. See also Washington *Post*, May 1, 1952, p. 1.

40. Stenographic Record of the Proceedings before the United States Court of Appeals, pp. 93–97.

41. Judges Prettyman, Bazelon, Fahy, and Washington formed the majority with Edgerton.

42. The minority consisted of Judges Stephens, Clark, Wilbur K. Miller, and Proctor. See Order–Filed April 30, 1952, U.S. House, *Steel Seizure Case*, p. 444. Judge Prettyman indicated that the debate in conference was very bitter. Prettyman Diary, May 1952, pp. 6–7, Box 1, Prettyman Papers, LC.

43. Washington *Post*, May 1, 1952, p. 2.

44. Application to Attach Condition to Stay in Order to Protect Appellees from Irreparable Injury—Filed May 1, 1952, U.S. House, *Steel Seizure Case*, pp. 444–45.

45. Stenographic Record of the Proceedings before the United States Court of Appeals, pp. 104–8. Judge Prettyman commented that Westwood was "truly magnificent." Prettyman Diary, May 1952, p. 7, Box 1, Prettyman Papers, LC.

46. Stenographic Record of the Proceedings before the United States Court of Appeals, p. 123.

47. *Ibid.*, p. 129. In fact, Truman opened his conference with the state-

ment: "I don't intend to answer any questions on the steel controversy." A wage increase was not announced. Truman did indicate during the meeting with the press, however, that he would abide by any court decision. He was not a dictator, he stated. The President's News Conference of May 1, 1952, U.S. President, *Public Papers 1952–1953*, pp. 306ff.

48. Stenographic Record of the Proceedings before the United States Court of Appeals, pp. 129–30.

49. *Ibid.*, p. 142. On the afternoon of May 1, however, Perlman dealt with the matter of a wage order. He sent the following note to Roger Putnam:

"I have your letter of April 28, 1952, transmitting a copy of a proposed 'Order No. 2' to be issued by the Secretary of Commerce, for the establishment of terms and conditions of employment in certain steel plants now in the possession of the Secretary of Commerce pursuant to Executive Order No. 10340. The proposed order, as your letter states, was drafted by you at the request of the Secretary of Commerce. . . .

I presume that you are aware that I stated to the Court of Appeals for the District of Columbia Circuit this morning that the Government would take no action with respect to terms and conditions of employment in the seized steel plants until the filing of a petition for certiorari in the Supreme Court by the Government. Such a petition will be filed prior to 4:30 P.M. tomorrow." Department of Justice Files, Washington, D.C.

Perlman apparently approved the wage order the following day and recommended that it be put into effect immediately. Charles Sawyer stated in a letter to the President, May 20, 1952, that "the Attorney General finally approved the recommendations of Mr. Putnam and suggested that I put them into effect immediately. In the light of the stay issued by the Supreme Court one day later [May 3, 1952], it is clear to me that such a course at that time would have been the height of folly." This letter is contained in Memorandum for Sawyer from the President, May 26, 1952, Box 136, President's Secretary's Files, Truman Papers, TL.

50. Stenographic Record of the Proceedings before the United States Court of Appeals, pp. 159–60. For a copy of the order, see U.S. House, *Steel Seizure Case*, p. 446. Judge Prettyman indicated that the vote in conference had been 4 to 4 and that he had cast the deciding vote. Prettyman voted against attaching conditions to the stay, because he believed that in the short period of time that the stay would be in effect the government would not change employment terms and because he did not wish to tie the President's hands without knowing all the facts. Prettyman Diary, May 1952, pp. 8–10, Box 1, Prettyman Papers, LC.

51. Opinion—Filed May 2, 1952, U.S. House, *Steel Seizure Case*, p. 448.

Holmes Baldridge, in three separate letters written to his niece for her use in a master's essay, stated that the court of appeals had decided the steel case on the merits, 5 to 4, thus reversing Judge Pine's ruling. Baldridge even went so far as to say that the Supreme Court had ignored the court of appeals action on the merits and had addressed itself only to Judge Pine's decision. Holmes Baldridge to Nancy Dixon Bradford, October 9, 1964, December 22, 1964, and undated, Box 1, Baldridge Papers, TL.

Ms. Bradford, in her master's essay, repeats many times Baldridge's error, citing his letter as authority. See "The 1952 Steel Seizure Case," unpublished M.A. thesis (Political Science), University of Southern California, January 1967, pp. 3, 126, 176.

52. 13 Wall. 623 (1871).

53. 341 U.S. 114 (1951).

54. 321 U.S. 414 (1944).

55. 13 Wall. 623, 627–28 (1871). See Opinion—Filed May 2, 1952, U.S. House, *Steel Seizure Case*, pp. 447–48.

56. 341 U.S. 114, 117 (1951). See Opinion—Filed May 2, 1952, U.S. House, *Steel Seizure Case*, p. 448.

57. *Yakus* v. *United States*, 321 U.S. 414, 440 (1944). See Opinion—Filed May 2, 1952, U.S. House, *Steel Seizure Case*, pp. 448–49.

58. Opinion—Filed May 2, 1952, U.S. House, *Steel Seizure Case*, p. 449. At the conclusion of the opinion was a statement that the minority judges dissented from the views expressed by the majority.

Chief Judge Stephens was highly exercised by the manner in which the majority had filed its opinion. To a friend, Stephens wrote: "The majority seemed to me to have been actuated throughout the case and in filing the 'statement' in the manner in which they filed it by something in the nature of confusion and bewilderment or emotionalism, rather than cool judicial considerations. This is the first instance in nearly 17 years that I have been on this court in which a majority did not submit to the minority in advance any proposed opinion or statement. This is the practice in all of the Circuit Courts of Appeals so far as I am informed." Harold M. Stephens to Frank E. Holman, Esq., May 16, 1952, Box 89, Harold M. Stephens Papers, LC.

To a conservative Michigan congressman Judge Stephens explained his views on the merits of the steel seizure: "Inherent power under our theory of Government rests only in the people. The Government has only such powers as are delegated by the Constitution. The 10th amendment carefully reserves to the people or the States powers not delegated. The duty of the President under the Constitution is to see that the laws are faithfully executed. In seizing the Steel Companies the President did not even purport to operate under a law, and there is no law which would authorize the seizure—the President had no 'inherent' power. His act was simply executive usurpation. I am frankly shocked that the majority of the members of this court, or indeed any member, takes any different view from the foregoing." Harold M. Stephens to Hon. Earl G. Michener, May 6, 1952, Box 89, Harold M. Stephens Papers, LC.

Stephens also confessed that he had mixed feelings about the steel case going directly from the district court to the Supreme Court. He thought that the courts of appeals should have a chance to "express themselves in cases of national importance as well as in others." He complained that the Supreme Court had taken away from the court of appeals the United Mine Workers case and the Nazi Saboteur case. On the other hand, Stephens confided, "I am somewhat reconciled to having the case taken out of this Court because I am apprehensive that if it had been decided here on the merits the division of the Court would have been the same as on the stay order question. And if perchance thereafter and before decision in the Supreme Court the case had become moot by settlement of the steel strike, there would then have been outstanding a Circuit Court of Appeals opinion sustaining the President's action." Stephens also noted that he thought that if the majority had had to write a "real opinion" it would have had a difficult time finding authoritative support for its position. To Stephens, the government's arguments in support of the seizure appeared to be based on expediency, not law. Harold M. Stephens to Golden W. Bell, Esq., May 7, 1952, Box 82, Harold M. Stephens Papers, LC.

Judge Fahy, one of the majority in the court of appeals, has indicated that he was not at all sure that the division on the merits would have been the same as on the application for a stay. Interview, March 10, 1972.

59. *New York Times*, May 2, 1952, p. 1. The six presidents invited to the White House were: Ben Moreell of Jones & Laughlin, Frank Purnell (chairman of the board, not the president) of Youngstown Sheet & Tube Co., Benjamin Fairless of U.S. Steel, Charles White of Republic, Arthur Homer of Bethlehem, and Clarence Randall of Inland.

Immediately after the court of appeals decision to deny the steel industry's application to attach conditions to the stay order, Secretary Sawyer delivered a statement to the press which indicated that he had no intention of moving on the wage order. Statement of Charles Sawyer, May 1, 1952, Doc. #37, "The 1952 Government Seizure of the Basic Steel Industry: Significant Documents," Department of Commerce Records.

60. *Newsweek*, May 12, 1952, p. 31; *New York Times*, May 3, 1952, p. 14.

61. See 28 *United States Code*, sec. 1254(1).

62. *New York Times*, May 3, 1952, p. 14.

63. No. 744—The Youngstown Sheet & Tube Co. et al., Republic Steel Corp., Armco Steel Corp. and Sheffield Steel Corp., Bethlehem Steel Co. et al., Jones & Laughlin Steel Corp., United States Steel Co., and E. J. Lavino & Co., Petitioners, v. Charles Sawyer, Respondent, Petition for a Writ of Certiorari to the United States Court of Appeals for the District of Columbia Circuit, U.S. House, *Steel Seizure Case*, pp. 461–74.

The steel industry's decision to file a petition gave the industry the right to open and close the oral argument in the Supreme Court. Although the petitioner (normally the losing party in the court below) usually had the right to open and close the argument, when cross petitions were filed Rule 28(1) of the Supreme Court provided that the plaintiff in the court below (the companies in the steel case) should be allowed to open and conclude the argument. Philip Perlman was so upset by the industry's move that he wrote to Chief Justice Vinson requesting that the Supreme Court depart from Rule 28(1) in this case and permit the government to begin and end the argument. Having heard about the solicitor general's letter, however, John W. Davis sent a note, upholding the industry's right to open and close the argument, to the Chief Justice on the same day. Davis pointed out that in *United States* v. *United Mine Workers,* 330 U.S. 258 (1947), the government did exactly what the steel companies had done in the present case: Having won in the district court, the government, which was the plaintiff in that court, filed a petition for certiorari as did the Mine Workers. Both petitions were granted and the government had the right to open and close the argument. The government did not question the application of Rule 28(1) in that case. In view of that precedent, Davis urged that no departure from Rule 28(1) be allowed in the steel case. John W. Davis to Hon. Fred M. Vinson, May 6, 1952, Pickering Papers, Washington, D.C.

On May 8, Chief Justice Vinson denied Perlman's request, for the reasons put forth by Davis. Note, Vinson to Philip Perlman, May 8, 1952, Department of Justice Files.

The rules of the Supreme Court have since been changed. Rule 44(2) now provides that "The appellant or petitioner shall be entitled to open and conclude the argument. But when there are cross-appeals or cross-writs of certiorari they shall be argued together as one case and in the time of one case, and the court will, by order seasonably made, advise the parties which one is to open and close." Robert L. Stern and Eugene Gressman, in their treatise on *Supreme Court Practice* (4th ed.; Washington, D.C.: 1969), p. 492, commented on the change in the rules: "At one time the Rules gave the plaintiff below the right to open and close. In several important cases [citing *Youngstown* and *United States* v. *United Mine Workers*], where the plaintiff had won in the district court but in which both parties petitioned for certiorari before judgment in the court of appeals, this provision resulted in giving the winning party below the right to open and close. The present language of Rule 44(2), enabling the Court in such cases to decide which party shall open and close, was designed to avoid the inflexibility of the old rule and the occasionally unreasonable consequences."

64. No. 745—Charles Sawyer, Secretary of Commerce, Petitioner v. The Youngstown Sheet & Tube Co. et al., Petition for a Writ of Certiorari to the United States Court of Appeals for the District of Columbia Circuit and Application for Stay, U.S. House, *Steel Seizure Case,* p. 513.

65. *Ibid.,* pp. 515–16.

66. No. 744—Memorandum on behalf of Respondent—Filed May 2, 1952, *ibid.,* p. 478.

67. No. 745—Memorandum on behalf of Respondents, *ibid.,* p. 548.

68. See Nos. 744 and 745, Brief for the United Steelworkers of America, CIO, as *Amicus Curiae* with regard to the Issuance of a Stay, *ibid.,* pp. 555–61.

69. Note, Sawyer to President Truman, May 2, 1952, Sawyer Papers. Sawyer never sent the letter because he received a phone call from Truman and discussed his views at that time. He told the President that he would put the letter in the files for the record.

Sawyer became increasingly disgusted with the way in which the steel situation was being handled. He complained to Secretary of the Treasury John Snyder and Secretary of Defense Robert Lovett that his position was intolerable. Although nominally in charge of the mills, he was never consulted about what measures should be taken and other people were constantly making statements about things that he would be doing without informing him about it first. He was ready to resign but for the fact that he loved his country and thought things might be even worse without him.

On the morning of May 3, Sawyer arrived at the White House early, spoke with the President and told him of his desire to resign. Truman asked Sawyer not to leave at this time because he was a steadying influence and the President depended upon his advice. Sawyer agreed to stay in the administration. Confidential Dictation, May 3, 1952, Sawyer Papers, Cincinnati, Ohio.

70. Remarks at a Meeting with Steel Industry and Labor Leaders, May 3, 1952, U.S. President, *Public Papers 1952–1953*, p. 317.

For a copy of Putnam's recommendations, see Note, Putnam to Sawyer, May 3, 1952, Doc. #41, "The 1952 Government Seizure of the Basic Steel Industry: Significant Documents," Department of Commerce Records.

71. Washington *Post*, May 4, 1952, pp. 1M, 14M; May 5, 1952, p. 2; *New York Times*, May 6, 1952, p. 22; Enarson, "Politics of an Emergency Dispute," in Industrial Relations Research Association, *Emergency Disputes and National Policy*, p. 64; Truman, *Memoirs*, 2:537–38; interview with David Stowe, January 9, 1973.

72. Justice Burton, with whom Justice Frankfurter concurred, voted to deny certiorari for the reasons stated in a memorandum filed by them: "The constitutional issue which is the subject of the appeal deserves for its solution all of the wisdom that our judicial process makes available. The need for soundness in the result outweighs the need for speed in reaching it. The Nation is entitled to the substantial value inherent in an intermediate consideration of the issue by the Court of Appeals. Little time will be lost and none will be wasted in seeking it. The time taken will be available also for constructive consideration by the parties of their own positions and responsibilities." Supreme Court of the United States, Journal of Saturday, May 3, 1952; U.S. House, *Steel Seizure Case*, p. 458.

73. *Ibid.*, p. 457. Justice Burton noted in his diary entry of April 30, 1952, that although he agreed with Judge Pine's decision that there was no authority for the steel seizure, he thought Pine should have issued a preliminary injunction along the line that U.S. Steel had requested, i.e., one preserving the status quo until the final disposition of the case. Enjoining the seizure was not in the public interest. Burton Diary, Box 2, Burton Papers, LC.

Burton also indicated that during the Supreme Court conference on May 3, the Chief Justice had received word that the President had threatened to increase wages on May 5 if no settlement were reached over the weekend. Vinson had not informed his colleagues of this news until after they had decided what action the Court would take in the steel case. Burton Diary, May 3, 1952, Box 2, Burton Papers, LC.

There was some speculation in the press that a conditional stay had been ordered by the Supreme Court because the Justices were piqued by the threat contained in Perlman's reply to the steel companies' petition for certiorari. Perlman had explicitly stated that if the Court ordered a ban on changes in wages and working conditions, another interruption in steel production was probable. *New York Times*, May 6, 1952, p. 22.

74. Washington *Post*, May 4, 1952, pp. 1M, 14M; May 5, 1952, p. 2; *New York Times*, May 6, 1952, p. 22; Truman, *Memoirs*, 2:537–38; interview with David Stowe, January 9, 1973.

75. See various memoranda detailing the Sunday meetings in Box 4, Stowe Papers, TL.

EIGHT ★ THE SUPREME COURT HEARS THE STEEL SEIZURE CASE

1. Five lawyers from the claims division and the solicitor general's office did the bulk of the work on the brief: Samuel Slade, Oscar Davis, James Morrisson, Benjamin Forman, and Herman Marcuse. Note, Robert L. Stern [attorney in the office of the solicitor general] to Prof. Douglas B. Maggs, May 16, 1952, Department of Justice Files, Washington, D.C. See also *New York Times*, May 11, 1952, p. 1E.

In Charles Murphy's opinion, "the Government's brief was really quite a good job in view of the fact that they had less than a week to prepare it." Note, Murphy to Douglas B. Maggs, May 13, 1952, Box 1, Murphy Papers, TL.

During the period in which the Justice Department prepared the brief and the oral argument and for a week thereafter, many letters were received from attorneys and other private citizens offering advice, quotes from authorities which upheld the government's position, and general support for the President. See Files on *Youngstown Sheet & Tube Co.* v. *Sawyer*, Department of Justice.

2. No. 745—Brief for Petitioner, U.S. House, *Steel Seizure Case*, p. 710.

3. *Ibid.*, pp. 733–34.

4. *Ibid.*, pp. 737–57. This whole section of the brief, from p. 733 to p. 757, is almost identical to an analysis I found in the Files of David D. Lloyd entitled "Summary of Position: The Emergency in which the President Acted," (Box 6, TL), which leads me to conclude that there was much more contact between the White House and the Justice Department in preparation for the Supreme Court argument than there had been at the district court level. Interviews with Justice Department attorneys who worked on the case also support this conclusion. Interview with Judge Oscar Davis, January 3, 1973; interview with Herman Marcuse, May 12, 1971.

Other White House papers that probably were used as background for the brief include: Memorandum to Assistant Attorney General Baldridge Re: Inapplicability of the Taft-Hartley Act to the Steel Dispute, undated, Box 4, Stowe Papers, TL; Summary of Argument, May 5, 1952, Box 6, Files of David D. Lloyd, TL; Draft, May 7, 1952, Box 6, Lloyd Files, TL.

5. No. 745—Brief for Petitioner, U.S. House, *Steel Seizure Case*, pp. 757–62.

6. The Defense Production Act of 1950 actually contained a procedure which allowed for immediate possession: the government was permitted to seize both real and personal property by the process of requisition; thereafter a suit for just

compensation could be brought by the property owner. As amended in 1951, however, the Defense Production Act prescribed a different procedure for the taking of real property: it would be condemned under the normal procedures used by the government when private property was acquired for public use, a more time-consuming process. Ironically, it was the Justice Department which had requested this amendment, because it thought in the long run it would be more convenient and efficient to follow the usual procedure for condemnation of realty. See *ibid.*, p. 766n.

7. 285 U.S. 95 (1932).

8. *Ibid.*, p. 104.

9. No. 745—Brief for Petitioner, U.S. House, *Steel Seizure Case*, p. 772. The brief also made the point that the Court of Claims might have jurisdiction of a just compensation suit by the companies even if no statute existed giving the President the power to take private property for defense purposes. See *ibid.*, pp. 772–77.

The plaintiffs' argument that the remedy in the Court of Claims would not be adequate for the kind of damages the companies would sustain had not been recognized by the Supreme Court, the government brief contended. In *Pewee Coal Co.* (341 U.S. 114 (1951)) the Court held specifically that money was adequate compensation for the type of damages alleged by the steel industry. *Ibid.*, pp. 780–82.

10. 285 U.S. 95, 104n (1932).

11. No. 745—Brief for Petitioner, U.S. House, *Steel Seizure Case*, p. 792. See also *ibid.*, pp. 783–93.

The government's brief noted that the steel industry's affidavits alleging irreparable injury had not been distributed to government counsel until the lunch recess of the first day of the hearing on the motion for preliminary injunction. The government, therefore, had not been able to offer specific contrary proof on all the injury issues raised by the plaintiffs. The government, accordingly, disagreed with the statement in Judge Pine's opinion that nothing further could be submitted at a trial on the facts that might change the case and observed that, if the case were remanded for trial, the government intended to put plaintiffs to their proof and would offer opposing evidence. *Ibid.*, p. 795n.

12. *Ibid.*, p. 795.

13. 321 U.S. 414, 440 (1944).

14. U.S. House, *Steel Seizure Case*, p. 75.

15. No. 745—Brief for Petitioner, *ibid.*, pp. 794–98.

16. *Ibid.*, p. 810.

17. *Ibid.*, pp. 810–11.

18. *Ibid.*, p. 814. The brief shied away from the more famous Lincoln pronouncements which indicated that in some situations it was necessary to go beyond the Constitution. For example, "Was it possible to lose the nation and yet preserve the Constitution? By general law, life and limb must be protected, yet often a limb must be amputated to save a life; but life is never wisely given to save

a limb. I felt that measures otherwise unconstitutional might become lawful by becoming indispensable to the preservation of the nation." (Note, Lincoln to A. G. Hodges, April 4, 1864, Nicolay and Hay, *Complete Works of Abraham Lincoln*, 10:66.) The Justice Department most emphatically renounced the use of such an interpretation of the President's powers, stressing instead that the President has always had the constitutional power to seize private property in times of emergency.

19. No. 745—Brief for Petitioner, U.S. House, *Steel Seizure Case*, pp. 815–16. The brief failed to point out that the Smith & Wesson seizure was also based on statutory power: Section 120 of the National Defense Act of 1916. See Clinton Rossiter, "The President and Labor Disputes," pp. 101–2; "Executive Commandeering of Strike-Bound Plants," pp. 287–88.

20. No. 745—Brief for Petitioner, U.S. House, *Steel Seizure Case*, p. 816n. The brief did not indicate, for good reason, the occasions on which Wilson contemplated seizure but was advised that it would be unconstitutional. In November, 1917, the President's mediation commission failed to settle a dispute at the Pacific Telephone & Telegraph Co., and a member of that commission advised President Wilson to seize the company. Wilson asked his Attorney General, Thomas W. Gregory, if the President had authority, under existing law, to take possession of Pacific Telephone & Telegraph. In an opinion, Gregory outlined the various statutes Congress had passed to give the power of eminent domain to the President for wartime purposes. Gregory then informed the President:

"Evidently none of these statutes, which constitute as I believe all the existing legislation of this character, is broad enough to cover the present case. No doubt as Commander-in-Chief of the army and navy you may, in time of war, seize upon those things which are presently necessary to equip, maintain, transport or manoeuvre either arm without waiting for express statutory authority; but the power is an extraordinary one and not to be pressed beyond the emergency which evokes it. So of the power of eminent domain itself; it draws from the necessity of the government, not from mere general convenience; and ceases when the governmental need has been supplied. Insofar as the service now rendered by the Pacific Telephone & Telegraph Co. is necessary to the government's purposes, Congress unquestionably could authorize the seizure of the property in order to secure its continuance: but until it has done so I think you can act only in the presence of some truly military emergency." Opinion of Attorney General T. W. Gregory, November 9, 1917, Unpublished Opinions of the Attorney General, File #188619, Department of Justice.

The government brief did mention the fact that Wilson had contemplated seizure of the Colorado coal mines in 1914 because of a long strike, but that seizure had not been effected. No reason was given in the brief for the decision not to seize. In a Justice Department document, however, which Perlman and his associates undoubtedly studied, reference is made to the decision not to seize the Colorado coal mines. The author of the memorandum, which is entitled "Constitutional Power of the President to Seize Property during the Period of an Emergency," noted that he could not find an official opinion of the Attorney General on the President's power to seize the Colorado mines, but that a *New*

York Times article, November 25, 1914, p. 12, revealed that President Wilson had informed callers who were interested in seizure that the Secretary of Labor had consulted the solicitor of his department and had decided that the seizure was "without warrant of law." William H. Rose to Hugh B. Cox, March 9, 1945, p. 18; File #146–49-103, Department of Justice.

21. Quoted in No. 745—Brief for Petitioner, U.S. House, *Steel Seizure Case*, p. 817n. The twelve seizures executed by Roosevelt were North American Aviation Plant, June 9, 1941; Federal Shipbuilding & Drydock Co., August 23, 1941; Air Associates, Inc., October 30, 1941; Grand River Dam Project, November 19, 1941; Toledo, Peoria & Western R.R. Co., March 21, 1942; Brewster Aeronautical Corp., April 18, 1942; General Cable Co., August 13, 1942; S. A. Woods Machine Co., August 19, 1942; Triumph Explosives, Inc., October 12, 1942; coal mines, May 1, 1943; American R.R. Co. of Puerto Rico, May 13, 1943; Howarth Pivoted Bearings Co., June 14, 1943. *Ibid.*, p. 817n.

The brief also noted the numerous other instances, during the depression and the war emergency, when President Roosevelt took actions grounded on his constitutional powers rather than on statutory law. *Ibid.*, p. 818n.

An attorney employed by the Justice Department at the time of the steel seizure commented that most of the government attorneys assumed Truman could seize property because FDR had seized property. Interview with Ellis Lyons, November 5, 1973.

22. *In re Debs*, 158 U.S. 564 (1895).

23. The brief duly noted that it was aware of the opinion of Attorney General Knox given to Roosevelt informing him that he had no power to effect such a seizure, but it pointed out that Knox's opinion rested on the contention that a coal strike in Pennsylvania was a matter of purely local, not federal, concern. No. 745—Brief for Petitioner, U.S. House, *Steel Seizure Case*, pp. 819, 819n. See also Roosevelt, *An Autobiography*, pp. 388–89.

24 *United States* v. *Midwest Oil Co.*, 236 U.S. 459 (1915).

25. Quoted in No. 745—Brief for Petitioner, U.S. House, *Steel Seizure Case*, p. 821.

26. Quoted *ibid.*

27. *Ibid.*, pp. 822–27.

28. Quoted *ibid.*, pp. 828–29n. For diverse views of the President's emergency powers and Congress' recognition of them, see Rossiter, "The President and Labor Disputes," pp. 104–5; "Executive Commandeering of Strike-Bound Plants," pp. 284–87; "The Executive Power in Foreign Affairs," in MacLean, *President and Congress: The Conflict of Powers*, p. 65.

29. No. 745—Brief for Petitioner, U.S. House, *Steel Seizure Case*, pp. 828–29. The quote is from *New York Trust Co.* v. *Eisner*, 256 U.S. 345, 349 (1921).

30. No. 745—Brief for Petitioner, U.S. House, *Steel Seizure Case*, p. 829. The quote from *Inland Waterways* is "even constitutional power, when the text is doubtful, may be established by usage." 309 U.S. 517, 525 (1940).

31. *United States* v. *Midwest Oil Co.*, 236 U.S. 459, 472–73 (1915), as quoted in No. 745—Brief for Petitioner; U.S. Congress, *Steel Seizure Case*, p. 829.

32. No. 745—Brief for Petitioner, U.S. House, *Steel Seizure Case*, pp. 831–32. Some of the Supreme Court cases which helped to mold this judicial doctrine and which the brief discussed at length were *Mitchell* v. *Harmony*, 13 How. 115 (1852); *United States* v. *Russell*, 13 Wall. 623 (1871); *United States* v. *Pacific Railroad*, 120 U.S. 227 (1887).

33. *United States* v. *Russell*, 13 Wall. 623, 627, as quoted in No. 745—Brief for Petitioner, U.S. House, *Steel Seizure Case*, pp. 830–31.

34. No. 745—Brief for Petitioner, U.S. House, *Steel Seizure Case*, p. 847.

35. *Alexander* v. *United States*, 39 Ct. Cls. 383, 396 (1904), as quoted in No. 745—Brief for Petitioner, U.S. House, *Steel Seizure Case*, p. 846.

36. *United States* v. *Pewee Coal Co., Inc.*, 341 U.S. 114 (1951), and *Dakota Coal Co.* v. *Fraser*, 283 Fed. 415 (D.N.D., 1919).

37. No. 745—Brief for Petitioner, U.S. House, *Steel Seizure Case*, p. 849.

38. 283 Fed. 415, 418, as quoted *ibid.*, p. 851.

39. *Ibid.*

40. Quoted *ibid.*, pp. 857–58.

41. 135 U.S. 1 (1890). In his opinion for the Court, Justice Miller asked, "Is this duty [the President's duty to see that the laws are faithfully executed] limited to the enforcement of acts of Congress or of treaties of the United States according to their *express terms*, or does it include the rights, duties and obligations growing out of the Constitution itself, our international relations, and all the protection implied by the nature of the government under the Constitution?" Miller thought the latter interpretation was correct. *Ibid.*, p. 64. See also Corwin, *The Twilight of the Supreme Court*, pp. 137–40.

42. 158 U.S. 564 (1895). The *Debs* case upheld the right of the President, despite the absence of statutory authorization, to petition the federal court for an injunction against striking railroad workers on the basis that they were interfering with interstate commerce. Justice Brewer's opinion for the Court noted: "Every government, entrusted, by the very terms of its being, with powers and duties to be exercised and discharged for the general welfare, has a right to apply to its own courts for any proper assistance in the exercise of the one and the discharge of the other. . . . While it is not the province of the Government to interfere in any mere matter of private controversy between individuals, or to use its granted powers to enforce the rights of one against another, yet, whenever the wrongs complained of are such as affect the public at large, and are in respect of matters which by the Constitution are entrusted to the care of the Nation and concerning which the Nation owes the duty to all the citizens of securing to them their common rights, then the mere fact that the Government has no pecuniary interest in the controversy is not sufficient to exclude it from the courts, or prevent it from taking measures therein to fully discharge those constitutional duties." *Ibid.*, pp.

584, 586. See also Berman, *Labor Disputes and the President of the United States*, pp. 13–35; Teller, "Government Seizure in Labor Disputes," p. 1044.

43. Quoted in No. 745—Brief for Petitioner, U.S. House, *Steel Seizure Case*, pp. 878–79.

44. *Ibid.*, p. 880. Despite the length of the government brief, 175 pages, it did not cover all points in which the Supreme Court was interested. The Justice Department and the White House received telephone calls from the Supreme Court asking for information from previous experience on how long it took a Taft-Hartley board of inquiry to make its report, what the results from Taft-Hartley "final offer" elections had been, and what allocation orders on steel had been issued by the National Production Authority. See Memorandum, Enarson to Murphy, May 12, 1952, Box 28, Murphy Files, TL.

45. The steel companies had decided to file one brief for the industry, although Armco and Sheffield Steel corporations elected to submit their own additional brief, which concentrated on the constitutional question, and E. J. Lavino & Co. filed a brief dealing with its particular situation. See U.S. House, *Steel Seizure Case*, pp. 885–947, 949–63.

46. S. Rept. No. 105, 80th Cong., 1st sess., 1947, as quoted in No. 744 and No. 745—Brief for Plaintiff Companies, U.S. House, *Steel Seizure Case*, p. 598.

47. Quoted *ibid.*, p. 599.

48. *Ibid.*, pp. 603–4.

49. In *Toledo, Peoria & Western R.R.* v. *Stover*, a district court had stated: "The executive department of our government cannot exceed the powers granted to it by the Constitution and the Congress, and if it does exercise a power not granted to it, or attempts to exercise a power not granted to it, or attempts to exercise a power in a manner not authorized by statutory enactment, such executive act is of no legal effect." 60 F. Supp. 587, 593 (S.D. Ill., 1945), as quoted *ibid.*, pp. 618–19.

50. 2 Cranch 170 (1804).

51. *Ibid.*, p. 177. See No. 744 and No. 745—Brief for Plaintiff Companies, U.S. House, *Steel Seizure Case*, pp. 622–23.

52. No. 744 and No. 745—Brief for Plaintiff Companies, U.S. House, *Steel Seizure Case*, pp. 624–30.

53. 218 U.S. 322, 336 (1910), as quoted *ibid.*, p. 635. Other cases cited by the industry brief as establishing the congressional power of eminent domain included *United States* v. *North American Transportation and Trading Co.*, 253 U.S. 330, 333 (1920); *United States* v. *Rauers*, 70 Fed. 748 (S.D. Ga. 1895); *Rindge Co.* v. *Los Angeles*, 262 U.S. 700, 709 (1923); *Bragg* v. *Weaver*, 251 U.S. 57, 58 (1919); *Chappell* v. *United States*, 160 U.S. 499, 510 (1896). See U.S. House, *Steel Seizure Case*, p. 636.

54. *Ibid.*, p. 645.

55. *Ibid.*, p. 595. The industry brief included arguments on the equitable

issues, but these were essentially the same as those presented by Theodore Kiendl in the district court. See chapter 6.

Two other briefs were submitted amicus curiae, one for the United Steelworkers of America, CIO, and one for the Brotherhood of Locomotive Engineers, Brotherhood of Locomotive Firemen and Enginemen, and Order of Railway Conductors. The Steelworkers' brief treated the questions in which the union had an interest. The major issue, from its point of view, was whether the Taft-Hartley Act could be imposed after the union had cooperated with the government and had agreed to take its dispute to the WSB. All the objectives of the Taft-Hartley Act had been fulfilled by April 8, the brief observed. If the Court decided that, regardless of the truth of that statement, the act should have been employed on April 8, no union would henceforth go before the Wage Stabilization Board; the union would be encouraged to get Taft-Hartley over with as soon as possible. The Steelworkers' brief also noted that the seizure order was not issued to settle a labor dispute. The dispute was between the government and the industry over prices. See *ibid.*, pp. 965–91.

The brief for the railroad brotherhoods argued that the steel seizure was unconstitutional, a point the brotherhoods intended to pursue in their own case opposing the railroad seizure. Their primary contention was that eminent domain was a legislative power. See *ibid*, pp. 997–1065.

56. Westin, *Anatomy of a Constitutional Law Case*, p. 112.

57. *New York Times*, May 9, 1952, p. 16; Memorandum to Secretaries, May 6, 1952, Box 83, Burton Papers, LC.

58. Interview with Stanley Temko, December 11, 1973. It was also decided that, should Davis need assistance or replacement during the argument, Bruce Bromley would take his place. *Ibid.*

59. *Ibid.* One of the problems with which Davis was most concerned was the *Midwest Oil* case. As solicitor general in 1914–1915, he had successfully taken a position in that case that was directly contrary to the one he would argue in the steel seizure case. Davis was certain that the Court would question him about *Midwest Oil*, and he pondered how to distinguish it from the steel seizure case. *Ibid.*

60. *New York Times*, May 13, 1952, p. 14; New York *Herald Tribune*, May 13, 1952, p. 1; *Wall Street Journal*, May 13, 1952, p. 2.

61. Harbaugh, *Lawyer's Lawyer*, p. 475.

62. New York *Herald Tribune*, May 13, 1952, p. 1.

63. *New York Times*, May 13, 1952, p. 14. Compare Holmes Baldridge's description of Davis' argument contained in a letter he sent to his niece. Baldridge asserted that Davis' presentation was emotional, that he was in tears most of the time, his voice trembled, and he frequently choked up. Note, Baldridge to Nancy Dixon Bradford, December 22, 1964, p. 3, Baldridge Papers, TL.

64. Transcript of the Stenographic Record of the Proceedings had before the Supreme Court of the United States commencing on Monday, May 12, 1952, pp. 22–23. I am indebted to Stanley Temko for allowing me to use his copy of the transcript.

I have referred to the transcript since it is the only complete reproduction of the argument. It is filled with errors, however. One of Davis' partners noted that Davis thought the transcript was so bad that he could not correctly reconstruct the argument from it. Note, Porter Chandler to John C. Gall, May 24, 1952, John Pickering Papers, Washington, D.C.

65. Transcript of the . . . Proceedings had before the Supreme Court of the United States, p. 23.

66. *Ibid.*, p. 34. At this point, Davis was interrupted by Justice Frankfurter who questioned him about the circumstances of the *Midwest Oil* case. Davis asked if he could finish his presentation and deal with that question afterwards, a suggestion to which Justice Frankfurter was receptive. *Ibid.*, pp. 34–35. This was the only question addressed to Davis during his argument. It is also noteworthy that Davis did not answer Justice Frankfurter's inquiry immediately. Only a supremely confident advocate, one who was aware of his standing with the Court, would venture to delay answering a Justice.

67. In the brief for the United States in *Midwest Oil*, Davis had stated: "While perfect flexibility is not to be expected in a Government of divided powers, and while division of power is one of the principal features of the Constitution, it is the plain duty of those who are called upon to draw the dividing lines to ascertain the essential, recognize the practical, and avoid a slavish formalism which can only serve to ossify the Government and reduce its efficiency without any compensating good. The function of making laws is peculiar to Congress, and the Executive can not exercise that function to any degree. But this is not to say that all of the subjects concerning which laws might be made are perforce removed from the possibility of Executive influence. The Executive may act upon things and upon men in many relations which have not, though they might have, been actually regulated by Congress. In other words, just as there are fields which are peculiar to Congress and fields which are peculiar to the Executive, so there are fields which are common to both, in the sense that the Executive may move within them until they shall have been occupied by legislative action. These are not the fields of legislative prerogative, but fields within which the law-making power may enter and dominate whatever it chooses. This situation results from the fact that the President is the active agent, not of Congress, but of the Nation. As such he performs the duties which the Constitution lays upon him immediately, and as such, also, he executes the laws and regulations adopted by Congress. He is the agent of the people of the United States, deriving all his powers from them and responsible directly to them. In no sense is he the agent of Congress. He obeys and executes the laws of Congress not because Congress is enthroned in authority over him, but because the Constitution directs him to do so." *United States* v. *Midwest Oil Co.*, Brief for the United States. Quoted in Schubert, *Constitutional Politics*, pp. 318–19.

68. "I should have answered Day," Davis later wrote, "that my lapse from virtue was due to the duty of defending Republican presidents." Quoted in Harbaugh, *Lawyer's Lawyer*, p. 101.

69. Transcript of the . . . Proceedings had before the Supreme Court of the United States, pp. 41–43.

70. *Ibid.*, p. 50. The correct quotation is: "In questions of power then let no more be heard of confidence in man, but bind him down from mischief by the claims of the Constitution." Kentucky Resolutions of November 16, 1798, in Commager, *Documents of American History,* p. 181.

Davis received many letters congratulating him on his performance and thanking him. See, for example, Note, Thomas Robins to John W. Davis, May 26, 1952, John W. Davis Papers, Yale University.

71. Transcript of the . . . Proceedings had before the Supreme Court of the United States, pp. 51–52.

72. *Ibid.*, pp. 56–58.

73. *Ibid.*, p. 88.

74. *Ibid.*, p. 116. The transcript indicates that it was Justice Minton who interrupted Perlman with these statements, but that appears to be a mistake. A partner of Davis', after reading the transcript, wrote that the reporter consistently confused Justices Burton and Minton and completely left out Justice Douglas' remark to Perlman that if Perlman's arguments as to the President's powers were sound there would be no need for Congress. See Note, Porter Chandler to John C. Gall, May 22, 1952, John Pickering Papers.

75. Transcript of the . . . Proceedings had before the Supreme Court of the United States, pp. 115–23; *New York Times*, May 13, 1952, p. 14; Cleveland *Plain Dealer*, May 13, 1952, p. 1.

On the evening of May 12, the White House held a reception to which Secretary Sawyer was invited. The President saw Sawyer privately for a few minutes and asked him how the Supreme Court argument was going. When Sawyer told Truman that Perlman had been hostilely questioned, the President declared that he would be "shocked, disappointed, and disturbed," if the Court ruled against him. Memorandum for Files, May 13, 1952, Sawyer Papers, Cincinnati, Ohio. See also Sawyer, *Concerns of a Conservative Democrat,* pp. 266–67.

76. This was the argument the steel industry attorneys most feared. Interview with Stanley Temko, January 16, 1974.

77. Transcript of the . . . Proceedings had before the Supreme Court of the United States, pp. 155–56. On May 15, 1952, the Washington *Post* ran an editorial which discussed the confusion created because Congress had not declared war. The editorial (p. 10) noted the Justices' reactions to Perlman's statement that this was wartime and concluded, "By fighting a war without having war declared the Administration has multiplied confusion and placed itself under unnatural handicaps."

78. For accounts of the second day of argument, see *New York Times*, May 14, 1952, p. 1; New York *Herald Tribune*, May 14, 1952, p. 1; Cleveland *Plain Dealer*, May 14, 1952, p. 1.

C. D. Williams, the solicitor of the Commerce Department, who attended the argument, reported to Secretary Sawyer that he had the impression that Perlman had done better than he had on the first day. Memorandum of telephone conversation with C. D. Williams, May 13, 1952, 2 P.M., Sawyer Papers, Cincinnati, Ohio.

Felix Frankfurter commented on the oral arguments: "What miserable reporting even the best of our papers do when there aren't any handouts. Précis
writing seems to be an unknown art with us. On the whole what you get from an
account of the argument like that in the *Steel* case are what to the reporters seem
quips or brushes between court and counsel. You rightly surmise, however, that
the S.G. was better the second day, though to me it is absurd that his argument
should have taken more than three hours. To me it is a constant surprise that
lawyers resent questioning from the Bench. Of course an English barrister should
be allowed to develop his argument and quote from the cases and all the rest of
the business, since the judges must take it all in by ear. But I see no excuse whatever for allowing them here to tell us by word of mouth what they have submitted
in print. With us the function of oral argument is that of supplementing the
printed briefs, by cross-examination as it were, of counsel on what is in the briefs.
Of course an artist at the business, like Davis, knows how to exercise his art and
holds the Court's attention by it. Incidentally, although he had two and a half
hours I don't think he used more than half of it. One of the differences between
good advocates and others is what R. L. Stevenson said was the difference between Homer and other poets: 'Homer knew what to omit.' " Felix Frankfurter to
Charles C. Burlingham, May 19, 1952, Box 36, Frankfurter Papers, LC.

79. Transcript of the . . . Proceedings had before the Supreme Court of the
United States, p. 160.

80. *Ibid.*, p. 163. Newspaper articles which mentioned Davis' rebuttal noted
that he spoke for only ten minutes and was asked no questions. Harbaugh characterized the brief rebuttal as "an extraordinary display of confidence." *Lawyer's
Lawyer*, p. 478.

Arthur Goldberg, in his oral presentation for the United Steelworkers of
America, concentrated almost totally on the Taft-Hartley Act. He tried to show
that every intention of Congress as enacted in the Taft-Hartley Act had been fulfilled by the President and the union before April 8, and that, on April 8, there
was no remedy intended by Congress available to the President. Goldberg chose
not to speak to the constitutional merits of the case. For accounts of his argument
see *New York Times*, May 14, 1952, p. 22; Cleveland *Plain Dealer*, May 14,
1952, p. 9; *CIO News*, May 19, 1952, p. 2.

Attorneys for the railroad brotherhoods argued on the inherent power question. On May 21, 1952, before their own case came up for argument, however,
the rail dispute was settled. See *New York Times*, May 22, 1952, p. 1.

Justice Burton, in his diary, rated the oral arguments as follows: Davis was
"excellent"; Perlman was "good"; Goldberg was "extremely excellent"; Clifford D.
O'Brien and Harold Heiss, attorneys for the railroad unions, were "good." See
Diary, May 12 and May 13, 1952, Box 2, Burton Papers, LC.

81. See, for example, *New York Times*, May 16, 1952, p. 9; *New York
Times*, May 26, 1952, p. 32; *Journal of Commerce*, May 19, 1952, p. 1. *Iron Age*
commented: "Inventories of steel users are so comfortable now that they might
spell trouble during the second half of the year. Throughout the steel crisis there

has been practically no panic on the part of consumers. Many of them have seemed almost blasé." May 8, 1952, p. 169.

82. U.S. Senate, Committee on Banking and Currency, *Defense Production Act Amendments of 1952, Hearings*, p. 2038.

Manly Fleischmann, administrator of the Defense Production Administration, testifying before the House Committee on Banking and Currency, claimed that the steel situation was good "unless we have a strike of long duration." When he was asked how long a strike could continue before its effects would be noticed in defense industries, Fleischmann replied that a strike beyond a week would harm the defense effort, especially the ammunition program. U.S. House, Committee on Banking and Currency, *Defense Production Act Amendments of 1952, Hearings*, pp. 286, 288–89.

83. U.S. Senate, Committee on Labor and Public Welfare, *National and Emergency Labor Disputes, Hearings*, p. 587.

84. *New York Times*, May 14, 1952, p. 1.

85. *Ibid.*, May 7, 1952, pp. 1, 13.

86. *Ibid.*, May 23, 1952, p. 14.

87. *Ibid.*, May 18, 1952, p. 1E; Lahey, "The Steelworkers State Their Case," *New Republic*, May 26, 1952, pp. 12–13.

88. *New York Times*, May 15, 1952, pp. 1, 20; May 18, 1952, p. 9E.

89. *New York Times*, May 14, 1952, p. 22; *Steel Labor*, June 1952, pp. 3, 13. The convention also passed a resolution condemning the majority of the nation's newspapers for prejudiced coverage of the steel dispute.

90. The President's News Conference, May 22, 1952, U.S. President, *Public Papers 1952–1953*, p. 362.

91. *Ibid.*

92. *Ibid.*, pp. 362–63.

93. *Ibid.*, p. 364. See also *New York Times*, May 23, 1952, pp. 1, 14. There were rumors in the press at this time that Truman had been informed of the Supreme Court's thoughts because of his friendship with several of the Justices. Chief Justice Vinson, President Truman, and their wives dined together before decision day, and this was duly noted in the newspapers. See, for example, *Washington Post*, May 27, 1952, p. 10.

Arthur Krock commented on the President's statements: "But . . . more interesting . . . was the implication the President left today that he knows the Supreme Court decision will not be based on his power to seize the steel industry as the central issue. Since Government counsel and the industry lawyers presented this as such, and the court has not made public its opinion, how does the President know that this issue will be subordinated to another? If he was guessing, that was an exercise of doubtful propriety. If he has had some reliable information in advance to the public, that will confirm a secret relationship which in the past has been suspected and criticized and will provide a footnote for history." *New York Times*, May 23, 1952, p. 20.

NINE ★ AWAITING THE DECISION

1. *Congressional Record*, 98:4151. See also statements by Senator Morse, *ibid.*, p. 4689, and Senator Bricker, *ibid.*, p. 4151.

2. *Ibid.*, 98:A3278.

3. Thomas Robins to John W. Davis, May 26, 1952, Davis Papers, Yale University. For similar views, see C. M. Lewis to Harold Burton, May 13, 1952, Box 83, Burton Papers, LC.

In his autobiography, Justice Douglas tells of a meeting with Sidney Weinberg, a Wall Street investment banker, just after the seizure case had been argued. Weinberg had a history of making remarks, in the presence of others, accusing Douglas of ruining the country. At this particular encounter, Weinberg, in a loud aside, declared, "The Court hasn't got the guts to hold it [the seizure] unconstitutional." *Go East, Young Man*, pp. 292–93.

4. Clarence Berdahl examined the new trends in government when he published *War Powers of the Executive in the United States*. To Berdahl one development was clear: World War I had increased the war powers of the President as compared with Congress. Concentration of power in the Executive appeared to be the necessary ingredient for successful crisis management. Berdahl did point out, however, that attention was paid "to constitutional forms in bringing about this necessary concentration of power" (p. 268).

5. *Twilight of the Supreme Court*, p. 133.

6. Corwin, *The President, Office and Powers 1787–1948*, (3d ed. rev.; New York: New York University Press, 1948), pp. 317–18.

7. *Ibid.*, p. 318. In the 1940s, Corwin viewed with alarm the tremendous expansion of presidential power and raised the question "whether the Presidency is a potential matrix of dictatorship." *Ibid.*, p. 353. He believed that the relationship between Congress and the President had to be changed to provide greater responsiveness in the legislative branch to public opinion, and greater accountability of the executive to the other branches of government. *Ibid.*, pp. 353–64. See also Zurcher, "The Presidency, Congress, and Separation of Powers," pp. 84–85.

8. Clinton Rossiter noted, even after the Taft-Hartley Act was enacted, that one of the President's prime responsibilities consisted in keeping the industrial peace and that this would continue to be a primary, time-consuming concern.

Rossiter classified the President's authority to deal with labor disputes as follows: "1) the power literally to 'keep the peace of the United States,' by instituting military action in strikes attended by violence and public disorder; 2) the power to remove obstructions to the flow of industrial production in time of war (or just before and after war) by establishing emergency agencies to settle labor disputes and by enforcing their decisions with seizure and other sanctions; 3) the power to intervene in disputes, not necessarily wartime or touched with violence, that constitute an economic national emergency." In Rossiter's opinion the first and second of these powers were "well established." "The President and Labor Disputes," pp. 94–95.

9. Rossiter, *The Supreme Court and the Commander in Chief*, p. 4. (The book was published first in 1951.)

Rossiter noted one great exception to this general statement, the Civil War case of *Ex parte Milligan*, in which the Court had been "lured into saying things about limits on the war powers that were simply not true." *Ibid.*, p. 7.

10. *Ibid.*, pp. 5–6. Rossiter cited *Duncan* v. *Kahanamoku* (327 U.S. 304 (1946)) and *Korematsu* v. *United States* (323 U.S. 214 (1944)) to illustrate the Court's problem. In *Duncan*, the Court was willing to assert, over vigorous dissent, that the judgment of the President and his military subordinates in regard to a state of emergency in Hawaii was wrong. *Korematsu* brought the opposite result, over equally vigorous dissent.

11. Rossiter, *The Supreme Court and the Commander in Chief*, p. 7.

12. *Ibid.*, pp. 67–68.

13. In *Constitutional Revolution, Ltd.* (Claremont, Calif.: 1941), p. 113, Corwin stated that "the principle of the Separation of Powers, so far as it has been thought to exclude the executive from the business of law-making," had become, in the hands of the New Deal Court, "otiose and superfluous." The Court had decided, Corwin surmised, to minimize "its own role in favor of the political forces of the country."

14. Pritchett, "The President and the Supreme Court," pp. 91–92.

15. *Ibid.*, p. 92. For similar views, see Sturm, "Emergencies and the Presidency," pp. 121–44.

Clinton Rossiter stated, in *The American Presidency*, which came out a few years after the *Steel Seizure* decision: "We delude ourselves cruelly if we count on the Court at all hopefully to save us from the consequences of most abuses of presidential power. The fact is that the Court has done more over the years to expand than to contract the authority of the Presidency." Rossiter cited the *Prize* cases, *In re Debs*, *Myers* v. *United States*, and *United States* v. *Curtiss-Wright Export Corp.* to support his view. *The American Presidency*, pp. 58–59. See also Frank, *Marble Palace*, p. vii, where Frank reviews people's expectations for the post–World War II Court.

One judge and legal scholar took strong exception to the opinion that the Court would not rule the seizure illegal. Arthur T. Vanderbilt, in 1952 chief justice of the supreme court of New Jersey, speaking after the seizure but before the case was decided, said: "The issues are so fundamental, so pressing, and so inescapable that it is inconceivable that the Court will attempt to dispose of them by resorting to the doctrine of judicial deference." The lecture is published in Vanderbilt, *The Doctrine of the Separation of Powers and Its Present-Day Significance*, p. 127.

16. For Black's career before he joined the Court, see Virginia Van Der Veer Hamilton, *Hugo Black*; Frank, *Mr. Justice Black*, pp. 3–108. For a brief overview of Black's entire career, see John P. Frank, "Hugo L. Black," in Friedman and Israel, *The Justices of the United States Supreme Court 1789–1969*, 3:2321–47. For Black's appointment to the Court, see Leuchtenburg, "A Klansman Joins the Court," pp. 1–31.

17. Prichett, *Roosevelt Court*, p. 258. See also Leuchtenburg, "A Klansman Joins the Court," pp. 22–23.

18. Pritchett, *Roosevelt Court*, pp. 208, 258; Westin, *Anatomy of a Constitutional Law Case*, p. 130.

19. Charles A. Reich, "The Living Constitution and the Court's Role," in Strickland, *Hugo Black and the Supreme Court*, p. 149.

In Paul Freund's words, Justice Black believed that judges must "follow the mandates of the Framers in their full and natural meaning, so that the powers granted are not constricted and the guaranties prescribed are not cabined, cribbed and confined." "Mr. Justice Black and the Judicial Function," in Freund, *On Law and Justice*, pp. 214–15.

20. 323 U.S. 214 (1944).

21. See Charlotte Williams, *Hugo L. Black*, pp. 165–66; Stephen P. Strickland, "Black on Balance," in Strickland, *Hugo Black and the Supreme Court*, pp. 249–50.

22. C. Herman Pritchett, "Stanley Reed," in Friedman and Israel, *The Justices of the United States Supreme Court*, 3:2389; Pritchett, *Roosevelt Court*, pp. 208, 260; Rodell, *Nine Men*, pp. 262, 268; Westin, *Anatomy of a Constitutional Law Case*, p. 130.

23. Fred Rodell was not convinced that Frankfurter really practiced judicial self-denial. Rodell called it "selective" judicial self-denial. *Nine Men*, p. 272.

24. Felix Frankfurter to Jack E. Brown, Esq., February 3, 1956, Box 149, Frankfurter Papers, LC.

25. After Justice Roberts' famous "switch in time," Frankfurter wrote to President Roosevelt: "And now, with the shift by Roberts, even a blind man ought to see that the Court is in politics, and understand how the Constitution is 'judicially' construed. It is a deep object lesson—a lurid demonstration—of the relation of men to the 'meaning' of the Constitution. This behavior . . . comes on top of the Hughes letter. That was a characteristic Hughes performance—part and parcel of that pretended withdrawal from considerations of policy, while trying to shape them, which is the core of the mischief of which the majority have so long been guilty." March 30, 1937, Box 243, Frankfurter Papers, LC.

26. 323 U.S. 214, 224 (1944).

27. *Ibid.*, p. 225.

28. 327 U.S. 304, 313, 335–37 (1946).

29. *Ibid.*, pp. 344–45.

30. Frankfurter, *The Public and Its Government*, p. 77. See also the section on separation of powers in James Willard Hurst, "Themes in United States Legal History," in Mendelson, *Felix Frankfurter*, pp. 206–13.

31. John P. Frank, "William O. Douglas," in Friedman and Israel, *The Justices of the United States Supreme Court*, 4:2454. For other accounts of Douglas' activities before he went on the Court, see Countryman, *Douglas of the Supreme Court*, pp. 9–14; Douglas, *Go East, Young Man*.

32. Douglas, *Go East, Young Man*, pp. 452–53.

33. Rodell, *Nine Men*, p. 276. In *Duncan* v. *Kahanamoku*, where the Court ruled against the government, Douglas was in the majority, but the case was decided in 1946, after the war had ended.

34. Rodell, *Nine Men*, pp. 279–80; Westin, *Anatomy of a Constitutional Law Case*, p. 131. Warner Gardner commented, in an assessment of Jackson's tenure as attorney for the government, "Nor can one escape the belief that the 1938–1941 revolution in the constitutional doctrine of the Supreme Court was made a great deal simpler, and much less noticeable, by Robert Jackson's uncommonly skillful advocacy." "Robert H. Jackson," p. 442.

35. U.S. Dept. of Justice, *Official Opinions of the Attorneys General*, 39:484–96. See too Philip Kurland, "Robert H. Jackson," in Friedman and Israel, *The Justices of the United States Supreme Court*, 4:2560–61. Corwin states that this transaction violated at least two statutes and was an example of presidential exercise of a power assigned by the Constitution to Congress. *The President*, p. 289.

36. Quoted in Kurland, "Robert H. Jackson," in Friedman and Israel, p. 2560.

37. Rodell, *Nine Men*, p. 262.

38. Pritchett, *Roosevelt Court*, pp. 260–61.

39. See Note, E. Barrett Prettyman, Jr., to Felix Frankfurter, October 13, 1955, Box 149, Frankfurter Papers, LC.

40. 320 U.S. 81 (1943).

41. See Jackson's dissent, 323 U.S. 214, 242–48.

42. Jackson, *The Struggle for Judicial Supremacy*, p. 114.

43. Westin, *Anatomy of a Constitutional Law Case*, p. 130.

44. Richard Kirkendall, "Fred M. Vinson," in Friedman and Israel, *The Justices of the United States Supreme Court*, 4:2649.

45. *Ibid.*, pp. 2640–41.

46. In 1946, the Court was split intellectually into two groups, the Black-Douglas-Murphy-Rutledge wing and the Frankfurter-Jackson faction, with the remaining Justices oscillating between the two to form a majority in any particular case. The split is commonly described in terms of liberal-conservative (or judicial activists as opposed to those who believed in restraint), but the labels are generally meaningless unless applied to a specific class of issues, e.g., civil liberties, at a specific time. The whole Court was "liberal," as compared to the Court of the 1930s, on the question of government regulation of the economy. See Pritchett, *Roosevelt Court*, for a study of the voting blocs in the Roosevelt Court.

The intellectual division in the Court had turned into a personality feud also, and Truman hoped that the force of Vinson's personality would have a restraining influence. Kirkendall, "Fred M. Vinson," in Friedman and Israel, pp. 2641–42.

47. Kirkendall, "Fred M. Vinson," in Friedman and Israel, p. 2648; Scigliano, *The Supreme Court and the Presidency*, p. 76.

For criticism of Vinson's performance, see Rodell, *Nine Men*, pp. 306–9, and Rodell, "Our Not So Supreme Supreme Court," *Look*, July 31, 1951, pp. 63–64.

48. Richard Kirkendall, "Harold Burton," in Friedman and Israel, *The Justices of the United States Supreme Court*, 4:2617–18; Westin, *Anatomy of a Constitutional Law Case*, p. 132.

49. Frank, *Marble Palace*, p. 68.

50. Quoted in Westin, *Anatomy of a Constitutional Law Case*, p. 132. Upon Truman's appointment of Clark to the Court, Harold Ickes remarked, "President Truman has not 'elevated' Tom C. Clark to the Supreme Court, he has degraded the Court." Quoted in Hamby, *Beyond the New Deal*, p. 337.

For a brief assessment of Clark's career see Richard Kirkendall, "Tom C. Clark," in Friedman and Israel, *The Justices of the United States Supreme Court*, 4:2665–77.

51. In early 1949, when the repeal of Taft-Hartley was being considered, the Justice Department had produced a memorandum entitled, "Inherent Executive Power to Deal with Emergencies Resulting from Labor Disputes in Vital Industries Affecting the Health, Safety, and Welfare of the Entire Nation" (February 7, 1949, Files of Holmes Baldridge, TL), which reviewed the Supreme Court's generous interpretation of the President's powers and advocated seizure as the most practicable solution to these problems (p. 14). The inherent powers of the President, the memorandum noted, would be sufficient authority in such emergencies (p. 15). Justice Jackson, in his concurring opinion, took notice of Clark's well-known support of the inherent powers of the President. See 343 U.S. 579, 649n17.

52. Swindler, *Court and Constitution in the Twentieth Century*, pp. 82–83. See also Leuchtenburg, "A Klansman Joins the Court," p. 29.

53. Richard Kirkendall, "Sherman Minton," in Friedman and Israel, *The Justices of the United States Supreme Court*, 4:2699–708; Westin, *Anatomy of a Constitutional Law Case*, p. 132. Fred Rodell called Minton "the least inept of Truman's four justices" ("Our Not So Supreme Supreme Court," p. 64). For Justice Black's opinion of Minton's work on the Court, see Note, Black to Gordon R. Owen, September 1, 1961, Special Correspondence, Box 6, Black Papers, LC.

54. In fact, on May 3, the day the Supreme Court granted certiorari, the industry and the union were negotiating at the White House and a settlement was assured—until the industry heard that the Court had taken the case. Fairless still wanted to settle but could not convince the rest of the industry. Interview with Arthur Goldberg, September 11, 1972.

Judge Charles Fahy of the United States Court of Appeals for the District of Columbia Circuit felt strongly that the Supreme Court should have returned the case to the court of appeals, to allow time for a settlement. Fahy believes the Supreme Court made an unnecessary decision on executive power. It is better to avoid big constitutional decisions, if possible, Fahy emphasized. Interview with Judge Fahy, March 10, 1972.

C. Herman Pritchett observed: "The doctrine of judicial self-restraint rigorously applied in this situation would call for upholding the President's action if at all possible, or, alternatively, for finding some reason for judicial nonintervention." *Civil Liberties and and the Vinson Court*, p. 207. See also pp. 206–13.

The classic statement of the Court's self-imposed restrictions on judicial review of constitutional questions was made by Justice Brandeis in a concurring opinion in *Ashwander* v. *Tennessee Valley Authority*, 297 U.S. 288, 346–48 (1936). See also McCloskey, *American Supreme Court*, pp. 20–21.

55. Freund, "The Year of the Steel Case," *Harvard Law Review*, pp. 90–91. Freund points out that this was not a procedure the Court was unfamiliar with. The Court had taken just this course when, on a motion for a temporary injunction, the constitutionality of the National Industrial Recovery Act was challenged as an improper delegation of power. The Court returned the case to the lower court with a statement that important constitutional questions need not be considered at that stage of the proceedings before the facts had been fully determined (*Wilshire Oil Co.* v. *United States*, 295 U.S. 100, 102 (1935)). *Ibid.*, pp. 89–90.

56. Weston, "Political Questions," pp. 331–32.

57. There are exceptions to this broad generalization, as the Court itself pointed out when it discussed the political question doctrine in *Baker* v. *Carr*, 369 U.S. 186 (1962). See too Black, *Perspectives in Constitutional Law*, pp. 14–18; U.S. Senate, *Constitution of the United States, Analysis and Interpretation*, pp. 546–49; Westin, *Anatomy of a Constitutional Law Case*, p. 23; Post, *The Supreme Court and Political Questions*.

58. Bickel, *The Least Dangerous Branch*, p. 184.

59. See McCloskey, *American Supreme Court*, pp. 191–92.

60. Corwin, "The Steel Seizure Case," p. 62. See too Scharpf, "Judicial Review and the Political Question," pp. 585, 587.

61. See Freund, "Year of the Steel Case," pp. 91–92, and Justice Douglas' opinion for the Court in *United States* v. *Causby*, where he construed the government's use of air space over private property to be a taking under the Constitution and entitled to just compensation. The purpose for which the air space was used was authorized by Congress, although the statute did not give specific authority to take private property in that manner. Douglas stated, "If there is a taking, the claim is 'founded upon the Constitution' and within the jurisdiction of the Court of Claims to hear and determine." Douglas cited *Hurley* v. *Kincaid* in support of this proposition. 328 U.S. 256, 267 (1946).

The court of appeals had implicitly recognized the validity of the *Hurley* v. *Kincaid* argument when it issued its memorandum on the granting of a stay of Judge Pine's injunction. See too "Comments on Recent Cases," pp. 592–93; Hasson, "The Steel Seizure Cases," p. 56.

62. 321 U.S. 414, 440 (1944).

63. 285 U.S. 95, 104n (1932).

64. See *United States* v. *Curtiss-Wright Export Corp.*, 299 U.S. 304 (1936);

Chicago and Southern Air Lines, Inc. v. *Waterman Steamship Corp.*, 333 U.S. 103 (1948); Westin, *Anatomy of a Constitutional Law Case*, p. 24.

65. See *In re Neagle, In re Debs,* and *Ex parte Quirin. Quirin,* the Nazi saboteurs case (317 U.S. 1 (1942)), involved the President's order to try the saboteurs before a military tribunal. Although Congress had not passed a statute specifically to that effect, the Court construed the President's order as in keeping with the general directives contained in the Articles of War enacted by Congress.

66. See, for example, the letter of Attorney General Francis Biddle to Congressman Robert F. Jones, May 29, 1942, defending certain federal agencies' power, pursuant to an executive order, to make or guarantee loans to contractors engaged in business operations necessary to the prosecution of the war. Biddle stated that although there existed no explicit statute authorizing these loans, it was to be assumed from other legislation that Congress intended the program of procurement of essential war materials to be carried out as expeditiously as possible and the President's executive order was issued with this purpose in mind. "It has been recognized throughout our constitutional history," Biddle asserted, "that action, not expressly authorized by Congress but incident and necessary to the faithful performance of duties imposed by Congress, must often be taken by the various branches of the Executive Arm of the Government." The Constitution vests in the President the executive powers which enable him to take whatever measures he deems necessary to fulfill the obligations imposed on him by Congress, Biddle continued. Moreover, as commander in chief, the President has the additional duty of ensuring that the armed forces of the United States are properly equipped for combat. From these constitutional sources the President derives the power to promulgate his executive order, Biddle concluded. See pp. 4–7, File #146-25-04, Department of Justice Records.

67. See Corwin, *The President,* pp. 147–49.

The Supreme Court reaffirmed the *Midwest Oil* doctrine in *Inland Waterways Corp.* v. *Young,* 309 U.S. 517 (1940). In that decision, Justice Frankfurter stated for the Court: "Illegality cannot attain legitimacy through practice. But when legality itself is in dispute—when Congress has spoken at best with ambiguous silence—a long continued practice pursued with the knowledge of the Comptroller of the Currency [in that particular case] is more persuasive than considerations of abstract conflict between such a practice and purposes attributed to Congress. . . . Even constitutional power, when the text is doubtful, may be established by usage." *Ibid.,* pp. 524–25. Black and Douglas joined Frankfurter in this opinion. Reed, the only other Justice on the Court in 1940 as well as in 1952, took no part in the case.

TEN ★ THE SUPREME COURT RULES

1. The Supreme Court, as an institution, has been throughout its history very secretive about the decision-making and opinion-writing processes in a particular case. Individual Justices differ, however, in degree of sensitivity to this issue. Justice Black, for one, ordered that his conference notes be burned after his death. For Black's views on the sacredness of the conference, see Frank, *Mr. Jus-*

tice Black, pp. 129–30. Justice Burton's papers, on the other hand, contain his conference sheets as well as his diary, which included some informative comments on the *Steel Seizure* case. His papers, in fact, constitute the only participant's account of the case I could find. Justice Black's Court papers, at the Library of Congress, and his general correspondence provided little information on the *Youngstown* case. Frankfurter's Supreme Court papers, at Harvard University Law School, and Douglas' at the Library of Congress, are closed. The papers of Justice Jackson are in the hands of Professor Philip Kurland, and access is restricted. Vinson's papers, located at the University of Kentucky, are closed for the period he was Chief Justice. Minton's papers, at the Truman Library, were unrevealing in regard to the *Steel Seizure* case, and Clark's and Reed's papers are unavailable.

2. The Justices held unofficial conferences with one another to discuss the seizure case, and it was often a topic for luncheon conversation. See Harold Burton's Diary, May 14 and May 15, 1952, Box 2, Burton Papers, LC.

3. *Ibid.*, May 10, 1952. See also June 2, when Burton wrote: "At one time I had found I might be the only one to upset the seizure."

4. *Ibid.*, May 15, 1952.

5. *Ibid.*, May 16, 1952. See also Conference Sheet for Friday, May 16, 1952, Box 235, Burton Papers, LC.

6. Westin, *Anatomy of a Constitutional Law Case*, pp. 126–28.

7. The Black opinion was circulated on May 28; the concurring opinions and the dissent made the rounds after that. See Justice Burton's Diary, May 28 and May 30, 1952, Box 2, Burton Papers, LC.

8. *Youngstown Sheet & Tube Co.* v. *Sawyer*, 343 U.S. 579, 589 (1952).

9. Sawyer, *Concerns of a Conservative Democrat*, pp. 268–69.

10. According to Justice Frankfurter, "An opinion of the Court is a controlling precedent for the relevant views expressed in such opinion. But when there is no opinion of the Court there is no precedent of views for the future." Frankfurter to Irving Dilliard, October 11, 1951, Box 52, Frankfurter Papers, LC.

11. 343 U.S. 579, 589. Sensitive to critical comment on the number of opinions in the *Steel Seizure* decision, Frankfurter wrote to one of his favorite correspondents:

"Which reminds me that I am also a bit puzzled that you should feel so strongly against multiplicity of opinions. As I said in my oral statement in the *Steel* case, there couldn't be a single opinion on behalf of all those who concurred in the result on issues such as those raised by the *Steel* case without covering up ambiguities and differences which ought not to be covered up and which only beget difficulties for the future. Why are we so squeamish about multiplicity of opinions? You know very well that that is the custom to this day in the courts of Great Britain and Australia and Canada. That was the custom in the beginning of this Court's history and Jefferson, in my view, was dead right in thinking that on these far-reaching constitutional and public law questions men should stand up and explain their position in their own way and not hide behind question-begging, evasive, litigation-breeding generalities." Frankfurter to Charles C. Burlingham, June 5, 1952, Box 36, Frankfurter Papers, LC.

A few days later Frankfurter again wrote to Burlingham:

"And now a word about concurring opinions. The Steel case is a good illustration of their necessity and desirability. As the senior on the majority side Black had the assignment of the case and assigned it to himself for very excellent reasons. But his views on the separation of powers are so rigid from my point of view, as well as from that of Bob Jackson's for instance, that we could not possibly agree to what he felt it necessary to write. And yet it was deemed desirable to have an opinion of the Court. That could be brought about only be agreeing with reservations and setting forth one's individual views. What's the matter with that? Why is our Bar less able to extract holdings from a multiplicity of opinions than the Bar of the other English-speaking nations?" June 8, 1952, Box 36, Frankfurter Papers, LC.

The Vinson Court had been consistently criticized for the number of opinions it produced on many of its decisions. Thomas Reed Powell noted that Vinson and his associates acted more like "a company of independent essay writers rather than . . . members of an official body wielding governmental power." Quoted in Swindler, *Court and Constitution in the Twentieth Century*, p. 182. See also Ralph F. Bischoff, "The Role of Official Precedents," in Cahn, *Supreme Court and Supreme Law*, p. 79; Frank, *Marble Palace*, pp. 123–29.

12. 343 U.S. 579, 584.

13. *Ibid.*, p. 585. Black ignored the *Pewee Coal Co.* case, in which the Court had measured damages resulting from a government seizure. Any damages arising from the seizure of the steel companies would presumably have been of a similar nature. See Nathanson, "The Supreme Court as a Unit of the National Government," pp. 342–43.

14. 343 U.S. 579, 585.

15. *Ibid.*, pp. 585–86.

16. *Ibid.*, p. 587.

17. *Ibid.*, p. 588.

18. *Ibid.*, p. 589.

19. 343 U.S. 579, 630.

20. *Ibid.*, pp. 631–32. In a footnote, Douglas accounted for wartime seizures by the military, which the Court had held did not require compensation, merely by stating that they were in a "different category." *Ibid.*, p. 632n.

21. 343 U.S. 579, 633–34. Douglas' final comment is a paraphrase of the argument presented in the industry's brief: "If arbitrary executive action to force a wage increase is lawful today, then arbitrary executive action to force a wage decrease, or longer hours, or anything else, will be equally lawful tomorrow." See U.S. House, *Steel Seizure Case*, p. 595.

22. 343 U.S. 579, 610.

23. *Ibid.*, p. 593.

24. This was the opposite of what the public expected in constitutional cases, Frankfurter declared. He noted in his opinion: "This eagerness to settle— preferably forever—a specific problem on the basis of the broadest possible constitutional pronouncements may not unfairly be called one of our minor national traits. An English observer of our scene has acutely described it: 'At the first sound of a new argument over the United States Constitution and its interpretation the hearts of Americans leap with a fearful joy. The blood stirs powerfully in their veins and a new lustre brightens their eyes. Like King Harry's men before Harfleur, they stand like greyhounds in the slips, straining upon the start.' *The Economist*, May 10, 1952, p. 370." 343 U.S. 579, 594.

25. *Ibid.*, p. 595.

26. 343 U.S. 579, 596.

27. *Ibid.*, pp. 598–602.

28. *Ibid.*, pp. 603–9. It is surprising that Justice Frankfurter, precise craftsman that he was, did not mention the power to seize private property given to the President in two statutes passed after the enactment of Taft-Hartley: the Selective Service Act of 1948 and the Defense Production Act of 1950. Although the seizure power granted in these acts had no specific reference to labor disputes, it was concerned with the ability of the government to obtain materials needed for the national defense. Neither provision expressly ruled out seizure where a work stoppage was responsible for the failure to produce the needed materials.

Frankfurter did state in his concurring opinion: "The Defense Production Act affords no ground for the suggestion that the 1947 denial to the President of seizure powers has been impliedly repealed and its legislative history contradicts such a suggestion." *Ibid.*, p. 607. Frankfurter is speaking only about Title V, however. Though it is true that the legislative history of Title V leaves that impression, the legislative history of Title II, Authority to Requisition (the 1951 amendments added the words "and condemn" to the title), which was not mentioned in Frankfurter's opinion, raises some doubts. The congressional committees dealing with the Defense Production Act were aware that Title II gave the President the power to seize plants, although administration spokesmen disclaimed any intention to use it for such a purpose. See U.S. Senate, Committee on Banking and Currency, *Defense Production Act of 1950, Hearings*, pp. 19–20, 23, 35–36, 183–84; U.S. House, Committee on Banking and Currency, *Defense Production Act of 1950, Hearings*, pp. 97–98.

These seizure provisions and their legislative history should have tempered Justice Frankfurter's conclusion that "To find authority so explicitly withheld is not merely to disregard in a particular instance the clear will of Congress. It is to disrespect the whole legislative process and the constitutional division of authority between President and Congress." 343 U.S. 579, 609.

29. 272 U.S. 52, 177, as quoted in 343 U.S. 579, 610. Frankfurter did not indicate that Holmes had written in dissent.

30. 343 U.S. 579, 610–11.

31. *Ibid.*, p. 613.

32. *Myers* v. *United States*, 272 U.S. 52, 293, as quoted by Frankfurter in 343 U.S. 579, 613–14. Again, Frankfurter neglected to say that Brandeis had dissented in *Myers*.

33. 343 U.S. 579, 635.

34. *Ibid.*, p. 637.

35. *Ibid.*, p. 640.

36. *Ibid.*, p. 641.

37. *Ibid.*, p. 644.

38. *Ibid.*, p. 645.

39. *Ibid.*, p. 646.

40. *Ibid.*, p. 655. A noted law professor described Jackson's concurrence as "a most brilliant exposition of 'undefined presidential powers' and their relation to legislation." See Jaffe, "Mr. Justice Jackson," p. 989n. For a different opinion see Petro, "The Supreme Court and the Steel Seizure," p. 454.

41. 343 U.S. 579, 655–57.

42. *Ibid.*, p. 660. It is interesting to note that, in preliminary drafts of his opinion, Justice Burton showed greater concern both about the emergency which existed ("The immediate crisis . . . is the threatened occurence [*sic*] of a nationwide strike affecting the entire steel industry. I accept the Government's conclusion that its occurence [*sic*] would imperil the national safety with particular relation to our national defense and to our military responsibilities in Korea and elsewhere") and about the possibility of a strike resulting from the Court's decision. In one such draft, Burton urged that the mandate of the Court be issued, as it usually was, in twenty-five days, and that during the interim between announcing its decision and issuing its order, the Court should continue in effect the stay it had granted when it took jurisdiction of the case. The parties then would be given another chance to compose their differences and avert a strike. See Case #744-745, Notes and Memos, Box 242, Burton Papers, LC.

43. 343 U.S. 579, 662.

44. *Ibid.*, pp. 664–66.

45. *Ibid.*, pp. 679–80.

46. Vinson quoted at length from John W. Davis' brief in the *Midwest Oil* case (*ibid.*, pp. 689–91), which supported the dissenters' view of presidential power. When the opinions were read in Court, Justice Frankfurter had commented in reference to this quotation: "I don't understand the relevance of quoting the brief of a lawyer in a case . . . with the well-known astigmatism of advocates." Quoted in the *New York Times*, June 3, 1952, p. 23.

47. 343 U.S. 579, 687–88. Clinton Rossiter commented: "In the light of the Debs and Neagle cases, it might easily be argued that there are no judicial limits to the President's real or alleged 'inherent' power to protect the peace of the United States." *The Supreme Court and the Commander in Chief*, p. 41.

48. 343 U.S. 579, 700.

49. *Ibid.*, pp. 695–96.

50. Quoted in the *New York Times*, June 3, 1952, p. 23.

51. 343 U.S. 579, 702–3.

52. The minority Justices did not think that the Taft-Hartley Act could have been invoked on April 8. Taft-Hartley was "a route parallel to, not connected with, the WSB procedure," Vinson noted. Taft-Hartley authorized a maximum delay of eighty days for a strike, whereas the steel walkout had already been postponed for ninety-nine days. *Ibid.*, p. 707.

53. *Ibid.*, pp. 703–4.

54. *Ibid.*, pp. 708–9.

55. *Ibid.*, p. 710.

56. Chief Judge Harold Stephens of the United States Court of Appeals for the District of Columbia Circuit observed that "The long opinion of the Chief Justice comes down in essence to the short and erroneous proposition that emergency creates power. I was not surprised at our friend, Reed, and not greatly surprised at Minton. I was agreeably surprised at Clark." Stephens to Wilson McCarthy, June 17, 1952, Box 82, Stephens Papers, LC.

57. Arthur Krock commented: "The suggestion has been made that the decision was received as it was because it was popular, and because five of the six justices in the majority held against the leader of their party and the executive power from which they had received their appointments. This doubtless explains why the applause has been so widespread." *New York Times*, June 5, 1952, p. 30.

Willard Hurst wrote: "The general approval of the Steel Seizure decision of 1952 expressed a different judgment [from that expressed in 1934–1936] of the balance between the need for action and the desirability of restraint." "Review and the Distribution of National Powers," in Cahn, *Supreme Court and Supreme Law*, p. 157.

58. June 3, 1952, p. 28. See also Chicago *Tribune*, June 4, 1952, p. 20; Cincinnati *Enquirer*, June 3, 1952, p. 4; Jackson (Tennessee) *Sun*, June 3, 1952, p. 4 (clipping in the John W. Davis Papers, Yale University); *New York Times*, June 3, 1952, p. 1.

59. June 3, 1952, p. 21. Arthur Krock observed in his column that he had heard numerous comments in Washington to the effect that it was better to suffer a strike than to expand the President's emergency powers during peacetime. See *New York Times*, June 3, 1952, p. 28.

60. June 3, 1952, p. 22.

61. June 3, 1952, p. 1. See also New York *Herald Tribune*, June 3, 1952, p. 22. For additional editorials appearing on that date praising the Court's decision, see Cleveland *Plain Dealer*, p. 14; *Wall Street Journal*, p. 10; *Washington News*, p. 22; Washington *Evening Star*, p. 12; Los Angeles *Times*, Part II, p. 4; Wheeling *Intelligencer*, p. 4; Pittsburgh *Post-Gazette*, p. 12.

62. June 3, 1952, p. 29.

63. *Congressional Record*, 98:6278.

64. *Ibid.*, p. 6289.

65. *Ibid.*, p. 6483.

66. *New York Times,* June 3, 1952, p. 23.

To his constituents, Lehman wrote: "The Taft-Hartley Act . . . gives scant promise of a basic solution. . . . Now that the Supreme Court has ruled the President's seizure illegal, the obligation is on Congress to enact a law which will specifically authorize the President to seize plants in industries in which a prolonged strike would critically affect the national security of the United States." Undated, Departmental File #19, Drawer 109, Lehman Papers, Columbia University.

67. Quoted in the Pittsburgh *Post-Gazette,* June 3, 1952, p. 6.

68. Quoted in the *New York Times,* June 3, 1952, p. 24.

69. *New York Times,* June 3, 1952, p. 24.

70. Quoted *ibid.,* p. 23.

71. Confidential Dictation, June 5, 1952, Sawyer Papers, Cincinnati, Ohio.

72. Felix Frankfurter wrote to Black, "That was a good party—Excellent, as far as your hostship . . . and your advice to your special guest." June 16, 1952, Special Correspondence, Box 5, Black Papers, LC.

73. Douglas, *Go East, Young Man,* p. 450. Douglas noted, however, that the good feeling did not last long. The Court usually paid an annual call on the President, Douglas reported. During the bitter days of the New Deal and the Court-packing plan, the call did not take place. It was resumed in 1939 and Truman had continued the custom. "But the year the Court held that Truman had acted unconstitutionally in seizing the steel mills during the Korean War, neither the White House nor the Court tried to arrange a meeting." *Ibid.,* p. 328.

74. Truman, *Memoirs,* 2:539. The *Steel Seizure* decision undoubtedly contributed to Truman's very negative feelings about Justice Clark. See Merle Miller, *Plain Speaking,* p. 225.

75. Virtually every law journal had something to say about the *Steel Seizure* case. See, for example, Banks, "Steel, Sawyer, and the Executive Power," pp. 467–537; H.L.C., "Constitutional Law—The President's Power of Eminent Domain," pp. 129–34; "Comments on Recent Cases," pp. 590–93; Eaton, "Constitutional Guarantees and the Steel Decision," pp. 220–33; Griffith, "Constitutional Law—Power of Executive to Seize Private Property," pp. 90–92; Grisanti, "Comment: Constitutional Law—Separation of Governmental Powers," pp. 460–63; Hearon, "Constitutional Law—Distribution of Governmental Powers and Functions," pp. 885–88; Lea, "The Steel Case," pp. 289–313; Richberg, "The Steel Seizure Cases," pp. 713–27; Saylor, "Court over the President," pp. 38–46.

76. See, for example, Freund, "Year of the Steel Case," p. 95; "The Supreme Court, 1951 Term," p. 104; Bernard Schwartz, *The Supreme Court,* pp. 71, 75–76.

77. See Corwin, *The President* (1957 ed.), p. 156; and Corwin, "Steel Seizure Case," p. 65.

78. Corwin, "Steel Seizure Case," p. 64. Corwin agreed with Justice Clark's approach to the case.

79. Corwin, *The President* (1957), p. 155. Chief Justice Vinson detailed this history in his dissent. 343 U.S. 579, 683–701.

80. See, for example, Mendelson, *Justices Black and Frankfurter*, pp. 10–14, Schubert, "The Steel Case," p. 66; Black, *Perspectives in Constitutional Law*, p. 62; Binkley, *The Man in the White House*, p. 121.

81. Freund, "Year of the Steel Case," p. 95.

82. *Ibid.*, pp. 94–95. Glendon Schubert commented: "One must go back sixteen years to the Butler case, the Carter case, and other decisions of the first term of the New Deal to find a majority of the Supreme Court using language of equal extravagance in a case of constitutional conflict with one of the other branches of the national government." "The Steel Case," p. 64.

83. 343 U.S. 579, 602. See pp. 598–604 for the whole argument. Glendon Schubert has commented that Frankfurter's opinion consisted of a "psychoanalysis of Congress, to divine that the Congress had in effect, although not expressly, disapproved the exercise of presidential seizure powers." See "The Steel Case," p. 67.

84. *Congressional Record*, 98:3965. Morse was referring to an amendment proposed to the Defense Production Act, but the principle is the same.

85. Interview with Judge Charles Fahy, March 10, 1972; Black, *Perspectives in Constitutional Law*, p. 62; Bickel, *Least Dangerous Branch*, pp. 215–16.

86. 330 U.S. 258 (1947).

87. *Ibid.*, pp. 282–83.

88. *Ibid.*, p. 282n. Justice Frankfurter concurred in the result in the *United Mine Workers* case, but on other grounds. He disagreed with the Court's treatment of the legislative history of the War Labor Disputes Act. *Ibid.*, p. 310.

89. *Congressional Record*, 93:3645.

90. *Ibid.*, pp. 3835–36; Frankfurter opinion, 343 U.S. 579, 600–601.

91. 343 U.S. 579, 604.

92. "Essential Principles of the Taft-Hartley Law, and Amendments Proposed by the Republican Minority," Statement of Hon. Robert A. Taft in Senate, May 4, 1949, p. 4, Box 614, Taft Papers. For background on this amendment see Aaron, "Amending the Taft-Hartley Act," pp. 330–31; David B. Truman, *The Congressional Party*, p. 23. For details of Sen. Taft's amendments to the Thomas Bill and the effect on the present Taft-Hartley Act, see analysis prepared by attorneys representing the AFL, CIO, and independent unions, p. 2, United Steelworkers of America, Papers, Pennsylvania State University; R. Alton Lee, *Truman and Taft-Hartley*, pp. 176–77; *Congressional Record*, 95:8717.

93. This does not imply that the decision striking down the seizure was incorrect, for the government did not rely in any way on statutory power.

94. 343 U.S. 579, 656–60, 639n8.

95. Clark, too, accepted Frankfurter's analysis of the legislative history of the

Taft-Hartley Act, but the crucial point of his opinion did not depend on it. 343 U.S. 579, 663.

96. *Ibid.*, p. 662.

97. "The Steel Seizure Case," p. 65.

98. Swisher, *Historic Decisions of the Supreme Court*, p. 147.

99. Schwartz, "Inherent Executive Power and the Steel Seizure Case," p. 478.

100. The opinions discussed the President's power to seize; they made no pretense of talking about Secretary Sawyer's power. Black's opinion for the Court stated: "The President's power, if any, to issue the order must stem either from an act of Congress or from the Constitution itself. There is no statute that expressly authorizes the President to take possession of property as *he* did here." [Emphasis added.] 343 U.S. 579, 585. Schwartz, "Inherent Executive Power and the Steel Seizure Case," pp. 479–80. See also Schwartz, *American Constitutional Law*, pp. 202–4; Fahy, "Judicial Review of Executive Action," pp. 712–13.

Judge Learned Hand disapproved of the Court's action in this respect. To Justice Frankfurter he wrote: "I think I should have taken the point that to enjoin the Secretary was to enjoin the President, save that nobody seems to agree with me." June 13, 1952, Box 64, Frankfurter Papers, LC.

101. Schubert, "The Steel Case," p. 74.

102. "Supreme Court, 1951 Term," pp. 103–4.

103. *Ibid.*, p. 103.

104. Kauper, "The Steel Seizure Case," pp. 181–82.

105. Schwartz, "Inherent Executive Power and the Steel Seizure Case," pp. 476–77.

106. Learned Hand to Felix Frankfurter, June 13, 1952, Box 64, Frankfurter Papers, LC. Dean Erwin Griswold of the Harvard Law School, future solicitor general of the United States, agreed with Judge Hand. Griswold wrote to Frankfurter: "My chief thought on the whole question is that the question should never have been presented to you. This seems to me to be one of those issues which it is perhaps best to leave in indefinite terms for as long as possible." June 2, 1952, Box 59, Frankfurter Papers, LC.

For more comments on the decision see Corwin, *The Constitution and What It Means Today*, p. 128; Corwin and Peltason, *Understanding the Constitution*, p. 72; Hyman, *The American President*, p. 307; McCloskey, *The American Supreme Court*, p. 190; Rankin and Dallmayr, *Freedom and Emergency Powers in the Cold War*, pp. 161–64; Schubert, *Constitutional Politics*, pp. 324–25; Smith and Cotter, *Powers of the President during Crises*, pp. 126, 134, 139; Swindler, *Court and Constitution in the Twentieth Century*, p. 182; Vanderbilt, *Doctrine of the Separation of Powers*, pp. 127–28n; Frank, "The Future of Presidential Seizure," p. 70; Monaghan, "Presidential War-Making," p. 22; Tanenhaus, "The Supreme Court and Presidential Power," pp. 106, 108, 113.

107. Schubert, *Presidency in the Courts*, p. 250.

108. Pritchett has written: "The plain fact of the matter is that the Supreme

Court of the United States always has been and, so long as it retains its present powers, always will be a political institution. The notion that the Court's justices are political eunuchs quietly going about their work of applying permanent canons of interpretation in the settlement of individual disputes with no relation to the province of government or the formulation of policy, has never corresponded with the facts. . . .

"The Supreme Court has been from the first an active participant in the power struggles of American politics. It could not be otherwise when the most important political questions must sooner or later be couched in the form of a lawsuit and presented to the Court for adjudication before the legality of governmental action is known. . . . Where the potential consequences of judicial decisions are so great, judges must be politicians in the higher sense of that much abused word." *Roosevelt Court*, p. 16. See also Jackson, *The Supreme Court in the American System of Government*, pp. 55–56; Schwartz, *The Supreme Court*, p. 141; Latham, "The Supreme Court as a Political Institution," p. 207.

109. Mason, *The Supreme Court from Taft to Warren*, pp. 186–87.

110. *Nine Men*, p. 316. See also Paul L. Murphy, *The Constitution in Crisis Times 1918–1969*, p. 290. The *Harvard Law Review* wrote: "The lasting importance of the decision would seem to rest with the political effect. . . . It is for this reason, perhaps, that the Court did not rest on the narrower grounds possible, but concerned itself with the larger and more popularly appreciated problems of power under the Constitution" ("Supreme Court, 1951 Term," p. 104).

111. Frankfurter, "Robert H. Jackson," p. 437. See too Francis Biddle to Felix Frankfurter, Dec. 10, 1962, Box 149, Frankfurter Papers, LC; Jackson's concurring opinion, 343 U.S. 579, 651–52; Frankfurter's opinion, 343 U.S. 579, 613.

112. 343 U.S. 579, 642.

113. Charles A. Miller, *The Supreme Court and the Uses of History*, p. 32; Rossiter, *The Supreme Court and the Commander in Chief*, p. 47.

114. Roche, "Executive Power and Domestic Emergency," p. 617.

115. 343 U.S. 579, 708.

116. After the Supreme Court's decision, President Truman asked James McGranery, the new Attorney General, for his opinion. McGranery said the matter "had not been presented as well as possible—the Court had not been given sufficient evidence of the emergency." Perlman, who was also present, replied that "there had indeed been such evidence before the Court" and he mentioned the affidavits filed by government officials and his own argument before the Court. This conversation is reported in Sawyer, *Concerns of a Conservative Democrat*, p. 271. See also Corwin and Peltason, *Understanding the Constitution*, p. 72; Smith and Cotter, *Powers of the President during Crises*, p. 139.

117. *Imperial Presidency*, p. 148. See too Roche, "Executive Power and Domestic Emergency," pp. 617–18; Jay W. Murphy, "Some Observations on the Steel Decision," *Alabama Law Review*, p. 230.

118. Interview with Charles Murphy, September 22, 1972; interview with

David Stowe, January 9, 1973. Union officials also knew that the need for steel was not pressing at that time. Interview with Arthur Goldberg, September 11, 1972; interview with Elliot Bredhoff, September 6, 1972. An attorney in the solicitor general's office at the time of the seizure told me that it was clear to Justice Department officials soon after the Supreme Court decision that the Commerce Department had exaggerated the disaster that would occur if steel production ceased. Interview with Judge Oscar Davis, January 3, 1973.

119. Memorandum, "Future Action in the Steel Case," Enarson and Kayle to Dr. Steelman, May 23, 1952, Box 4, Stowe Papers, TL.

120. Black, *The People and the Court*, pp. 51–55.

121. Justice Frankfurter stated this in his letter to Charles C. Burlingham, June 8, 1952, Box 36, Frankfurter Papers, LC.

In fact, less than two months before, the Court had found circumstances in which the President could exercise legislative power. See *Madsen v. Kinsella*, 343 U.S. 341 (1952). Only Justice Black dissented.

122. *Myers v. United States* (Justice Brandeis dissenting), 272 U.S. 52, 293 (1926), quoted in 343 U.S. 579, at 613 and 629.

ELEVEN ★ THE CONSTITUTIONAL SIGNIFICANCE OF THE YOUNGSTOWN DECISION

1. As Willard Hurst has noted: "The more significant the collisions of power, the more likely—if only because top-level conflicts spell unusually complex policy issues—that the circumstances will be highly individual. Perhaps this is especially true where action of the Chief Executive comes into question. How far may the recent Steel Seizure decision have important limiting effects in the future, and how far may its very importance in its day limit its practical compulsion on events of another day? How significant for the future was the Dred Scott case except with regard to slavery?" "Review and the Distribution of National Powers," in Cahn, *Supreme Court and Supreme Law*, p. 147.

2. *Youngstown* has value as a precedent in other areas of the law too. In equity, for example, *Youngstown* has been cited as authority for the propriety of determining the merits of a case on a motion for preliminary injunction (see *The Delaware & Hudson Railway Company, et al. v. United Transportation Union*, 450 F.2d 603, 620 (1971)), and for the proposition that an injunction is the proper remedy when a government official acts without statutory authority and that the government will not be liable for damages resulting from such an unauthorized act (see *Central Eureka Mining Co. v. United States*, 138 F. Supp. 281, 297, 312 (1956) and 357 U.S. 155, 166n (1958)).

Youngstown has also become an established precedent cited when a party wants to petition the Supreme Court for certiorari before judgment in the court of appeals. See for example *Griffin v. School Board*, 375 U.S. 391, 392 (1964), and *United States v. Nixon*, 418 U.S. 683, 686 (1974).

3. 347 U.S. 483 (1954).

4. 349 U.S. 331 (1955). The case was one of several involving the legality of actions of the Loyalty Review Board. In *Peters*, the constitutional issue concerned

whether the petitioner had been denied an opportunity to confront and cross-examine his secret accusers. But the Court decided it could determine the case without ruling on the constitutional question. *Ibid.*, p. 338. Justice Black concurred, but he had this to say: "I agree that it is generally better for this Court not to decide constitutional questions in cases which can be adequately disposed of on non-constitutional grounds. . . . But this generally accepted practice should not be treated as though it were an inflexible rule to be inexorably followed under all circumstances . . . [cites *Youngstown*]. Here, as in the *Youngstown* case, I think it would be better judicial practice to reach and decide the constitutional issues, although I agree with the Court that the Presidential Order can justifiably be construed as denying the Loyalty Review Board the power exercised in this case. For this reason I join the opinion of the Court. But I wish it distinctly understood that I have grave doubt as to whether the Presidential Order has been authorized by any act of Congress. . . . These orders look more like legislation to me than properly authorized regulations to carry out a clear and explicit command of Congress. . . . The Constitution does not confer lawmaking power on the President [cites *Youngstown*]." *Ibid.*, pp. 349–50. See too Kalven, "Upon Rereading Mr. Justice Black on the First Amendment," p. 437.

5. Bickel, *Least Dangerous Branch*, pp. 189–94.

6. 328 U.S. 549 (1946). The case was heard by only seven Justices and was decided 4 to 3. But only Justices Reed and Burton joined Frankfurter's opinion, so it was not an opinion of the Court.

7. *Ibid.*, p. 552.

8. *Ibid.*, p. 553.

9. *Ibid.*, p. 556.

10. 369 U.S. 186 (1962).

11. *Ibid.*, p. 217.

12. *Ibid.*, p. 246n.

13. *Powell* v. *McCormack*, 266 F. Supp. 354, 359 (1967).

14. 395 U.S. 486, 548 (1969).

15. *Ibid.*, p. 549. Judge Hart, in his opinion, had distinguished the *Youngstown* case on its facts: It did not involve a separation-of-powers issue between the legislature and the judiciary. 266 F. Supp. 354, 357 (1967). See also *Doe* v. *McMillan* (412 U.S. 306 (1973)), where the Supreme Court took jurisdiction of an issue which required an interpretation of the scope of congressional immunity under the Speech or Debate clause of the Constitution.

16. 357 U.S. 116 (1958).

17. 381 U.S. 1 (1965).

18. 357 U.S. 116, 128 (1958).

19. 381 U.S. 1, 12–13 (1965). Other grounds asserted for the propriety of the Secretary's embargo on travel to Cuba were that the Department of State has consistently imposed area restrictions (pp. 8–11); that restriction on travel does not abridge constitutional rights (pp. 13–18); that liberty can be inhibited through due

process of law (p. 14) and that the restriction was justified by national security considerations (pp. 14–15).

20. *Ibid.*, pp. 20–21.

21. *Ibid.*, p. 30. A case which involved a similar question of power—whether the President could obtain "voluntary" steel import restrictions from foreign producers when Congress is assigned the power to regulate foreign trade—was decided in October, 1974, by the U.S. Court of Appeals for the D.C. Circuit in favor of the President (*Consumers Union of U.S., Inc.*, v. *Henry A. Kissinger, et al.*, 506 F.2d 136 (1974)). The majority held that the action of the President in setting up the restriction agreements was valid because it was a matter left to the President's discretion in the conduct of foreign affairs and because it was not preempted by Congress' statutory scheme for the conduct of foreign trade. Judge Harold Leventhal dissented, and the influence of *Youngstown* is evident throughout his opinion. Leventhal wrote that "Congress has enacted a comprehensive statutory scheme that identifies the Executive's role in shaping the terms of foreign trade. In doing so Congress has followed a well-established practice of combining broad executive discretion with procedural protection for those who may be affected by its exercise." The Executive's negotiation of the voluntary import restrictions on steel did not conform to the procedures outlined by Congress. "Whatever may be the President's prerogative when Congress has not spoken, it is untenable, in my view," Leventhal maintained, "to assert a continuing inherent Executive authority to negotiate import restraints outside and in disregard of a consistent and comprehensive Congressional pattern for protective procedures."

22. *Massachusetts* v. *Laird*, 400 U.S. 886 (1970).

23. 400 U.S. 886, 894. Douglas was rejecting the Court's tradition of respect for the President's prerogatives in the conduct of foreign relations.

Louis Henkin recently examined the basis for this "respect" shown by the Court and concluded that there was not a firm foundation in precedent for the Court's refusal to consider the constitutionality of the Vietnam War. See "Viet-Nam in the Courts of the United States," pp. 284–89. See also Henkin, *Foreign Affairs and the Constitution*, pp. 208–16.

24. 400 U.S. 886, 897.

25. *Ibid.*, p. 898.

26. *Ibid.*, pp. 896, 899.

27. Washington *Evening Star*, April 14, 1970, p. 3.

28. 317 F. Supp. 715, 729 (1970).

29. 327 F. Supp. 378, 381 (1971).

30. *Atlee* v. *Laird*, 347 F. Supp. 689, 701–2 (1972).

31. *Ibid.*, p. 710.

32. *DaCosta* v. *Laird*, 405 U.S. 979, 980 (1972).

33. Comparing the Court's action in the *Steel Seizure* case with its inaction in the Indochina War cases, one observer drew this conclusion: "Seizure of the steel industry is riskier for the President than waging war in Asia. Moreover, executive action which affects the well-established rights of powerful individuals in

this country is more likely to spark political backlash than action that impinges upon the more inchoate rights of less powerful persons. Since Congress is unlikely to bring an action against the President for infringement of its right to participate in war decisions, the plaintiff in the hypothetical case would probably be a serviceman seeking to avoid participation in the conflict. . . . Unlike the property rights at issue in *Youngstown*, a draftee or reservist's right to avoid involvement in an unauthorized war is not an interest that our legal system has traditionally recognized as worthy of protection. Further, the draftee's political power is minuscule compared to that of the steel magnate." W. Taylor Reveley III, "Presidential War-making: Constitutional Prerogative or Usurpation?" in U.S. Senate, Committee on Foreign Relations, *Documents Relating to the War Power of Congress*, p. 218n.

The Court exercised similar restraint when the question of the President's power to order bombings in Cambodia arose. Justice Douglas reiterated the lesson of the *Steel Seizure* case to no avail. See *Holtzman* v. *Schlesinger*, 414 U.S. 1304, 1308, 1312, 1316, 1318 (1973).

For a discussion of the Vietnam War in the courts, see Schlesinger, *Imperial Presidency*, pp. 288–95.

34. *New York Times* v. *United States*, 403 U.S. 713, 714 (1971).

35. Opinion of Justice Thurgood Marshall, *ibid.*, p. 742. See also Justice Byron White's opinion, pp. 733–40. The Espionage Act of 1917 dealt with the problem of unauthorized disclosure of information. In formulating that act, Congress had rejected a provision which would have given the President the power to make regulations proscribing the publication of certain kinds of information on national defense.

36. *United States* v. *Sinclair*, 321 F. Supp. 1074, 1080 (1971).

37. *United States* v. *U.S. District Court for the Eastern District of Michigan*, 444 F.2d 651, 660–61 (1971).

38. 18 U.S.C. §2511 (3).

39. Justice Powell pointed out that the Court was not deciding the scope of the President's wiretapping powers in regard to the activities of foreign countries, either within or without the United States. 407 U.S. 297, 308, 321–22.

40. By a series of cases, the Supreme Court had brought the protection of private speech from unreasonable surveillance under Fourth Amendment safeguards. *Ibid.*, p. 313.

41. *Ibid.*, p. 317.

42. *Ibid.*, pp. 318–21. Douglas, the only Justice to write a concurring opinion, pointed out the large number of judicially unapproved wiretaps used by the Nixon administration and declared: "When the Executive attempts to excuse these tactics as essential to its defense against internal subversion, we are obliged to remind it, without apology, of this Court's long commitment to the preservation of the Bill of Rights from the corrosive environment of precisely such expedients." Here Douglas cited *Youngstown* and other cases where the Court had upheld individual rights. *Ibid.*, pp. 331–32.

43. William Rehnquist (then assistant attorney general, Office of Legal

Counsel) to deputy counsel to the President, December 19, 1969, quoted in Schlesinger, *Imperial Presidency*, p. 237.

44. For a discussion of Nixon's impoundment actions, see Schlesinger, *Imperial Presidency*, pp. 235–40, 397–400.

45. 367 F. Supp. 686 (1973).

46. *Ibid.*, p. 698.

47. *Local 2677, the American Federation of Government Employees* v. *Phillips*, 358 F. Supp. 60, 76–77 (1973). See also Schlesinger, *Imperial Presidency*, pp. 241–42.

48. 360 F. Supp. 1 (1973).

49. 487 F.2d 700 (1973).

50. Letter of July 25, 1973, quoted in Memorandum in Support of an Order to Produce Documents or Objects in Response to the Subpoena, *In re Grand Jury Subpoena Duces Tecum Issued to Richard M. Nixon, or Any Subordinate Officer, Official, or Employee with Custody or Control of Certain Documents or Objects*, Misc. No. 47-73, U.S. District Court for the District of Columbia, p. 1.

51. Brief in Opposition, *In re Grand Jury Subpoena*, Misc. No. 47-73, U.S. District Court for the District of Columbia, p. 2.

52. *Ibid.*, p. 3.

53. *Ibid.*, p. 24.

54. *Ibid.*, p. 4.

55. *Ibid.*, p. 7. Compare Holmes Baldridge's oral argument before Judge Pine in the *Steel Seizure* case.

56. Brief in Opposition, *In re Grand Jury Subpoena*, pp. 29–30.

57. *In re Grand Jury Subpoena Duces Tecum Issued to Richard M. Nixon*, 360 F. Supp. 1, 5, 6 (1973).

58. *Ibid.*, p. 8.

59. *Ibid.* The special prosecutor noted in his brief that, in the *Steel Seizure* case, Judge Holtzoff had declined to issue a temporary restraining order because an act of the President was involved and the courts could not enjoin the President. But Judge Pine, and after him the Supreme Court, had not hesitated to strike down the President's executive order. See Memorandum in Support, *In re Grand Jury Subpoena*, p. 25n.

60. Quoted in Memorandum in Support, *In re Grand Jury Subpoena*, p. 23.

61. 360 F. Supp. 1, 8–9. See discussion in Memorandum in Support, *In re Grand Jury Subpoena*, pp. 23–25. See also editorial, *New York Times*, August 15, 1973, p. 32.

62. *Youngstown Sheet & Tube Co.* v. *Sawyer*, 343 U.S. 579, 635, as quoted in 360 F. Supp. 1, 9.

63. 360 F. Supp. 1, 9.

64. *Ibid.*, p. 14.

65. 487 F. 2d 700 (1973). The court of appeals did modify Judge Sirica's order, however, by establishing a specific procedure for *in camera* inspection.

66. 487 F. 2d 700, 709.

67. *Ibid.*

68. 4 Wall. 475, 498 (1867).

69. 6 Wall. 50 (1867).

70. 487 F.2d 700, 712n. See, for example, *National Treasury Employees Union* v. *Nixon*, 492 F.2d 587, 606–8, 613–14 (1974). After pointing out that *Mississippi* v. *Johnson* specifically left unanswered the question of whether a court can order the President to perform a ministerial act (a duty imposed by law involving no discretion), the court of appeals in this case answered the question in the affirmative. The court discussed at length the interpretation of *Youngstown* put forth in *Nixon* v. *Sirica* and held that the President is subject to judicial process. Executive acts had always been reviewed by the courts, and if no relief could be obtained from a subordinate executive branch official, the President could be sued, the court of appeals ruled. 492 F.2d 587, 614–15.

71. 418 U.S. 683 (1974).

72. The President's counsel revealed in court that two of the subpoenaed conversations had not been recorded and a third had an inexplicable eighteen-and-one-half-minute gap.

73. *United States* v. *Mitchell, et al.*, D.D.C. Crim. No. 74-110.

74. Opinion and order of May 20, 1974, 377 F. Supp. 1326 (1974). Judge Sirica also considered and rejected a motion by the President to expunge the grand jury's action in naming him a coconspirator in the Watergate cover-up and to enjoin everyone but the President and his counsel from ever disclosing this information. When the case came before the Supreme Court, this issue was moot because the press had already reported the grand jury's finding. See Chief Justice Burger's opinion., *United States* v. *Nixon*. The district court also ruled that the special prosecutor had met the requirements for a subpoena *duces tecum* specified in the Federal Rules of Criminal Procedure (17(c)): the materials subpoenaed were relevant and evidentiary in nature.

75. *United States* v. *Nixon*, 418 U.S. 683, 686n (1974).

76. Only eight Justices participated in the decision because Justice Rehnquist recused himself.

77. The issues were technically different, in that *In re Grand Jury Subpoena* and *Nixon* v. *Sirica* involved the right of the grand jury to every man's evidence and *United States* v. *Nixon* concerned whether the President could withhold evidence needed by the prosecution in the trial, on criminal charges, of the President's own associates. The separation-of-powers argument put forward by the President was the same in both cases: the Chief Executive was the absolute and sole judge of his own privilege, and he could not be reached by compulsory process of the courts.

78. Before reaching the privilege question, Chief Justice Burger concluded that the Supreme Court had proper jurisdiction of the case since the district court order was a final, appealable order for review of which the Court could grant certiorari before judgment in the court of appeals. The Chief Justice also held that the case represented a justiciable question, notwithstanding the fact that it concerned a dispute between a subordinate and a superior officer of the executive branch. And the Court also ruled that the special prosecutor had satisfied all the requirements of Federal Rule of Criminal Procedure 17(c).

79. 1 Cranch 137 (1803).

80. 369 U.S. 186, 211 (1962).

81. *United States* v. *Nixon*, 418 U.S. 683, 704–5 (1974).

82. *Ibid.*, p. 707.

83. *Ibid.*

84. *Ibid.*

85. *Ibid.*, p. 713.

86. *Ibid.*, p. 714.

EPILOGUE

1. Corwin commented: "In the Steel Seizure case . . . the Court in 1952 professed to believe that it had found in the principle of the Separation of Powers a judicially enforceable concept in restraint of Presidential emergency power, but . . . this was an empty gesture. If nature abhors a vacuum, so does an era of crisis, and a vacuum is all that Judicial Review has to offer in such a situation." *The Constitution and What It Means Today*, p. vi.

Nathaniel Nathanson, many years after the seizure decision, speculated about Justice Frankfurter's position: "I have sometimes wondered whether, if the division of the Court had been close enough to make his own vote decisive, the Justice might not have been content to dispose of it upon that ground [adequate remedy at law]. Perhaps as he watched the long strike which followed the Court's decision, without Congress lifting a finger to provide a solution as the Justice in his closing paragraph hopefully anticipated it would, he too may have wondered whether this may not have been another instance for the application of his oft-repeated principle: 'A basic rule is the duty of the Court not to pass on a constitutional issue at all, however narrowly it may be confined, if the case may, as a matter of intellectual honesty, be decided without even considering delicate problems of power under the Constitution.' " "Separation of Powers: The Justice Revisits His Own Casebook," in Mendelson, *Felix Frankfurter*, pp. 17–18.

2. Sawyer, *Concerns of a Conservative Democrat*, pp. 269–71.

3. *Ibid.*, p. 272. According to Sawyer, Truman "said that he would consider it [invoking Taft-Hartley], although he was a very stubborn man and did not easily change his mind." Confidential Dictation, June 5, 1952, Sawyer Papers, Cincinnati, Ohio.

4. Sawyer, *Concerns of a Conservative Democrat*, p. 273.

5. Special Message to the Congress on the Steel Strike, June 10, 1952, in U.S. President, *Public Papers 1952–1953*, p. 412.

6. *Ibid.*, p. 413. For a detailed analysis of the reasons for and against using Taft-Hartley see "Should the National Emergency Provision of the Taft-Hartley Act be Invoked in the Current Steel Industry Dispute?" Box 5, Enarson Papers, TL.

7. The amendments were offered by Senators Morse, Maybank, and Monroney. For the debate on these amendments, see *Congressional Record*, 98:6900 ff.

8. The amendment passed 49 to 30. *Ibid.*, p. 6926.

9. *Ibid.*, p. 8050; McConnell, *Steel Seizure of 1952*, p. 50.

10. Congress renewed the Defense Production Act with amendments that forbade the Wage Stabilization Board to handle labor disputes in industries important to the defense program and further weakened the inflation control measures contained in the previous acts. *Congressional Record*, 98:8409, 8522–39. Truman signed the bill on June 30, 1952, and issued a statement excoriating Congress for the debilitating changes it had made in the law. See Statement by the President on the Defense Production Act Amendments, July 1, 1952, U.S. President, *Public Papers 1952–1953*, pp. 453–54.

The House Committee on Education and Labor had sent its report on the WSB to the House on June 17, 1952. Among its conclusions were:

"In recommending that the parties agree to some kind of union shop in the Steel case, the Douglas Aircraft Co. case, and the Boeing Airplane Co. case the Wage Stabilization Board failed to respect the national labor policy as set forth in the Labor-Management Relations Act.

"The existence of the Wage Stabilization Board with jurisdiction in dispute cases in itself reduces the effectiveness of the Federal Mediation and Conciliation Service and interferes with collective bargaining.

"The Wage Stabilization Board has adopted policies and made decisions and recommendations inconsistent with the intent of Congress with respect to stabilization and in contravention of the public interest.

"The policies and regulations of the Wage Stabilization Board have been unusually liberal in an emergency period. As applied to 'petition cases' they have been extremely liberal and as interpreted and modified in dispute cases they have been highly unstabilizing and inflationary.

"Recommendations by the Board in disputes cases have been made without consideration of their effect on prices and have thereby contributed to the failure of the stabilization program to hold the line against inflation." U.S. House, Committee on Education and Labor, *Investigation of the Wage Stabilization Board*, pp. 12, 15, 18, 23.

A minority of the committee believed that the WSB had done a creditable job, except in the steel case, and that the board's actions had not conflicted with national labor policy as expressed in the laws passed by Congress. U.S. House, Committee on Education and Labor, *Investigation of the Wage Stabilization Board*, p. 28.

11. See, for example, the President's News Conference, June 19, 1952, U.S. President, *Public Papers 1952-1953*, p. 435. The decision not to use Taft-Hartley was made by the President, and he stuck with it despite advice to the contrary from his advisers. Interview with David Stowe, January 9, 1973. See also Memorandum, Enarson to Steelman, June 24, 1952, OF 407B, Truman Papers, TL.

12. June 4, 1952, p. 12. See too *New York Times*, June 3, 1952, p. 24.

13. National Production Authority, The Office of Labor, "Government Action in the Steel Dispute of 1952," December 1, 1952, p. 11, copy in Government Documents Steel Dispute, 1952, United Steelworkers of America Papers, Pennsylvania State University.

The same study noted that the actual effects of the strike "were not too serious; the economy functioned, the defense program went forward, and within a few months after the termination of the strike any lost ground in most important fields had been largely recovered." *Ibid.*, p. 12.

Though a few analysts describe the effects of the 1952 strike as disastrous (see, for example, Irving Bernstein, "The Economic Impact of Strikes in Key Industries," in Industrial Relations Research Association, *Emergency Disputes and National Policy*, pp. 38-42), many think its consequences have been exaggerated (see, for example, Livernash, *Collective Bargaining in the Basic Steel Industry*, p. 7). For differing views see Ching, *Review and Reflection*, p. 108; Statement by Secretary Sawyer, August 25, 1952, Department of Commerce Press Release, Decentralized Files of Various Directors of OPS, RG 295, NA; National Production Authority, Office of Labor, "The Impact of the 1952 Steel Strike," copy in Box 1, John C. Houston Papers, TL; U.S. Department of Defense, *Semiannual Report of the Secretary of Defense, January 1 to June 30, 1952*, p. 41; Memorandum, Leo R. Werts to Secretary Tobin, July 25, 1952, 1952 Defense Manpower Administration, Secretary Tobin, General Subject File 1952, NA.

A shortage of ammunition, reported in the summer of 1952, was blamed on the steel strike (see Harry S. Truman, *Memoirs*, 2:540), but sources in the Labor Department stated that this was inaccurate: "Reports that the steel strike has resulted in a shortage of ammunition are being treated with skepticism in informed quarters. The shortage refers to only certain types of heavy ammunition which require less than 1 percent of the total steel output, and the lack of steel can be remedied quickly. The rationing of ammunition, both in training and in actual combat is a standard procedure. If the ammunition program is seriously behind schedule—the report was that 35 percent of the 1952 output, or about four months' production had been lost—the major explanation probably lies in the delayed receipt of certain machine tools and the delay in establishing and staffing plant facilities, rather than the steel strike, which lasted less than two months." Memorandum, Ewan Clague to Secretary Tobin, August 6, 1952, 1952 Labor Statistics—General Subject File, Secretary Tobin, NA.

14. For an account of the negotiations, see Benjamin M. Selekman, Sylvia K. Selekman, and Stephen H. Fuller, *Problems in Labor Relations* (2d ed.; New York: McGraw-Hill, 1958), pp. 493-98.

15. McConnell, *Steel Seizure of 1952*, pp. 50-51.

16. The stumbling block was said to be the union shop agreement. Not only were the companies generally opposed to the union shop (although union shops existed in the mines and railroads owned by some of the companies), but their determination to resist it was strengthened by the government's attempts, through the WSB, to impose it on them. See Harbison and Spencer, "Politics of Collective Bargaining," pp. 716, 716n. Another analyst, however, states that the companies had agreed to the Bethlehem formula and that the last two weeks of the strike were spent negotiating on incentives and the fair-day's-work issue. Livernash, *Collective Bargaining in the Basic Steel Industry*, p. 107.

According to Arthur Goldberg, it was the union shop issue which prolonged the strike. Bethlehem Steel chose to compromise on that issue in order to end the strike, for that company was motivated purely by economic factors. But the other companies were willing to bear some economic losses in an attempt to uphold their principles. Interview, September 11, 1972.

Moreover, in view of the rhetoric of their public relations campaign, it seems logical that the companies would hold out as long as they could on that issue. Their theme had been that the union shop itself, and especially government imposition of it, was un-American. The industry had to put up a good fight on this issue if it was to maintain its public image.

17. McConnell, *Steel Seizure of 1952*, p. 51.

18. Interview with David Stowe, January 9, 1973.

19. The $5.65 figure mentioned by R. Alton Lee in *Truman and Taft-Hartley* (p. 220) is incorrect. He cited Truman's *Memoirs* as his source. But on July 25, 1952, John Steelman had written to the President: "In accordance with our understanding earlier today, I have issued a directive to the Economic Stabilization Administrator ordering him, acting through the Office of Price Stabilization, to grant a price increase for steel. This directive was issued because, as you know, the steel industry would not settle the steel strike without the price increase. . . . These increases on top of the $2.84 Capehart allowance give steel a total increase of $5.20 per ton for carbon steel." Memorandum, OF Box 342, Truman Papers, TL. $5.20 was the figure used in other stabilization agency correspondence also. See, for example, Roger Putnam to Ellis Arnall, July 25, 1952, Decentralized Files of Various Directors of OPS, OPS, RG 295, NA. See too Selekman, Selekman, and Fuller, *Problems in Labor Relations*, p. 497; McConnell, *Steel Seizure of 1952*, p. 51.

20. Selekman, Selekman, and Fuller, *Problems in Labor Relations*, p. 498. According to Livernash, the 1952 union security provision virtually established a union shop. After 1952, "well over 90 percent of all production workers in the industry joined the union, compared with approximately 80 percent in the previous 5 years." *Collective Bargaining in the Basic Steel Industry*, p. 107.

21. Putnam to Arnall, July 25, 1952, Decentralized Files of Various Directors of OPS, OPS, RG 295, NA.

22. The Chairman's Letter, in *U.S. Steel Quarterly*, August, 1952, pp. 1–2, in Decentralized Files of Various Directors of OPS, OPS, RG 295, NA.

23. Fairless neglected to say that the steel industry, though publicly attack-

ing the union-government position on the union shop, actually refused to settle before the seizure because it was not satisfied with the price increase offered by the government.

24. The Chairman's Letter, p. 2.

25. Quoted in McConnell, *Steel Seizure of 1952*, p. 52.

26. See chapter 7.

27. See R. Alton Lee, *Truman and Taft-Hartley*, p. 220.

28. Labor Plank, 1952, copy in Box 5, Kroll Papers, LC.

29. Interview with Arthur Goldberg, April 20, 1975.

30. Tugwell, *The Enlargement of the Presidency*, p. 55.

31. In an interview, Leon Ulman, deputy assistant attorney general in the Office of Legal Counsel, Department of Justice, told me that it is hard to assess the influence of *Youngstown* on the advice his office gives the President. Attorneys in the office do not often cite the case, but it is always in the back of their minds. Questions of inherent power do not come up frequently, Ulman stated. The legality of an action may not be clear, especially when it involves national security, he continued. The President will be advised of the legal status of the measures he is contemplating, but he is not obliged to follow the advice. Telephone interview, October 16, 1974.

bibliography

PRIMARY SOURCES

Manuscript Collections

American Iron and Steel Institute Papers. American Iron and Steel Institute Library, New York, N.Y.

Holmes Baldridge Papers. TL, Independence, Mo.

Bernard M. Baruch Papers. Princeton University, Princeton, N.J.

George H. Bender Papers. The Western Reserve Historical Society, Cleveland, Ohio.

Hugo L. Black Papers. Library of Congress, Washington, D.C.

Nancy Dixon Bradford. "The 1952 Steel Seizure Case." M.A. thesis, University of Southern California, January 1967.

John Brophy Papers. Catholic University of America, Washington, D.C.

Harold Burton Papers. Library of Congress, Washington, D.C.

Emmanuel Celler Papers. Library of Congress, Washington, D.C.

John D. Clark Papers. TL, Independence, Mo.

Tom Connally Papers. Library of Congress, Washington, D.C.

John W. Davis Papers. Yale University, New Haven, Conn.

George Elsey Papers. TL, Independence, Mo.

Harold L. Enarson Papers. TL, Independence, Mo.

Felix Frankfurter Papers. Library of Congress, Washington, D.C.

General Historical Documents Collection. TL, Independence, Mo.

John C. Houston Papers. TL, Independence, Mo.

Arthur Krock Papers. Princeton University, Princeton, N.J.

Jack Kroll Papers. Library of Congress, Washington, D.C.

William Langer Papers. Chester Fritz Library, University of North Dakota, Grand Forks, N.Dak.

Herbert H. Lehman Papers. Columbia University, New York, N.Y.

David D. Lloyd Files. TL, Independence, Mo.

Sherman Minton Papers. TL, Independence, Mo.

Charles S. Murphy Files. TL, Independence, Mo.

Philip Murray Papers. Catholic University of America, Washington, D.C.

Richard E. Neustadt. "Notes on the White House Staff under President Truman," June 1953. TL, Independence, Mo.

Irving Perlmeter. Oral History Interview. TL, Independence, Mo.

John H. Pickering Papers. Washington, D.C.

E. Barrett Prettyman Papers. Library of Congress, Washington, D.C.

Charles Sawyer Papers. Cincinnati, Ohio.

Harold M. Stephens Papers. Library of Congress, Washington, D.C.

David H. Stowe Papers. TL, Independence, Mo.

James L. Sundquist. Oral History Interview. TL, Independence, Mo.

Robert A. Taft Papers. Library of Congress, Washington, D.C.

Harry S. Truman Papers. Official File. TL, Independence, Mo.

Harry S. Truman. President's Personal Files. TL, Independence, Mo.

U.S. Department of Commerce. Records. "The 1952 Seizure of the Basic Steel Industry," by Clarence H. Osthagen. Washington, D.C., July 1952.

U.S. Department of Justice. Memorandum of the Attorney General Commenting on the Report on Executive Order 9439 (Montgomery Ward and Co. Inc.) Made by the Majority of a Subcommittee of the Senate Committee on the Judiciary. June 12, 1944. Department of Justice Library, Washington, D.C.

—— Statement of the Attorney General Before the Committee of the House of Representatives Appointed Pursuant to House Resolution 521. May 24, 1944. Department of Justice Library, Washington, D.C.

—— *Youngstown Sheet & Tube Co. v. Sawyer* Files, Washington, D.C.

U.S. Department of Labor. Papers. Office of the Secretary, Maurice J. Tobin General Subject File, 1952. National Archives, Washington, D.C.

U.S. Economic Stabilization Agency. Records. National Archives, Washington, D.C.

U.S. Office of Price Stabilization. Records. National Archives, Washington, D.C.

U.S. Senate. Committee on Banking and Currency. Records. 81st Cong., 2d sess., 1950; 82d Cong., 1st sess., 1951; 82d Cong., 2d sess., 1952. National Archives, Washington, D.C.

—— Committee on the Judiciary. Records. 82d Cong., 2d sess., 1952. National Archives, Washington, D.C.

United Steelworkers of America. Papers. Labor Archives, Pennsylvania State University Library, University Park.

U.S. Government Publications

U.S. Department of Defense. *Semiannual Report of the Secretary of Defense: January 1 to June 30, 1952.* Washington, D.C., 1952.

U.S. Department of Justice. *Official Opinions of the Attorneys General of the United States Advising the President and Heads of Departments in Relation to Their Official Duties.* Vol. 39. Washington, D.C., 1971.

U.S. House of Representatives. Committee on Banking and Currency. *Defense Production Act of 1950.* H. Rept. 2759, 81st Cong., 2d sess., 1950.

—— Committee on Banking and Currency. *Defense Production Act of 1950. Hearings,* on H.R. 9176, 81st Cong., 2d sess., 1950.

—— Committee on Banking and Currency. *Defense Production Act Amendments of 1951.* H. Rept. 639, 82d Cong., 1st sess., 1951.

—— Committee on Banking and Currency. *Defense Production Act Amendments of 1951. Hearings.* 82d Cong., 1st sess., 1951.

—— Committee on Banking and Currency. *Defense Production Act Amendments of 1952. Hearings.* 82d Cong., 2d sess., 1952.

—— Committee on Conference. *Defense Production Act of 1950.* Conference Report. H. Rept. 3042, 81st Cong., 2d sess., 1950.

—— Committee on Education and Labor. *Investigation of the Wage Stabilization Board. Hearings,* pursuant to H. Res. 532. 82d Cong., 2d sess., 1952.

—— Committee on Education and Labor. *Investigation of the Wage Stabilization Board.* H. Rept. 2190, 82d Cong., 2d sess., 1952.

—— Committee on Foreign Affairs. *Background Information on Korea.* H. Rept. 2495. 81st Cong., 2d sess., 1950.

—— Committee on Foreign Affairs. *Background Information on the Use of United States Armed Forces in Foreign Countries.* H. Rept. 127, 82d Cong., 1st sess., 1951.

—— Committee on Government Operations. *Executive Orders and Proclamations: A Study of a Use of Presidential Powers.* Committee Print. 85th Cong., 1st sess., 1957.

—— Committee on Government Operations. Charles J., Zinn. *Extent of the Control of the Executive by the Congress of the United States.* 87th Cong., 2d sess., 1962.

—— Committee on Rules. *Directing the Committee on Education and Labor to Conduct an Investigation of the Wage Stabilization Board.* H. Rept. 1667, 82d Cong., 2d sess., 1952.

—— *The Steel Seizure Case.* H. Doc. 534, 82d Cong., 2d sess., 1952.

U.S. National Labor Relations Board. *Legislative History of the Labor Management Relations Act, 1947.* Washington, D.C.: 1948.

U.S. Office of Temporary Controls. Office of Price Administration. *Studies in Industrial Price Control,* 1947.

U.S. President. *Public Papers of the Presidents of the United States. Harry S. Truman, 1951.* Washington, D.C.: 1965.

—— *Public Papers of the Presidents of the United States. Harry S. Truman, 1952–1953.* Washington, D.C.: 1966.

U.S. Senate. Committee on Banking and Currency. *A Bill to Amend and Extend the Defense Production Act of 1950 and the Housing and Rent Act of 1947, As Amended.* S. Rept. 470, 82d Cong., 1st sess., 1951.

—— Committee on Banking and Currency. *Defense Production Act of 1950. Hearings,* on S. 3936, 81st Cong., 2d sess., 1950.

—— Committee on Banking and Currency. *Defense Production Act of 1950.* S. Rept. 2250, 81st Cong., 2d sess., 1950.

—— Committee on Banking and Currency. *Defense Production Act Amendments of 1951. Hearings.* 82d Cong., 1st sess., 1951.

—— Committee on Banking and Currency. *Defense Production Act Amendments of 1952. Hearings.* 82d Cong., 2d sess., 1952.

—— Committee on Foreign Relations. *Documents Relating to the War Power of Congress, the President's Authority as Commander-in-Chief, and the War in Indochina.* 91st Cong., 2d sess., 1970.

—— Committee on Foreign Relations. *National Commitments.* S. Rept. 91-129. 91st Cong., 1st sess., 1969.

—— Committee on Foreign Relations. *Powers of the President to Send the Armed Forces Outside the United States.* 82d Cong., 1st sess., 1951.

—— Committee on Foreign Relations. *War Powers.* S. Rept. 92-606, 92d Cong., 2d sess., 1972.

—— Committee on Labor and Public Welfare. *Disputes Functions of the Wage Stabilization Board. Hearings.* 82d Cong., 2d sess., 1952.

—— Committee on Labor and Public Welfare. *The Dispute Functions of the Wage Stabilization Board, 1951, together with the Individual Views of Mr. Taft.* S. Rept. 1037. 82d Cong., 1st sess., 1951.

—— Committee on Labor and Public Welfare. *Federal Labor Relations Act of 1947.* S. Rept. 105. 80th Cong., 1st sess., 1947.

—— Committee on Labor and Public Welfare. *National Emergency Labor Disputes Act.* S. Rept. 2073, 82d Cong., 2d sess., 1952.

—— Committee on Labor and Public Welfare. *National and Emergency Labor Disputes. Hearings.* 82d Cong., 2d sess., 1952.

—— Committee on Labor and Public Welfare. *Staff Report on WSB Recommendations in Steel Dispute.* S. Doc. 122, 82d Cong., 2d sess., 1952.

—— *The Constitution of the United States of America Annotated.* S. Doc. 170. 82d Cong., 2d sess., 1953.

Interviews

Elliot Bredhoff	Arthur Goldberg	Charles Sawyer
Bruce Bromley	Ellis Lyons	David H. Stowe
Oscar Davis	Herman Marcuse	Stanley Temko
Charles Fahy	Charles S. Murphy	Leon Ulman
	John Pickering	

SECONDARY SOURCES

Books

Allen, Robert S. and William V. Shannon. *The Truman Merry-Go-Round.* New York: Vanguard, 1950.

American Iron and Steel Institute. *The Making of Steel.* New York, 1950.

Anderson, Patrick. *The President's Men: White House Assistants of Franklin D. Roosevelt, Harry S. Truman, Dwight D. Eisenhower, John F. Kennedy, and Lyndon B. Johnson.* New York: Doubleday, 1968.

Baker, Liva. *Felix Frankfurter*. New York: Coward-McCann, 1969.

Berdahl, Clarence A. *War Powers of the Executive in the United States*. Urbana: University of Illinois Press, 1921.

Berg, Robert Louis. "Presidential Power and the Royal Prerogative." Ph.D. dissertation, University of Minnesota, 1958.

Berger, Raoul. *Congress v. The Supreme Court*. Cambridge, Mass.: Harvard University Press, 1969.

Berman, Edward. *Labor Disputes and the President of the United States*. New York: AMS Press, 1968; 1st ed. 1924.

Bernstein, Barton J., ed. *Politics and Policies of the Truman Administration*. Chicago: Watts, 1970.

Bernstein, Barton J. and Allen J. Matusow. *The Truman Administration: A Documentary History*. New York: Harper & Row, 1966. Paperback: 1968.

Bernstein, Barton J. and Allen J. Matusow, eds. *Twentieth-Century America: Recent Interpretations*. 2d ed. New York: Harcourt, Brace, 1972.

Bickel, Alexander M. *The Least Dangerous Branch: The Supreme Court at the Bar of Politics*. Indianapolis and New York: Bobbs-Merrill, 1962.

Binkley, Wilfred E. *The Man in the White House*. Rev. ed. New York: Harper, 1964.

Black, Charles L., Jr. *The People and the Court: Judicial Review in a Democracy*. New York: Macmillan, 1960.

—— *Perspectives in Constitutional Law*. Englewood Cliffs, N.J.: Prentice-Hall, 1970.

Blackman, John L., Jr. *Presidential Seizure in Labor Disputes*. Cambridge, Mass.: Harvard University Press, 1967.

Brody, David. "The Rise and Decline of Welfare Capitalism." In John Braeman, Robert H. Bremner, and David Brody, eds. *Change and Continuity in Twentieth-Century America: The 1920s*. Columbus: Ohio State University Press, 1968.

Brooks, Robert R. R. *As Steel Goes, . . . : Unionism in a Basic Industry*. New Haven, Conn.: Yale University Press, 1940.

Broude, Henry W. *Steel Decisions and the National Economy*. New Haven, Conn.: Yale University Press, 1963.

Burns, James MacGregor. *The Deadlock of Democracy: Four-Party Politics in America*. Englewood Cliffs, N.J.: Prentice-Hall, 1963.

Cabe, John Carl, Jr. "Governmental Intervention in Labor Disputes from 1945–1952." Ph.D. dissertation, University of Illinois, 1952.

Cahill, Fred V., Jr. *Judicial Legislation: A Study in American Legal Theory*. New York: Ronald, 1952.

Cahn, N. Edmond, ed. *Supreme Court and Supreme Law*. Bloomington: Indiana University Press, 1954.

Calkins, Fay. *The CIO and the Democratic Party*. Chicago: University of Chicago Press, 1952.

Caridi, Ronald J. *The Korean War and American Politics: The Republican Party as a Case Study*. Philadelphia: University of Pennsylvania Press, 1968.

Chamberlain, Neil W., Frank C. Pierson, and Theresa Wolfson, eds. *A De-*

cade of Industrial Relations Research 1946–1956. New York: Industrial Relations
Research Associates, 1958.

Chandler, Lester V. and Donald H. Wallace, eds. *Economic Mobilization
and Stabilization.* New York: Holt, 1951.

Ching, Cyrus S. *Review and Reflection: A Half-Century of Labor Relations.*
New York: Forbes, 1953.

Collins, J. Lawton. *War in Peacetime: The History and Lessons of Korea.*
Boston: Houghton Mifflin, 1969.

Congressional Quarterly Service. *Politics in America 1945–1964.* Washington, D.C., 1965.

Connally, Thomas. *My Name Is Tom Connally.* New York: Crowell, 1954.

Cornwell, Elmer E., Jr. *Presidential Leadership of Public Opinion.* Bloomington: Indiana University Press, 1965.

Corwin, Edward S. *The Constitution and What It Means Today.* Princeton,
N.J.: Princeton University Press, 1954.

—— *Constitutional Revolution, Ltd.* Claremont, Calif.: Claremont College
Administration Office, 1941.

—— *Court Over Constitution: A Study of Judicial Review as an Instrument
of Popular Government.* Princeton, N.J.: Princeton University Press, 1938.

—— *The Doctrine of Judicial Review: Its Legal and Historical Basis and
Other Essays.* Gloucester, Mass.: Peter Smith, 1963; 1st ed. 1914.

—— *The President: Office and Powers.* New York: New York University
Press, 1957.

—— *The Twilight of the Supreme Court. A History of Our Constitutional
Theory.* New York: Archon Books, 1970; 1st ed. 1934.

Corwin, E. S. and J. W. Peltason. *Understanding the Constitution.* New
York: Dryden, 1970; 1st ed. 1958.

Countryman, Vern, ed. *Douglas of the Supreme Court: A Selection of His
Opinions.* New York: Doubleday, 1959.

Crosskey, William Winslow. *Politics and the Constitution in the History of
the United States.* 2 vols. Chicago: University of Chicago Press, 1953.

Daniels, Jonathan. *The Man of Independence.* Port Washington, N.Y.: Kennikat, 1971.

Desmond, Charles S., Paul A. Freund, Potter Stewart, and Lord Shawcross. *Mr. Justice Jackson: Four Lectures in his Honor.* New York and London:
Columbia University Press, 1969.

Director, Aaron, ed. *Defense, Controls, and Inflation: A Conference Sponsored by the University of Chicago Law School.* Chicago: University of Chicago
Press, 1952.

Douglas, William O. *Go East, Young Man: The Early Years. The Autobiography of William O. Douglas.* New York: Random House, 1974.

—— *We the Judges.* New York: Doubleday, 1956.

Elman, P., ed. *Of Law and Men.* New York: Archon Books, 1956.

Fenton, John M. *In Your Opinion. . . .* Boston: Little, Brown, 1960.

Finer, Herman. *The Presidency: Crisis and Regeneration.* Chicago: University of Chicago Press, 1960.

Flash, Edward S., Jr. *Economic Advice and Presidential Leadership: The Council of Economic Advisers*. New York: Columbia University Press, 1965.

Frank, John P. *Marble Palace: The Supreme Court in American Life*. Westport, Conn.: Greenwood, 1972; 1st ed. 1958.

—— *Mr. Justice Black: The Man and His Opinions*. New York: Knopf, 1949.

Frankfurter, Felix. *The Public and Its Government*. New Haven, Conn.: Yale University Press, 1930.

Frankfurter, Felix and James M. Landis. *The Business of the Supreme Court: A Study in the Federal Judicial Sytem*. New York: Macmillan, 1928.

Freund, Paul A. *On Law and Justice*. Cambridge, Mass.: Harvard University Press, 1968.

—— *On Understanding the Supreme Court*. Boston: Little, Brown, 1950.

Friedman, Leon and Fred L. Israel, eds. *The Justices of the United States Supreme Court 1789–1969: Their Lives and Major Opinions*. Vols. 3 and 4. New York: Bowker, 1969.

Gallup, George H. *The Gallup Poll: Public Opinion 1935–1971*. 3 vols. New York: Random House, 1972.

Gerhart, Eugene C. *America's Advocate: Robert H. Jackson*. Indianapolis: Bobbs-Merrill, 1958.

Goldberg, Arthur J. *AFL–CIO Labor United*. New York: McGraw-Hill, 1956.

Goldman, Eric F. *The Crucial Decade—and After: America, 1945–60*. New York: Random House, 1960.

Greenstone, J. David. *Labor in American Politics*. New York: Knopf, 1969.

Gregory, Charles O. *Labor and the Law*. New York: Norton, 1961.

Gwyn, W. B. *The Meaning of the Separation of Powers: An Analysis of the Doctrine from its Origin to the Adoption of the United States Constitution*. New Orleans: Tulane University Press, 1965.

Hamby, Alonzo L. *Beyond the New Deal: Harry S. Truman and American Liberalism*. New York: Columbia University Press, 1973.

Hamilton, Virginia Van Der Veer. *Hugo Black: The Alabama Years*. Baton Rouge: Louisiana State University Press, 1972.

Harbaugh, William H. *Lawyer's Lawyer: The Life of John W. Davis*. New York: Oxford University Press, 1973.

Harris, Seymour E. *The Economics of Mobilization and Inflation*. New York: Norton, 1951; reprinted 1968.

Hartmann, Susan M. *Truman and the 80th Congress*. Columbia: University of Missouri Press, 1971.

Henkin, Louis. *Foreign Affairs and the Constitution*. Mineola, N.Y.: Foundation Press, 1972.

Hirschfield, Robert S. "Crisis Government in America." Ph.D. dissertation, New York University, 1958.

Hofstadter, Richard. *The American Political Tradition*. New York: Random House, 1959.

Hogan, William T., S.J. *Economic History of the Iron and Steel Industry in the United States.* Vols. 3 and 4. Lexington, Mass.: Lexington Books, 1971.

Hughes, Charles Evans. *The Supreme Court of the United States.* New York: Columbia University Press, 1928.

Huitt, Ralph K. and Robert L. Peabody. *Congress: Two Decades of Analysis.* New York: Harper & Row, 1969.

Hyman, Sidney. *The American President.* New York: Harper, 1954.

Industrial Relations Research Association. *Emergency Disputes and National Policy.* Irving Bernstein, Harold L. Enarson, and R. W. Fleming, eds. New York, 1955.

—— *Proceedings of the Third Annual Meeting.* Chicago, Ill., December 28–29, 1950, 1951.

—— *Proceedings of the Fourth Annual Meeting.* Boston, Mass. December 28–29, 1951, 1952.

Jackson, Robert. *The Struggle for Judicial Supremacy.* New York: Knopf, 1941.

—— *The Supreme Court in the American System of Government.* Cambridge, Mass.: Harvard University Press, 1955.

Kampelman, Max M. "Labor in Politics." In Industrial Relations Research Association, *Interpreting the Labor Movement.* Champaign: University of Illinois Press, December 1952.

Kariel, Henry S. *The Decline of American Pluralism.* Stanford, Calif.: Stanford University Press, 1961.

Kelly, Alfred H. and Winfred A. Harbison. *The American Constitution: Its Origins and Development.* Rev. ed. New York: Norton, 1955.

Kirkendall, Richard S. "Election of 1948." In Arthur M. Schlesinger, Jr., Fred L. Israel, and William P. Hansen, eds. *History of American Presidential Elections, 1789–1968.* 4 vols. New York: McGraw-Hill, 1971.

—— *The Truman Period as a Research Field.* Columbia: University of Missouri Press, 1967.

Koenig, Louis W. *The Chief Executive.* New York: Harcourt Brace, 1964.

Koenig, Louis W., ed. *The Truman Administration: Its Principles and Practice.* New York: New York University Press, 1956.

Konefsky, Samuel J., ed. *The Constitutional World of Mr. Justice Frankfurter.* New York: Hafner, 1949.

Kurland, Philip B. *Mr. Justice Frankfurter and the Constitution.* Chicago: University of Chicago Press, 1971.

Latham, Earl. *The Group Basis of Politics: A Study in Basing-Point Legislation.* Ithaca, N.Y.: Cornell University Press, 1952.

Lee, R. Alton. *Truman and Taft-Hartley: A Question of Mandate.* Lexington: University of Kentucky Press, 1966.

Leuchtenburg, William E. "The Case of the Contentious Commissioner: *Humphrey's Executor* v. *U.S.*" In Harold M. Hyman and Leonard Levy, eds. *Freedom and Reform: Essays in Honor of Henry Steele Commager.* New York: Harper, 1967.

—— "The Constitutional Revolution of 1937." In Victor Hoar, ed. *The Great Depression: Essays and Memoirs from Canada and the United States.* Vancouver, Toronto, and Montreal: Copp Clark, 1969.

Livernash, Edward R. *Collective Bargaining in the Basic Steel Industry.* Washington, D.C.: U.S. Department of Labor, 1961.

Lofgren, Charles A. "Congress and the Korean Conflict." Ph.D. dissertation, Stanford University, 1966.

Lowi, Theodore J. *Legislative Politics U.S.A.* Boston: Little, Brown, 1962.

—— *The Politics of Disorder.* New York: Basic Books, 1971.

Lubell, Samuel. *The Future of American Politics.* New York: Harper & Row, 1955.

McCabe, David A. "Union Policies As to the Area of Collective Bargaining." In Industrial Relations Research Association, *Interpreting the Labor Movement.* Champaign: University of Illinois Press, December 1952.

McCloskey, Robert G. *The American Supreme Court.* Chicago: University of Chicago Press, 1960.

McClure, Arthur F. *The Truman Administration and the Problems of Postwar Labor, 1945–1948.* Rutherford, Madison, Teaneck, N.J.: Fairleigh Dickinson University Press, 1969.

McConnell, Grant. *The Steel Seizure of 1952.* Inter-University Case Program, Case Series No. 52. University: University of Alabama Press, 1960.

MacLean, Joan Coyne, ed. *President and Congress: The Conflict of Powers.* New York: Wilson, 1955.

MacLeish, Archibald, and E. F. Prichard, Jr., eds. *Law and Politics: Occasional Papers of Felix Frankfurter 1913–1938.* New York: Harcourt, 1939.

McNaughton, Frank and Walter Hehmeyer. *Harry Truman—President.* New York: McGraw-Hill, 1948.

Mason, Alpheus Thomas. *The Supreme Court from Taft to Warren.* Baton Rouge: Louisiana State University Press, 1968.

Mason, Alpheus T. and William M. Beaney. *The Supreme Court in a Free Society.* New York: Norton, 1968.

Mendelson, Wallace. *Felix Frankfurter: The Judge.* New York: Reynal, 1964.

—— *Justices Black and Frankfurter: Conflict in the Court.* Chicago: University of Chicago Press, 1966.

Mendelson, Wallace, ed. *Felix Frankfurter: A Tribute.* New York: Reynal, 1964.

Miller, Charles A. *The Supreme Court and the Uses of History.* Cambridge, Mass.: Harvard University Press, 1969.

Miller, Merle. *Plain Speaking: An Oral Biography of Harry S. Truman.* New York: Putnam, 1973.

Millis, Harry A. and Emily Clark Brown. *From the Wagner Act to Taft-Hartley: A Study of National Labor Policy and Labor Relations.* Chicago: University of Chicago Press, 1950.

Mills, C. Wright. *The New Men of Power: America's Labor Leaders.* New York: Harcourt, 1948.

Murphy, Paul L. *The Constitution in Crisis Times 1918–1969.* New York: Harper & Row, 1972.

Murphy, Walter F. *Congress and the Court: A Case Study in the American Political Process.* Chicago: University of Chicago Press, 1962.

Murphy, Walter F. and C. Herman Pritchett. *Courts, Judges, and Politics: An Introduction to the Judicial Process*. New York: Random House, 1961.

Neustadt, Richard E. *Presidential Power: The Politics of Leadership*. New York: Wiley, 1960.

Paige, Glenn D. *The Korean Decision*. New York: Free Press, 1968.

Peltason, Jack W. *Federal Courts in the Political Process*. New York: Random House, 1955.

Perlman, Mark. "Labor in Eclipse." In John Braeman, Robert H. Bremner, and David Brody, eds. *Change and Continuity in Twentieth-Century America: The 1920s*. Columbus: Ohio State University Press, 1968.

Phillips, Cabell. *The Truman Presidency: The History of a Triumphant Succession*. New York: Macmillan, 1966.

Phillips, Harlan B., ed. *Felix Frankfurter Reminisces*. New York: Reynal, 1960.

Post, Charles G., Jr. *The Supreme Court and Political Questions*. New York: DaCapo Press, 1969.

Pritchett, C. Herman. *Civil Liberties and the Vinson Court*. Chicago: University of Chicago Press, 1954.

—— *The Roosevelt Court: A Study in Judicial Politics and Values 1937–1947*. New York: Quadrangle, 1948.

Randall, Clarence B. *Over My Shoulder: A Reminiscence*. Boston: Little, Brown, 1956.

Rankin, Robert S. and Winfried R. Dallmayr. *Freedom and Emergency Powers in the Cold War*. New York: Appleton, Century, Crofts, 1964.

Rayback, Joseph G. *A History of American Labor*. New York: Free Press, 1966.

Rees, David. *Korea: The Limited War*. New York: St. Martin's Press, 1964.

Rodell, Fred. *Nine Men: A Political History of the Supreme Court from 1790 to 1955*. New York: Random House, 1955.

Roosevelt, Theodore. *An Autobiography*. New York: Scribner, 1913.

Roper, Elmo. *You and Your Leaders: Their Actions and Your Reactions 1936–1956*. New York: Morrow, 1957.

Rossiter, Clinton. *The American Presidency*. New York: Harcourt Brace, 1960.

—— *Constitutional Dictatorship: Crisis Government in the Modern Democracies*. New York: Harcourt Brace, 1963.

—— *The Supreme Court and the Commander in Chief*. New York: DaCapo Press, 1970; 1st ed. 1951.

Rostow, Eugene V. *The Sovereign Prerogative: The Supreme Court and the Quest for Law*. New Haven, Conn.: Yale University Press, 1962.

Rothe, Anna and Evelyn Lohr. *Current Biography: Who's News and Why 1952*. New York: Wilson, 1953.

Sawyer, Charles. *Concerns of a Conservative Democrat*. Carbondale and Edwardsville: Southern Illinois University Press, 1968.

Schaefer, Arthur M. "Presidential Intervention in Labor Disputes during the Truman Administration: A History and Analysis of Experience." Ph.D. dissertation, University of Pennsylvania, 1967.

Schlesinger, Arthur M., Jr. *The Imperial Presidency*. Boston: Houghton Mif-flin, 1973.

—— *The Vital Center: The Politics of Freedom*. Cambridge, Mass.: Harvard University Press, 1949.

Schroeder, Gertrude G. *The Growth of the Major Steel Companies, 1900–1950*. Baltimore: Johns Hopkins University Press, 1953.

Schubert, Glendon A., Jr. *Constitutional Politics: The Political Behavior of Supreme Court Justices and the Constitutional Policies That They Make*. New York: Holt, 1960.

—— *The Judicial Mind: The Attitudes and Ideologies of Supreme Court Justices 1946–1965*. Evanston: Northwestern University Press, 1965.

—— *The Presidency in the Courts*. Minneapolis: University of Minnesota Press, 1957.

Schwartz, Bernard. *American Constitutional Law*. Westport, Conn.: Greenwood, 1969; 1st ed. 1955.

—— *The Supreme Court*. New York: Ronald, 1957.

Scigliano, Robert. *The Supreme Court and the Presidency*. New York: Free Press, 1971.

Seidman, Joel. *American Labor from Defense to Reconversion*. Chicago: University of Chicago Press, 1953.

Shapiro, Martin. *Law and Politics in the Supreme Court*. London: Macmillan, 1964.

Small, Norman J. *Some Presidential Interpretations of the Presidency*. Baltimore: Johns Hopkins University Press, 1932.

Smith, J. Malcolm and Cornelius P. Cotter. *Powers of the President during Crises*. Washington, D.C.: Public Affairs Press, 1960.

Smith, Merriman. *A President Is Many Men*. New York: Harper & Row, 1948.

Steinberg, Alfred. *The Man from Missouri: The Life and Times of Harry S. Truman*. New York: Putnam, 1962.

Strickland, Stephen Parks, ed. *Hugo Black and the Supreme Court: A Symposium*. Indianapolis, Kansas City, New York: Bobbs-Merrill, 1967.

Sutherland, Arthur E. *Constitutionalism in America*. 2 vols. Waltham, Mass.: Blaisdell, 1964.

Sweeney, Vincent D. *The United Steelworkers of America: Twenty Years Later 1936–1956*. 1957.

Swindler, William F. *Court and Constitution in the Twentieth Century: The New Legality 1932–1968*. Indianapolis and New York: Bobbs-Merrill, 1970.

Swisher, Carl B. *Historic Decisions of the Supreme Court*. 2d ed. New York: Van Nostrand, 1969.

Taft, Philip. "Theories of the Labor Movement." In Industrial Relations Research Association. *Interpreting the Labor Movement*. Champaign: University of Illinois Press, 1952.

Taft, William Howard. *Our Chief Magistrate and His Powers*. New York: Columbia University Press, 1916.

Tate, Jaunita D. "Philip Murray As a Labor Leader." Ph.D. dissertation, New York University, 1962.

Thomas, Helen Shirley. *Felix Frankfurter: Scholar on the Bench.* Baltimore: Johns Hopkins University Press, 1960.

Truman, David B. *The Congressional Party: A Case Study.* New York: Wiley, 1959.

——— *The Governmental Process: Political Interests and Public Opinion.* 2d ed. New York: Knopf, 1971.

Truman, Harry S. *Memoirs.* New York: Doubleday, 1955–1956.

——— *Mr. Citizen.* New York: Geis, 1960.

——— *Mr. President.* William Hillman, ed. New York: Farrar, Straus, 1952.

——— *The Truman Program.* Washington, D.C.: Public Affairs Press, 1948–1949.

Truman, Margaret. *Harry S. Truman.* New York: Morrow, 1973.

Tugwell, Rexford G. *The Enlargement of the Presidency.* Garden City, N.Y.: Doubleday, 1960.

Twiss, Benjamin R. *Lawyers and the Constitution: How Laissez Faire Came to the Supreme Court.* Princeton, N.J.: Princeton University Press, 1942.

Urofsky, Melvin I. *Big Steel and the Wilson Administration: A Study in Business-Government Relations.* Columbus: Ohio State University Press, 1969.

Vanderbilt, Arthur T. *The Doctrine of the Separation of Powers and Its Present-Day Significance.* Lincoln: University of Nebraska Press, 1953.

Vatter, Harold G. *The U.S. Economy in the 1950's: An Economic History.* New York: Norton, 1963.

Walker, Charles R. *Steeltown.* New York: Harper & Row, 1950.

Warne, Colston E., ed. *Labo.* in Postwar America.* Vol. 2 of *Yearbook of American Labor.* New York: Remsen, 1949.

Warren, Charles. *The Supreme Court in United States History.* 2 vols. Rev. ed. Boston: Little, Brown, 1935.

Wechsler, Herbert. *Principles, Politics, and Fundamental Law.* Cambridge, Mass.: Harvard University Press, 1961.

Westin, Alan F. *The Anatomy of a Constitutional Law Case: Youngstown Sheet and Tube Co. v. Sawyer; The Steel Seizure Decision.* New York: Macmillan, 1958.

Wiener, Frederick B. *Uses and Abuses of Legal History: A Practitioner's View.* London, 1962.

Williams, Charlotte. *Hugo L. Black: A Study in the Judicial Process.* Baltimore: Johns Hopkins University Press, 1950.

Periodicals

Aaron, Benjamin. "Amending the Taft-Hartley Act: A Decade of Frustration," *Industrial and Labor Relations Review* 11(April 1958):327–38.

Atkinson, David N. and Dale A. Neuman. "Toward a Cost Theory of Judicial Alignments: The Case of the Truman Bloc," *Midwest Journal of Political Science* 13(May 1969):271–83.

Auerbach, Jerold S. "The La Follette Committee: Labor and Civil Liberties in the New Deal," *Journal of American History* 51(December 1964):435–59.

Banks, Robert F. "Steel, Sawyer, and the Executive Power," *University of Pittsburgh Law Review* 14(Summer 1953):467–537.

Bernstein, Barton. "The Presidency under Truman," *Yale Political* 3(Fall 1964):8, 9, 24.

Binkley, Wilfred E. "The President and Congress," *Journal of Politics* 11(February 1949):65–79.

Brown, Everett S. "The Term of Office of the President," *American Political Science Review* 41(June 1947):447–52.

C., H. L. "Constitutional Law—the President's Power of Eminent Domain in the Absence of Enabling Legislation," *Brooklyn Law Review* 19(December 1952):129–34.

Cater, Douglass. "The President and the Press," *Annals of the American Academy of Political and Social Science* 307(September 1956):55–65.

Ching, Cyrus S. "Some Instruments of Industrial Peace," *Annals of the American Academy of Political and Social Science* 274(March 1951):179–84.

"Coddled Criminal," *New Republic*, April 17, 1971, pp. 10–11.

Commager, Henry Steele. "The Misuse of Power," *New Republic*, April 17, 1971, pp. 17–21.

"Comments on Recent Cases: Constitutional Law—President's Power to Seize Private Property—Adequacy of Remedy at Law," *Iowa Law Review* 37(Summer 1952):590–93.

Dahl, Robert A. "Decision-Making in a Democracy: The Role of the Supreme Court as a National Policy-Maker," *Journal of Public Law* 6(Fall 1957):279–95.

Davis, Kenneth Culp. "Ripeness of Governmental Action for Judicial Review," *Harvard Law Review* 68(May 1955):1122–53, and 68(June 1955):1326–73.

Day, Rutherford. "The Federal Government's Power to Requisition under the Defense Production Act of 1950," *Georgetown Law Journal* 41(November 1952):18–39.

Deutsch, Eberhard P. "The Legality of the United States Position in Vietnam," *American Bar Association Journal* 52(May 1966):436–42.

Durham, James A. "Congress, the Constitution and Crosskey," *Indiana Law Journal* 29(Spring 1954):355–66.

Eaton, William D. "Constitutional Guarantees and the Steel Decision," *Rocky Mountain Law Review* 25(February 1953):220–33.

"Executive Commandeering of Strike-Bound Plants," *Yale Law Journal* 51(December 1941):282–98.

Fahy, Charles. "Judicial Review of Executive Action," *Georgetown Law Journal* 50(Summer 1962):709–32.

Fairman, Charles. "The President as Commander-in-Chief," *Journal of Politics* 11(February 1949):145–70.

—— "Robert H. Jackson, 1892–1954: Associate Justice of the Supreme Court," *Columbia Law Review* 55 (April 1955):445–87.

Fellman, David. "Constitutional Law in 1951–52," *American Political Science Review* 47(March 1953):126–70.

Frank, John P. "The Future of Presidential Seizure," *Fortune*, July 1952, pp. 70–71, 168.

Frankfurter, Felix. "Mr. Justice Jackson," *Harvard Law Review* 68(April 1955):937–39.

—— "Robert H. Jackson: 1892–1954: Foreword," *Columbia Law Review* 55(April 1955):435–37.

Franklin, Mitchell. "War Power of the President: An Historical Justification of Mr. Roosevelt's Message of September 7, 1942," *Tulane Law Review* 17(November 1942):217–55.

Freund, Paul A. "Mr. Justice Black and the Judicial Function," *UCLA Law Review* 14(January 1967):467–74.

—— "The Year of the Steel Case," *Harvard Law Review* 66(November 1952):89–97.

Galenson, Walter. "The Unionization of the American Steel Industry," *International Review of Social History* 1(1956):8–40.

Gallup, George. "How Labor Votes," *Annals of the American Academy of Political and Social Science* 274 (March 1951):123–24.

Gardner, Warner W. "Robert H. Jackson: 1892–1954: Government Attorney," *Columbia Law Review* 55(April 1955):438–44.

Gilmore, Eugene A. "War Power—Executive Power and the Constitution," *Iowa Law Review* 29(March 1944):463–80.

Goebel, Julius, Jr. "Constitutional History and Constitutional Law," *Columbia Law Review* 38(April 1938):555–77.

Green, William. "The Taft-Hartley Act: A Critical View," *Annals of the American Academy of Political and Social Science* 274(March 1951):200–205.

Griffith, J. T. "Constitutional Law—Power of Executive to Seize Private Property," *Georgia Bar Journal* 15(1952–53):90–92.

Grisanti, Eugene P. "Comment: Constitutional Law—Separation of Governmental Powers—'Inherent' Power of the President to Seize an Industry for Prevention of Strikes," *Boston University Law Review* 32(November 1952):460–63.

Hamby, Alonzo L. "The Liberals, Truman, and FDR as Symbol and Myth," *Journal of American History* 56(March 1970):859–67.

—— "The Vital Center, the Fair Deal, and the Quest for a Liberal Political Economy," *American Historical Review* 77(June 1972):653–78.

Hamilton, Walton H. and George D. Braden. "The Special Competence of the Supreme Court," *Yale Law Journal* 50(June 1941):1319–75.

Hammond, Mary K. "The Steel Strike of 1952." *Current History* 23(November, 1952):285–90.

Harbison, Frederick H. and Robert C. Spencer. "The Politics of Collective Bargaining: The Postwar Record in Steel," *American Political Science Review* 48(September 1954):705–20.

Hasson, Bernard J., Jr. "The Steel Seizure Cases," *Georgetown Law Journal* 41(November 1952):45–63.

Hearon, Robert J., Jr. "Constitutional Law—Distribution of Governmental Powers and Functions—Power of the President to Seize an Industry to Prevent Strikes," *Texas Law Review* 30(October 1952):885–88.

Henkin, Louis. "Viet-Nam in the Courts of the United States: 'Political Questions,' " *American Journal of International Law* 63(April 1969):284–89.

Hyman, Sidney. "The Art of the Presidency," *Annals of the American Academy of Political and Social Science* 307(September 1956):1–9.

Jackson, Robert H. "The Task of Maintaining Our Liberties: The Role of the Judiciary," *American Bar Association Journal* 39(November 1953):961–65.

Jaffe, Louis L. "An Essay on Delegation of Legislative Power," *Columbia Law Review* 47(April 1947):359–76, and 47(May 1947):561–93.

—— "Mr. Justice Jackson," *Harvard Law Review* 68(April 1955):940–98.

Janeway, Eliot. "Mobilizing the Economy: Old Errors in a New Crisis," *Yale Review* 40(1950–51):201–19.

Kalven, Harry, Jr. "Upon Rereading Mr. Justice Black on the First Amendment," *UCLA Law Review* 14(January 1967):428–53.

Kaufmann, John H. "The Problem of Coordinating Price and Wage Programs in 1950–1953," *Indiana Law Journal* 29(Summer 1954):499–537, and 30(Fall 1954):18–58.

Kauper, Paul G. "The Steel Seizure Case: Congress, the President and the Supreme Court," *Michigan Law Review* 51(December 1952):141–82.

Kleiler, Frank M. "Presidential Seizures in Labor Disputes," *Industrial and Labor Relations Review* 6(July 1953):447–56.

Koenig, Louis W. "The Man and the Institution," *Annals of American Academy of Political and Social Science* 307(September 1956):10–14.

Kroll, Jack. "Labor's Political Role," *Annals of American Academy of Political and Social Science* 274(March 1951):118–22.

Laski, Harold J. "The American President and Foreign Relations," *Journal of Politics* 11(February 1949):171–205.

Latham, Earl. "The Supreme Court as a Political Institution," *Minnesota Law Review* 31(February 1947):205–31.

Lea, L. B. "The Steel Case: Presidential Seizure of Private Industry," *Northwestern University Law Review* 47(July–August 1952):289–313.

Lee, Frederic P. "The Origins of Judicial Control of Federal Executive Action," *Georgetown Law Journal* 36(March 1948):287–309.

Leiserson, Avery. "Organized Labor as a Pressure Group," *Annals of the American Academy of Political and Social Science* 274(March 1951):108–17.

Lester, Richard A. "Wage Troubles," *Yale Review* 41(Autumn 1951):54–65.

Letizia, Donald J. "Inherent Power of the President to Seize Property: 'Steel Seizure Cases,' " *The Catholic University of America Law Review* 3(January 1953):27–32.

Leuchtenburg, William E. "A Klansman Joins the Court: The Appointment of Hugo L. Black," *University of Chicago Law Review* 41(Fall 1973):1–31.

Leviero, Anthony. "How the President Makes Decisions," *New York Times Magazine*, October 8, 1950, pp. 14–15, 62, 64–65.

Lofgren, Charles A. "War-Making under the Constitution: The Original Understanding," *Yale Law Journal* 81(March 1972):672–702.

McCoy, Jean R. "Growing Executive Power and the Constitution," *Syracuse Law Review* 4(Fall 1952):109–17.

Mendelson, Wallace. "Mr. Justice Frankfurter and the Process of Judicial Review," *University of Pennsylvania Law Review* 103(December 1954):295–320.

Monaghan, Henry P. "Presidential War-Making," *Boston University Law Review* 50(Spring 1970):19–33.

Murphy, Jay W. "Some Observations on the Steel Decision," *Alabama Law Review* 4(Spring 1952):214–31.

Nathanson, Nathaniel L. "The Supreme Court as a Unit of the National Government: Herein of Separation of Powers and Political Questions," *Journal of Public Law* 6(Fall 1957):331–62.

Neustadt, Richard. "Congress and the Fair Deal: A Legislative Balance Sheet," *Public Policy* 5(1954):351–81.

––––– "The Presidency at Mid-Century," *Law and Contemporary Problems* 21(Autumn 1956):609–45.

Nulton, William C. "Steel Seizure Case—A Light in the Dark Continent," *Journal of the Bar Association of the State of Kansas* 21(February 1953):242–49.

O'Brien, William, S. J. "Mr. Justice Reed and Democratic Pluralism," *Georgetown Law Journal* 45(Spring 1957):364–87.

Perlman, Selig. "The Basic Philosophy of the American Labor Movement," *Annals of the American Academy of Political and Social Science* 274(March 1951):57–63.

Petro, Sylvester. "The Supreme Court and the Steel Seizure," *Labor Law Journal* 3(July 1952):451–54, 494–95.

Pomper, Gerald. "Labor and Congress: The Repeal of Taft-Hartley," *Labor History* 2(Fall 1961):323–43.

Pritchett, C. Herman. "The President and the Supreme Court," *Journal of Politics* 11(February 1949):80–92.

Rehnquist, William H., Robert B. McKay, John R. Stevenson, and Abram Chayes. "Hammarskjöld Forum: Expansion of the Vietnam War into Cambodia—The Legal Issues," *New York University Law Review* 45(June 1970):628–78.

Reilly, Gerard D. "The Legislative History of the Taft-Hartley Act," *George Washington Law Review* 29(December 1960):285–300.

Richberg, Donald. "The Steel Seizure Cases," *Virginia Law Review* 38(October 1952):713–27.

Riddick, Floyd M. "The Eighty-Second Congress: First Session," *Western Political Quarterly* 5(March 1952):94–108.

––––– "The Eighty-Second Congress: Second Session," *Western Political Quarterly* 5(December 1952):619–34.

Roche, John P. "Executive Power and Domestic Emergency: The Quest for Prerogative," *Western Political Quarterly* 5(December 1952):592–618.

Rodell, Fred. "For Every Justice, Judicial Deference is a Sometime Thing," *Georgetown Law Journal* 50(Summer 1962):700–708.

––––– "Our Not So Supreme Supreme Court," *Look*, July 31, 1951, pp. 60–64.

Root-Tilden Scholars, New York University School of Law. "The War in Southeast Asia: A Legal Position Paper," *New York University Law Review* 45(June 1970):695–726.

Rossiter, Clinton L. "The President and Labor Disputes," *Journal of Politics* 11(February 1949):93–120.

Rovere, Richard H. "Labor's Political Machine," *Harper's Magazine,* June 1945, pp. 592–601.

Rutledge, Ivan C. "Justice Black and Labor Law," *UCLA Law Review* 14(January 1967):501–23.

Saylor, J. R. "Court over the President," *Texas Law Review* 31(November 1952):38–46.

Scharpf, Fritz W. "Judicial Review and the Political Question: A Functional Analysis," *Yale Law Journal* 75(March 1966):517–97.

Schubert, Glendon A., Jr. "The Steel Case: Presidential Responsibility and Judicial Irresponsibility," *Western Political Quarterly* 6(March 1953):61–77.

Schwartz, Bernard. "Executive Power and the Disappearance of Law," *Labor Law Journal* 3(June 1952):423–32.

—— "Inherent Executive Power and the Steel Seizure Case: A Landmark in American Constitutional Law," *Canadian Bar Review* 30(May 1952):466–82.

Schwartz, Marvin and Edwin M. Zimmerman. "Review of Recent Decisions of the United States [Supreme] Court," *The Record of the Association of the Bar of the City of New York* 7(October 1952):390–98.

Seligman, Daniel. "The Last Days of Charlie Wilson," *Fortune,* June 1952, pp. 85–86, 222, 224.

Slichter, Sumner H. "Revision of the Taft-Hartley Act," *Quarterly Journal of Economics* 67(May 1953):149–80.

Steelman, John R. and H. Dewayne Kreager. "The Executive Office as Administrative Coordinator," *Law and Contemporary Problems* 21(Autumn 1956):688–709.

Sturm, Albert L. "Emergencies and the Presidency," *Journal of Politics* 11(February 1949):121–44.

"The Supreme Court, 1951 Term," *Harvard Law Review* 66(November 1952):99–104.

Taft, Robert A. "The Taft-Hartley Act: A Favorable View," *Annals of the American Academy of Political and Social Science* 274(March 1951):195–99.

"The Talk of the Town," *New Yorker,* May 16, 1970, pp. 31–33.

Tanenhaus, Joseph. "The Supreme Court and Presidential Power," *Annals of the American Academy of Political and Social Science* 307(September 1956):106–13.

Taylor, George W. "National Labor Policy," *Annals of the American Academy of Political and Social Science* 274(March 1951):185–94.

Taylor, Telford. "Robert H. Jackson: 1892–1954: The Nuremberg Trials," *Columbia Law Review* 55(April 1955):488–525.

Teller, Ludwig. "Government Seizure in Labor Disputes," *Harvard Law Review* 60(September 1947):1017–59.

Tyler, Gus. "The Presidency and Labor," *Annals of the American Academy of Political and Social Science* 307(September 1956):82–91.

Wakstein, Allen M. "The Origins of the Open-Shop Movement, 1919–1920," *Journal of American History* 51(December 1964):460–75.

Weston, Melville F. "Political Questions," *Harvard Law Review* 38(January 1925):296–333.

Williams, Arthur M. "The Impact of the Steel Seizures upon the Theory of Inherent Sovereign Powers of the Federal Government," *South Carolina Law Quarterly* 5(September 1952):5–32.

Williams, Jerre. "The Steel Seizure: A Legal Analysis of a Political Controversy," *Journal of Public Law* 2(1953):29–40.

Willis, Paul G. and George L. Willis. "The Politics of the Twenty-Second Amendment," *Western Political Quarterly* 5(September 1952):469–82.

Wilmerding, Lucius, Jr. "The President and the Law," *Political Science Quarterly* 67(September 1952):321–38.

Winger, William J. "Constitutional Power of President to Take Temporary Action in an Emergency," *Illinois Bar Journal* 41(December 1952):166–71.

Woolsey, Mark H. "The Steel Seizure Litigation," *Case and Comment* 58(January–February 1953):20, 22.

Yabroff, Bernard and Daniel P. Willis, Jr. "Federal Seizures in Labor-Management Disputes, 1917–52," *Monthly Labor Review* 76(June 1953):611–16.

Zurcher, Arnold J. "The Presidency, Congress, and Separation of Powers: A Reappraisal." *Western Political Quarterly* 3(March 1950):75–97.

Newspapers and Magazines

American Federationist
Business Week
Chicago *Sun-Times*
Chicago *Tribune*
Christian Science Monitor
Cincinnati *Enquirer*
CIO News
Cleveland *Plain Dealer*
Daily Metal Reporter
Fortune
Iron Age
Los Angeles *Times*
Nation
New Republic
Newsweek
New York *Daily News*
New York *Herald Tribune*

New York *Post*
New York Times
Pittsburgh *Post-Gazette*
Pittsburgh *Press*
St. Louis *Post-Dispatch*
Saturday Evening Post
Steel
Steel Labor
Time
U.S. News & World Report
Wall Street Journal
Washington *Evening Star*
Washington *News*
Washington *Post*
Washington *Times-Herald*
Wheeling *Intelligencer*

index

Fahy, Charles, 137, 318n58, 336n54, 346n100
Fairless, Benjamin, 58-59, 73, 147, 252-54
Federal Mediation and Conciliation Service, 59, 269n27
Federal Tort Claims Act, 106, 112-13
Feinsinger, Nathan, 24, 61, 67, 97, 296n81; Wilson's attempt to settle dispute and, 70, 71
Fenton, John M., 272n69, 70, 273n77
Ferguson, Homer, 91
Fifth Amendment: due process requirement, 117; preliminary injunction motion and, 117-18; see also Just compensation for seizure
Flash, Edward S., Jr., 264n17, 18, 265n22, 25, 267n68, 269n17, 20
Fleischmann, Manly, 330n82
Fourth Amendment, electronic surveillance and, 237, 238
Fowler, Henry H., 74, 249
Frank, John P., 190, 333n15, 16, 334n31, 336n49, 338n1, 346n106
Frankfurter, Felix, 167, 171, 173, 224, 330n78, 338n67; background of, 184-86; Youngstown opinion, 201-5, 216-21, 226, 227; Colegrove opinion, 229
Freund, Paul, 217, 334n19, 337n55, 344n76, 345n81
Frick, Henry Clay, 41
Fringe-benefits issue, 62, 65, 67, 68
Fuller, Stephen H., 356n14, 357n19, 20

Galenson, Walter, 275n33, 38, 40, 41
Gallup Poll, 92, 130
Galpin, Stephen K., 276n51
Gardner, Warner, 335n34
Garrett, Sylvester, 266n48
Gary, Elbert, 45, 47
George, Walter, 130
Georgia v. Stanton, 245
Geyelin, Philip, 296n84
Girdler, Tom, 51
Goldberg, Arthur, 70, 285n96, 287n114, 289n13, 330n80; personal background of, 55; on steel dispute, 61; seizure of mills and, 79; on Taft-

Hartley Act, 214, 330n80; on passport restrictions, 232
Grace, Eugene, 44
Great Depression, 47-48
Grede, William J., 293n47
Greenstone, J. David, 268n9
Gregory, Charles O., 298n5
Gregory, Thomas W., 323n20
Gressman, Eugene, 319n63
Griffith, J. T., 344n75
Grisanti, Eugene P., 344n75
Griswold, Erwin, 346n106

Hall, Max, 287n109
Hamby, Alonzo, L., 264n16, 19, 267n69, 268n1, 271n50, 273n80-82
Hamilton, Virginia Van Der Veer, 333n16
Hand, Learned, 222, 346n100, 106
Harbaugh, William H., 301n29, 308n92, 327n61, 330n80
Harbison, Frederick H., 357n16
Harlan, John Marshall, 233
Harris, Seymour E., 264n17, 268n75
Harsch, Joseph C., 31-32
Hart, George L., Jr., 231
Hartmann, Susan M., 268n3
Harvard Law Review, 221
Hasson, Bernard J., Jr., 337n61
Hayes, Rutherford, 156
Hearon, Robert J., Jr., 344n75
Heiss, Harold, 330n80
Henderson, James McI., 290n13
Henkin, Louis, 350n23
Hirabayashi v. United States, 188
Hogan, William T., 275n30, 41, 276n45, 48
Holmes, Oliver Wendell, 158
Holtzoff, Alexander, 103-4, 106-8, 112, 114, 290n13
Homestead plant, strike at (1892), 41-42
Hooe v. United States, 166, 303n54
House v. Mayes, 124
Hughes, Charles Evans, 107, 300n23
Humphrey, Hubert, 96, 125
Humphrey's Executor v. United States, 124
Hunt, Lester, 295n72
Hurley, Patrick J., 151
Hurley v. Kincaid, 151-52, 173, 193